Technological Revolutions in Europe

To the memory of Professor J.R. Harris

(1923–1997)

Professor of Economic History at the University of Birmingham,

1970–1990

Technological Revolutions in Europe

Historical Perspectives

Edited by

Maxine Berg
Reader in History
University of Warwick

and

Kristine Bruland
Professor of Economic History
University of Oslo

Edward Elgar
Cheltenham, UK • Northampton, MA, USA

Published by
Edward Elgar Publishing Limited
8 Lansdown Place
Cheltenham
Glos GL50 2HU
UK

Edward Elgar Publishing, Inc.
6 Market Street
Northampton
Massachusetts 01060
USA

A catalogue record for this book
is available from the British Library

Library of Congress Cataloguing in Publication Data

Technological revolutions in Europe: historical perspectives / edited
by Maxine Berg and Kristine Bruland.
Includes index.
1. Technological innovations—Europe—History. 2. Technology—
Social aspects—Europe—History. I. Berg, Maxine, 1950– .
II. Bruland, Kristine.
T173.8.T415 1998
306.3'094—dc21 97–50056
CIP

ISBN 1 85898 681 8

Printed and bound in Great Britain by MPG Books Ltd, Bodmin, Cornwall

Contents

PART 3: INVENTORS AND PRODUCTS

PART 4: LEARNING, SKILLS AND BOUNDARIES

PART 5: LAW AND INSTITUTIONS

Tables

Figures

Contributors

Edward Anderson holds an MA in Economics from the University of East Anglia, England. He is currently working as a Research Assistant to Nick von Tunzelmann and Adrian Wood, and is based at the Institute for Development Studies, University of Sussex, England.

Maxine Berg is Reader in History at the University of Warwick, England, the Director of the Warwick Eighteenth-Century Centre and a Fellow of the Society for the History of Technology. She is the author of *The Age of Manufactures* (2nd edition, 1994), and the editor of *Markets and Manufacture in Early Industrial Europe* (1991).

Kristine Bruland is Professor of Economic History at the University of Oslo, Norway. She is the author of *British Technology and European Industrialisation. The Norwegian Textile Industry in the Mid-Nineteenth Century* (1989), and the editor of *Technology Transfer and Scandinavian Industrialisation* (1991).

François Caron is Professor of History at the University of Paris IV (Sorbonne), France. He is the author of *An Economic History of Modern France* (1979), and of *Histoire des Chemins de Fer en France, 1740–1883*, vol. 1 (1997).

Chris Evans is Lecturer in History in the School of Humanities and Social Sciences at the University of Glamorgan, Wales. He is the author of *The Labyrinth of Flames: Work and Social Conflict in Early Industrial Merthyr Tydfil* (1993), and the co-editor, with P.J. Corfield, of *Youth and Revolution in the 1790s: Letters of William Pattisson, Thomas Amyot and Henry Crabb Robinson* (1996).

Robert Fox is Professor of the History of Science at the University of Oxford, England. He is the author of *Science, Industry and the Social Order of Post-Revolutionary France* (1995), the co-editor, with Anna Guagnini, of *Education, Technology and Industrial Performance in Europe 1850–1939* (1993), and the editor of *Technological Change: Methods and Themes in the History of Technology* (1996).

Trevor Griffiths is Lecturer in Economic and Social History at the University of Edinburgh. He is the author of 'Work, Class and Community: Social Identities and Political Change in the Lancashire Coalfield, 1910–1939', in A. Campbell, N. Fishman and D. Howell (eds), *Miners, Unions and Politics, 1910–47* (1996). With P. O'Brien and P. Hunt, he is the author of 'Theories of Technological Progress and the British Textile Industry from Kay to Cartwright', *Revista de Historia Economica*, **14** (1996), and of 'Inventive Activity in the British Textile Industry, 1700–1800', *Journal of Economic History*, **52** (1992).

Philip Hunt is Research Associate at the Institute of Historical Research, University of London, England. He is the author of *Corpus Christi College, Oxford: Biographical Register, 1880–1974* (1988). With P. O'Brien and T. Griffiths, he is the author of 'Technological Change during the First Industrial Revolution: The Paradigm Case of Textiles', in R. Fox (ed.), *Technological Change: Methods and Themes in the History of Technology* (1996), and of 'The Emergence and Consolidation of Excises in the English Fiscal System before the Glorious Revolution', *British Tax Review*, **1** (1997).

Margaret C. Jacob is Professor of the History of Science at the University of Pennsylvania, U.S.A. She is the author of *The Cultural Meaning of the Scientific Revolution* (1988), of *Living the Enlightenment* (1991) and of *Science and the Making of the Industrial West* (1997).

Alf O. Johansson is Professor of Economic History at the University of Oslo, Norway. He is the author of *Arbetarrorelsen och Taylorismen. Olofstrom 1895–1925* (The Labour Movement and Taylorism. Olofstrom 1895–1925) (1990), and of 'Taylorism and the Rise of Organized Labour: United States and Sweden', in Bo Gustafsson (ed.), *Power and Economic Institutions. Reinterpretations in Economic History* (1991). With J. Melling, he is the author of 'The Roots of Consensus: Bargaining Attitudes and Political Commitment among Swedish and British Workers, *c.* 1920–1950', *Economic and Industrial Democracy*, **16** (1995).

B. Zorina Khan is Professor of Economics at Bowdoin College, U.S.A. She is the author of 'Married Women's Property Laws and Female Commercial Activity: Evidence from United States Patent Records, 1790–1895', *Journal of Economic History*, **55** (1995). With Kenneth Sokoloff, she is the co-author of '"Schemes of Practical Utility": Entrepreneurship and Innovation Among "Great Inventors" in the United States, 1790–1868', *Journal of Economic History*, **53** (1993).

David S. Landes is Professor Emeritus of History and Economics at Harvard University, U.S.A. He is the author of *The Unbound Prometheus* (1969), and of *A Revolution in Time* (1983); his new book, *The Wealth and Progress of Nations*, will be published in 1998.

Christine MacLeod is Senior Lecturer in Economic and Social History at the University of Bristol, England. She is the author of *Inventing the Industrial Revolution: the English Patent System, 1660–1800* (1988), and of 'Strategies for Innovation: the Diffusion of New Technology in Nineteenth-Century British Industry', *Economic History Review*, **45** (1992). She is currently writing a book on inventors in the nineteenth century.

Joel Mokyr is the Robert H. Strotz Professor of Arts and Sciences and Professor of Economics and History at Northwestern University, U.S.A. He is the author of *Why Ireland Starved* (1985), and of *The Lever of Riches* (1990). He is currently working on a book on evolution and technology, to be entitled *Neither Chance nor Necessity: Evolutionary Models in Economic History.*

Patrick O'Brien is Professor of Economic History and Director of the Institute of Historical Research of the University of London, England. He is the author, with C. Keyder, of *Economic Growth in Britain and France: Two Paths to the Twentieth Century* (1978), and of 'Path Dependency in Britain and France', *Economic History Review*, **49** (1996).

Göran Rydén is Research Fellow in the Institute of Economic History of the University of Uppsala, Sweden. He is the author of *Hammarlag och Hushåll. Om Relationen Mellan Smidesarbetet och Smedshushållen vid Tore Petrés Brukskomplex 1830–1850* (Hammer Crews and Households. On the Relationship between Labour and the Households of Iron Workers at Tore Petrés Complex of Ironworks 1830–1850) (1991), and, with M. Ågren, of *Ironmaking in Sweden and Russia. A Survey of the Social Organization of Production before 1900* (1993).

Francis Sejersted is Professor of Economic History and Director of the Centre for Technology and Culture at the University of Oslo, Norway. He is the author of *Den Vanskelige Frihet, 1814–1851* (The Difficult Freedom – a general history of Norway) (1978), and of *Demokratisk Kapitalisme* (Democratic Capitalism) (1993). He is the editor of *Vekst gjennom Krise* (Growth through Depression) (1982).

Kenneth L. Sokoloff is Professor of Economic History at the University of California, Los Angeles, and at the National Bureau of Economic Research, U.S.A. He is the author of 'Inventive Activity in Early Industrial America: Evidence from Patent Records 1790–1846', *Journal of Economic History*, **48** (1988). He is the co-author, with B.Z. Khan, of 'The Democratization of Invention during Early Industrialization: Evidence from the United States, 1790–1846', *Journal of Economic History*, **50** (1990) and, with N.R. Lamoreaux, of 'Long-Term Change in the Organization of Inventive Activity', *Proceedings of the National Academy of Sciences*, **93** (1996).

Nick von Tunzelmann is Professor of the Economics of Science and Technology in the Science Policy Research Unit at the University of Sussex, England. He is the author of *Steam Power and British Industrialization to 1860* (1978), and of *Technology and Industrial Progress. The Foundations of Economic Growth* (1995).

Preface

This volume on Technological Revolutions in Europe arises from a symposium held in Oslo in May 1996, organized by the editors. The symposium aimed to bring together recent research on the importance of political, cultural and institutional factors in the economic history of technological change. Issues were discussed pertaining to the evolution of technologies across boundaries and borders, and to the different histories and experiences of both larger and smaller European economies, reaching out to the U.S.A. and eastern Asia for comparative purposes. The symposium could not cover all of Europe, and the book focusses mainly on Northern Europe.

A key theme addressed in this book is the relationship between technological change, learning and the social context. European economic growth and technological change relied heavily on a range of cultural and institutional factors which proved to be integral to the dynamics of growth or to the constraints which generated inertia. The central rôle of learning and skill in this process was demonstrated some years ago, notably by Peter Mathias and John R. Harris. Mathias's work on human embodied skills and mobility prefigured some of the key concepts in the economics of technology, in particular the rôle of tacit knowledge and movements of people in the diffusion of technological capabilities. John Harris's work on industrial espionage and on skills and technology in provincial manufacturing regions was likewise a source of inspiration. In this book the exploration has been extended into the cultural and institutional frameworks for the creation of such skills and capabilities in different national contexts.

The editors are grateful to Peter Mathias, who supervised us both, and who attended the symposium and wrote a summary which helped to shape the framework of this book.

This book is dedicated to the memory of John R. Harris, whose work has proved to be such a generous and inspiring source of knowledge, debate and research.

Maxine Berg, University of Warwick
Kristine Bruland, University of Oslo

Acknowledgements

The editors and publisher gratefully acknowledge the generous financial support of the Norwegian Research Council in the publication of this book. The editors also wish to express their gratitude to several institutions which together funded the 1996 symposium in Oslo on 'Technological Revolutions in Europe', from which this book arose. We thank the Faculty of Arts, the Department of History, the E.S.S.T. unit (Education in Society, Science and Technology), and the Centre for Technology and Culture, at the University of Oslo; the British Council, Oslo; B.P. Norge a/s; and the S.T.E.P. group (Studies in Technology, Innovation and Economic Policy), Oslo. We are grateful to Dr Frances White for assistance with the preparation of the book.

INTRODUCTION

1. Culture, Institutions and Technological Transitions

Maxine Berg and Kristine Bruland

INTRODUCTION

In seeking to understand and explain trajectories of economic growth, economic and technological historians are increasingly turning to the cultural and institutional contexts for invention and innovation. The contributions to this book explore these dimensions of economic change.[1] Organizational innovation and the continuous emergence and use of new technologies are clearly central to the processes of growth and change which characterize the industrial and post-industrial eras. But how should we conceptualize and understand innovation itself? In this area, much has changed in recent years. The technological determinism of early histories of industrialization, the old certainties over the benefits of technological change, and the traditional enquiries into the economic theory behind invention, technological diffusion and retardation have all been called into question. Most now acknowledge the need for a political economy of technological change, in the sense of an account which extends the parameters of analysis into institutional and social contexts.[2] There is increasing recognition also of the need for greater emphasis on the varieties of experience across countries and regions at different points in time.[3] It is important to develop and extend such approaches: although economists have tried to endogenize technological change, that is to make its explanation a part of the theory of economic growth rather than an outside unexplained factor, their accounts of innovation are sketchy at best. They are certainly not conceptually rich enough to understand either the springs of invention or the complexity of the processes of innovation and diffusion.[4]

A post-modern turn to a cultural and institutional history of Europe's technological revolution provides new directions for explanation of the dynamics of innovation and economic growth. This does not imply a return to old sociological explanations steeped in technological determinism.

Rather, it means opening up histories of innovation to analysis of the cultures of debate and scientific discourse, and to an understanding of the institutional structures of the 'learning economies' of Europe during the eighteenth and nineteenth centuries. From a historiographical point of view, this means that where once histories of science and technology went their separate ways, historians are now once again bringing them together.[5] Where political frameworks and legal institutions had been abandoned to the older concerns of the historical school, and customary codes had been left to the social historians, now economic historians are once again acknowledging their place in explaining technological transitions.[6] Above all, the cultural matrix, the speech communities and the priorities placed upon knowledge and learning are now seen to have provided new dimensions of creativity, motivation and practice for both the vital breakthroughs and the general transition to industrialization. Against this background, the chapters in this book address culture and institutions as integral factors, not simply as external parameters, in the dynamics of growth and the constraints governing inertia. General and comparative surveys of Europe and Asia and Europe and America, and new case studies focussed on northern Europe, provide some broader approaches to the history of technological revolutions in Europe.

EXOGENOUS AND ENDOGENOUS EXPLANATIONS

What are our more familiar explanations for European successes and failures, and for Southern European and Asian retardation in growth and technological dynamism? Explanations of technological innovation are generally cast in terms of exogenous and endogenous factors, and of supply-side and demand-side causation. Technology, institutions and taste have traditionally been treated as exogenous, the 'black boxes' of historical explanation.[7] Spontaneous advances of science, basic technological research and learning belong to these exogenous factors. The separation or exclusion of these elements of innovation has been particularly sharp in mainstream economics. In neo-classical economic theory, technological innovation was seen primarily in terms of process innovation, as a 'one-off' event which changed the production function, increasing productivity and real incomes. This approach has well-known limitations. In one sense, in the neo-classical account there is no explanation of technological change at all. Growth in productivity came about endogenously through capital formation, but in empirical analysis capital formation was found to explain only a fraction of long-term productivity growth. The rest was ascribed to a 'residual' of technical progress and a variety of other factors, left unexplained and uninvestigated. This gave us no economic or historical analysis of the determinants and direction of innovation itself.[8] Key features of innovation processes are in consequence absent from the analysis. In comparative static analysis, for example, the inherent continuity of innovation that came with the Industrial Revolution was missing from the story. Likewise the structuring of technological development, its organization into 'paradigms' of dominant technological approaches and their trajectories, is also absent.

Economic historians have been much more comfortable with what is now referred to as endogenous technological change. Early work treated this primarily in terms of demand-side and supply-side inducements. Landes[9] and Schmookler[10] addressed the issue of factors determining the pace and direction of innovation, arguing the importance of demand in influencing the level and direction of inventive activity. Growing demand provided investment opportunities; investment growth in effect signalled to inventors and engineers the possibilities of profit from product and process innovations. But the relevant signals were not just market-driven: Rosenberg suggested a range of 'focussing devices' – such as imbalances in the development of complementary technologies, or labour conflict, or resource limitations, which encouraged engineers to address and overcome key bottlenecks.[11] Familiar to economic historians as the 'challenge and response' model of technological change, these demand factors and focussing devices became associated with a specific chronology of the great textile inventions of the eighteenth and early nineteenth centuries. Recent investigation of patent and historiographical sources by Griffiths, Hunt and O'Brien raises questions over basic assumptions about intentions which lie behind this model.[12]

In recent years, however, supply-side theories have been ascendant. Rosenberg, followed by Mokyr, argued the case of supply-side effects; they held that prior to industrialization the existing stock of knowledge allowed for very little further invention; shifts in demand soon exhausted its modest potential. Because of this, and because of the intellectual or technical interrelatedness of many inventions, the supply of scientific and technological knowledge became a crucial input to industrialization.

Mokyr has deployed this supply-side case in an evolutionary framework: macro-inventions, large and sudden changes in technology, operated like random exogenous shocks, and were followed by micro-inventions, the improvement and development of 'learning by doing' innovations. The success of supply-side innovations was, however, dependent upon a large skilled labour force with a technical advantage in mechanical and metal-working skills. Much has been pinned on the existence in Britain of an élite core of mechanically-skilled workers used to working in metals and to adapting coal-using techniques.[13] This explanation had the intellectual attraction of honing the search for explanatory power down to a few key industries, even processes, and isolating the skills and technical know-how necessary to implement the still random exogenous technological breakthroughs. But this explanation is also circular: the key technologists emerged from training on the job, but we still have no social and economic explanation for why a critical minimum supply of technologists and craftsmen emerged from these jobs and in Britain.[14]

The supply-side explanations have been enhanced by evolutionary theories of technological change. Businesses have been seen as organizations carrying out search and selection processes, as in evolutionary biology and genetics. The genetic characteristics of industrial organizations resemble the technological 'paradigms' of firms; technologies develop along certain trajectories. Opportunities are paradigm-bound, so that there is a gradual exhaustion of technological change along particular trajectories. New paradigms then emerge, bringing with them new breakthroughs or macro-inventions.[15]

If change is all a matter of the impact of macro-inventions, then models of critical mass and clustering come into play. The critical mass means that 'positive feedback' and 'virtuous circles' will result, but the clustering of new technologies appearing in this way still needs to be explained. That clustering has hardly been investigated; there are simple models of agent interaction; that is, 'invention breeds invention'.[16] McCloskey sums up where this gets us to. Rapid advance in applied technologies was a North European and North American phenomenon; it was not due either to routine invention or to dumb luck. The macro-inventions came to the 'alert' ('Gutenberg just did it'); they are the hardest inventions to pin down to an economic explanation.[17]

That explanation must go back to the cultural contexts within which clusterings of invention occur, as well as to demand-side features which have been dismissed too readily. The markets for and expansion of European consumer-goods industries were closely linked to concentrated innovation in the finishing trades and to product innovation. In large economies, as Sokoloff found for early American industrialization, domestic competition, a large core of relatively small firms, and the emergence of a large middling social stratum prosperous enough to demand consumer decencies, provided the competitive market. Where this industrial population also had a working knowledge of current technology, ordinary citizens with common skills and knowledge could respond to market opportunities. In such cases demand factors and the expansion of markets could indeed stimulate increasing invention.[18] The prior existence of this consumer-led innovation helped to create a wider climate of invention. Where these features did not exist, the state or key élite groups might provide a distinctive technological leadership. This was one element shaping the relative success of some 'follower' economies.

CULTURE AND LEARNING

McCloskey has opened up a way to bring cultural history into the realm of explanation for the technological revolution in Europe. Through discussion of the economics of persuasion,[19] she has connected the economic history of technological change to the cultural historians' concept of the public sphere. She has identified the vital part played by 'speech communities', the 'rhetorical environment that makes it possible for inventors to be heard'.[20] Pointing the way to more comparative work, she asks why the 'wave of gadgets' was confined to North-West Europe, and why it did not emerge in Italy, Spain and China. The argument is that a culture of disagreement and of debate in North-West Europe created the climate for investors, inventors and customers. Riches and discourse grew together in Holland, Scotland, England, Belgium and the United States. Just as Adam Smith attributed the propensity to truck, barter and trade to the faculties of reason and speech, McCloskey concludes, 'we need an account of the age of industrialisation that admits into the tale ... a combination of reason and of speech, the economic historian's calculations and the social historian's sensibilities'.[21]

What were the direct ramifications of this culture of debate and difference for scientific and technological advance? Despite technological and scientific precocity in the Arab world and China, it was Europe which took the first initiative to technological breakthrough. Many multicultural reasons, and a series of contingent events, blocked the transition to cumulative technological progress in the East. Ultimately, as Landes argues, whatever the cultural limitations of European knowledge systems and aspirations, the priority which they gave to materialism and mechanical aptitude made the difference to economic and technological improvement. Republican reformers in France from the 1790s put in place a new applied and technical education which became the model for the rest of Europe.[22]

Nineteenth-century perspectives on invention and science were steeped in a culture of heroes. The connections between science and technology, as MacLeod shows in the case of James Watt, were believed to have been made by a few men of genius.[23] A culture of science and technology was, however, very much a part of the mentality of the educated and mercantile classes of Britain, much of northern Europe and some parts of southern Europe during the eighteenth century. A cultural matrix was not just based on ideas, but encoded in objects available to people or invented by them, and displayed in mechanisms and mathematical formulae. Scientific cultures developed across enlightened Europe were, however, very differently integrated into incentives to invent and into basic technologies. Jacob attributes much to the different receptions of Newtonianism in Britain, France and Holland. In Britain, to a much higher degree than in the rest of

Europe, science was closely integrated with civil and mechanical engineering. The scientists and the engineers were part of the establishment; their interactions with each other and with other businessmen and the élite were institutionalized in scientific societies, in the Royal Society of Arts, and in a consumer culture of scientific and technological novelty and entertainment. Different traditions – in France, where engineers were concentrated in the military and saw themselves as employees of the state; and in Holland, where Newtonian science remained confined to the universities – limited informal contact and the social value placed on solving practical technical problems.[24]

Nevertheless, the culture of science and technology flourished among the educated classes: O'Brien, Griffiths and Hunt have connected a small number of the great engineers and chemists of the industrial revolution with a background of scientific education at Scottish, Dutch, French and German universities. Many of the textile inventors were, however, less formally educated and from the middle ranks; they were part of a more widespread culture of interest in natural philosophy and technology. Direct connections between scientific theory and specific technological processes are limited to a few examples, such as the increasing rôle of chemical knowledge, beginning with the manufacture of alkalis, soaps, soda, bleaching agents and glass. Chlorine and its uses in chemical bleaching was discovered by the Swedish chemist Carl Wilhelm Scheele in 1774, but it was turned into a bleaching powder by a Scottish bleacher, Charles Tennant. Scientific and experimental methods, including time-keeping and precise measurement, however, became pervasive in many areas of industry, commerce and agriculture.[25]

The rhetoric of intention has also to be matched to the cultures of practical implementation. Scientists, engineers and technological leaders might talk amongst themselves, but they could not always communicate with those who implemented technologies: the skilled workers, the craftsmen and the scarce and gifted technicians. There was, as Fox tells us, a variety of scientific cultures. Practising dyers and colourists shared information and recipes, but their discourse was impenetrable to the academic chemist.[26] Many have now argued the decisive case, made many years ago by J.R. Harris, for the unarticulated 'on the job' skills and know-how of craftsmen working in precision engineering, in the adaptation of coal-fired techniques and in chemical processes in pottery, glass-making and other industries. It is important to remember that innovation and economic change were not confined to radical breakthroughs, and that they were widespread across activities and sectors; the implication is that learning and knowledge creation – in a variety of forms – occurred across many non-obvious aspects of economic life. But this is our new black box; we know very little indeed about the speech communities, literary and graphic cultures of the skilled workforces who now carry the weight of explanation for Europe's technological revolutions.[27]

What part was played by the learning cultures of the regions of North-West Europe? This aspect of the culture of technology has recently been labelled the 'learning economy'.[28] The advent of industrialization made learning strategic in the ability to produce and market products with changed technical and performance characteristics. The connection between industrialization and the skills on which learning is based has been ambiguous. It has sometimes been argued that skill-intensive mechanical engineering was matched by the deskilling of the workers using the machines. Yet, as Bruland shows, skills remained central to the key breakthroughs in a wide range of technologies, which often built on pre-existing craft skills. The key to reinforcing the accumulation of knowledge lies, it is argued, in education focussed on stimulating and protecting technological and institutional diversity.

Economic historians have debated the difference made to technological progress by learning. Jacob argues that early training in basic science and numeracy made the difference in England. But others are sceptical. The contents of syllabuses, modes of transmission, and enhancement of knowledge and motivation structures associated with educational traditions may be critical. But the work still remains to be done in relation to skilled artisans and innovators, and not just for intellectual élites. Much is made of the unacademic skills of craftsmen – all learned in the apprentice system.[29] Doing not talking was the way of iron puddlers and other iron tradesmen. That 'doing' was, however, as Evans and Rydén show, a complex and sophisticated knowledge, conveyed through kinship, dynasticism and close-knit cultural groups grounded in immigrant communities.[30] We know little of what literacy and numeracy children had gained before they were appprenticed, and what they were taught afterwards. A fluid social structure in England included many from the middling classes in the ranks of craftsmen and tradesmen. Literacy rates were high among these groups. Overall literacy for men in England was 60 per cent by 1800; for women it was 40 per cent. Male literacy was concentrated among the gentry, professionals, government officials, retail traders, and skilled tradesmen. Literacy among women in the London needle trades had reached 70 per cent by the early eighteenth century; for those in domestic service it was less than 40 per cent. Neither did apprenticeship mean the end of formal learning. Matthew Boulton took parish apprentices, and those with aptitude were instructed in literacy, numeracy, drawing and design.[31]

Low literacy levels have been correlated with economic retardation in Southern Europe for the eighteenth and nineteenth centuries. Spain, Italy and Portugal all had literacy rates far below those of North European countries until the 1920s and even later. Debate continues over whether literacy was required for skills in communication, or whether working skills are learned orally and by observation. But literacy provided the capacity to expand skills, to move between trades and to acquire flexibility. It has also been linked to improved reasoning and learning abilities.[32]

INSTITUTIONS

If different cultures of technology and the learning economy come to the forefront in the analysis of invention and innovation in eighteenth- and nineteenth-century Europe, then politics, institutions and the law are also important factors behind the dynamic of technological change. The introduction of institutions into the framework of economic and historical analysis in recent years derives in large part from the work and approach of North. He argues that institutions should be seen in terms of cultural norms, written rules and unwritten codes of conduct which provide the framework within which organizations such as firms function. There are lower transaction costs when institutions are efficient, generated by savings on acquiring information or by reductions of losses due to imperfect monitoring. Thus the historical development of the culture and institutions will effectively determine the outcome of any economic activity in a community. North accepts that economic growth is 'path-dependent', that is, once a development path is set on a particular course, the network externalities, the learning process of organizations, and the historically derived perception of issues reinforce this course. Equally, an initial set of institutions that provide disincentives to productive activity will create organizations and interests with a stake in those constraints that provide the disincentives.[33]

David and Mokyr work out in detail what this 'path-dependence' entails. Drawing on evolutionary models of technological systems, they analyse the positive and negative feedback effects of knowledge on innovation. History, as Mokyr points out, is one of the main determinants of technological outcomes. Individual technologies are tied into technological systems; change in any one part of the system has huge hurdles to overcome. This path-dependence explains much of the technological inertia of the period. This implies a 'systems' approach to technology: technological systems include political, cultural and ideological factors, and the introduction of new technologies entails political debate and rhetoric.[34] These technological systems with their political and legal ramifications could block major leaps of technology because a shift in technology could demand corresponding shifts in organizational and commercial structures. This was the case in Swedish attempts to introduce steam-powered sawmilling in the nineteenth century, as Johansson shows. Sejersted demonstrates that barriers could be overcome by effective government intervention, as in Norway's experience with the same transition in the 1860s. The development of Europe's railway networks is a classic case of the impact of cultural inheritance and state regulation, producing diversity rather than standardization of a network technology, here illustrated by Caron's study of the French railway system.[35]

Path-dependence is not, however, historical determinism – it leaves openings for historical contingency, surprise and the vital 'few' inventors.[36] Gaps in the diffusion of knowledge left room for the amateur who did not know what to expect; equally, technological blockages might be resolved in the process of the growing sophistication of the wider technological system. Those gifted amateurs had their own priorities, often little understood in economic histories of technological change. Some might be directed by an 'inner quest'. Few in the large group of British textile inventors studied by O'Brien, Griffiths and Hunt admitted to seeking only pecuniary gain; they sought solutions to technical problems, and perceived their inventive activity as a part of the wider culture of natural philosophy, mechanics, automata and technological fantasy.[37] Many inventors, as Berg and Griffiths, Hunt and O'Brien show, also saw their inventions as part of the wider consumer society – they took out patents for new products and for quality improvements. Griffiths, Hunt and O'Brien point out that just when innovations were needed in the preparatory and spinning stages of textile production, most patents were actually taken out for finishing techniques and new products. Berg argues that through connections in the designing, processing and marketing of sets of domestic and ornamental consumer wares, they entered the cultural sphere of taste and aesthetics. New kinds of information flows generated new ways of thinking.[38]

DIFFUSION OF TECHNOLOGIES AND CULTURAL FRAMEWORKS ACROSS BOUNDARIES

Institutions and cultural frameworks set the terms for the diffusion of technologies across regional and national boundaries. Yet institutions and cultures, as revealed in the different patent systems of Britain, France and the United States discussed by Khan and Sokoloff, either prompted or slowed down the speed of these information flows.[39] The development of new technologies on the Continent during early industrialization is partly a story of precocious and aborted developments, and partly one of international transfers of technologies, mainly from Britain. In both cases the contexts for successes and failures can be found in the cultures of technology, in the differences in learning processes, and in political and institutional constraints. The economy of Europe during the eighteenth century functioned, as Sidney Pollard has argued, as an economy of regions.[40] Technological experiment was widespread across export processing and finishing trades, the textile industries and ironworks and mining ventures, not just in North Europe, but in South Europe as well. The technological high noon of Holland was based in its finishing trades of the seventeenth and eighteenth centuries. The Dutch took charge of and applied up-to-date techniques in the bleaching of German, French and Flemish linen, and prepared Italian silk. Holland was a front-runner in the development of textile machinery, copper stills, presses, saws, shipbuilding innovations, and scientific and consumer novelties such as microscopes, firefighting engines, street lights and new methods of processing food. But these technologies were developed in trade-dependent manufactures, and lost momentum with the rise of tariff barriers in eighteenth-century Europe.[41]

Calico-printing works, involving large-scale concentrated manufactures deploying new technologies patented continuously over the eighteenth century, spread rapidly from Augsburg and Geneva to Silesia and Barcelona. German and Swiss towns like Bremen, Frankfurt, Hamburg, Neuchâtel, Lausanne, Geneva and Basel developed dense concentrations in the new industry. Barcelona's works were many and large-scale from an early stage, for they incorporated not just printing but also weaving. There was a direct route in several cases, especially in Holland and Switzerland, but also in Barcelona, to a mechanized, factory-based cotton industry.[42] State policy often lay behind these initiatives. Calico-printing from the mid-eighteenth century was regarded as a basic industry alongside wool and silk production, which was often the object of mercantilist import-substituting programmes.

Italy, too, during the course of the seventeenth and eighteenth centuries, developed the most advanced silk-throwing machinery, which led to the introduction of new Piedmontese mills for reeling and spinning silk yarn. While the new mills flourished around Turin, their success elsewhere was patchy. Workers' resistance and popular hostility in the south of Italy broke the success of the Piedmontese mills in Caserta, and destroyed the initiatives to introduce cotton factories in the region around Salerno. In the north, in Bergamo and Bologna, the rigidities of guild organization and fiscal and mercantile policies hindered adaptation to the new technology.[43]

There was thus no inevitability to the technological retardation of southern Europe which became its byword in the nineteenth and early twentieth centuries. This retardation was so striking that von Tunzelmann has recently compared two regions Scandinavia and South East Europe which looked much the same in the mid-nineteenth century but completely different fifty years later. 'The former had managed to develop successful industrial learning strategies, underpinned by social, educational and similar changes, while the latter had not.'[44] What happened between the eighteenth-century technological precursors and the nineteenth-century retardation was the shock to the Mediterranean economies of fifty years of war and imperial domination. The capture of Malta in 1800 by the British changed the dynamics of the Catalan textile industry by removing its traditional source of yarn. In Italy, the development of industry in Lombardy and the Veneto was hindered by heavy import duties on machinery imposed by the Austrian empire, and designed to protect Austrian industries.[45]

By the nineteenth century there was no lack of information flows among Europe's economies; indeed, there was an international community of savants in the eighteenth century, and highly-skilled workers and technicians were extremely mobile across national frontiers. Even after the scientific discoveries which underlay industrial processes in the nineteenth century, technologies still required skilled workers to implement them. Technology transfer between Britain and France was based in industrial espionage aimed at gaining key workmen as well as machines. Iron making, machinery, copper, glass and ceramics all recorded failures of successful transplantation where the knowledge base and the key technical workers were absent. But even transplanted workers were not enough, for these had to be able to adapt what they knew to unfamiliar surroundings. The experience of the Norwegian cotton industry is a case in point.[46] Imitation had to be adjusted to local environments.

The crucial factor in developing successful technology transfers in the nineteenth century was the development of an indigenous technology. This meant establishing a mechanical and electrical engineering industry. The technologies to start these might be imported – indeed engineering exports from Britain to Scandinavia were whole packages, including training schemes, long-term contracts with British workmen, repair and maintenance arrangements and even distribution systems. In the nineteenth century, the sectors which imported new technologies were spread widely across the economy, and these then shaped both the technological trajectories and the learning capabilities of the regions concerned. Germany developed its own engineering industry, and in both Scandinavia and Germany the emergence of indigenous learning capacity was crucial. Widespread com-pulsory primary education was followed by a practical curriculum in secondary schools, and by new universities and tertiary education in applied science.[47]

The evolution and international transfer of skills had complex economic effects. The general links between technology, trade patterns, skills and wage formation have rarely been studied in an integrated way. Von Tunzelmann and Anderson show that these effects were sector- and nation-specific. From the late nineteenth century on, wage differentials between skilled and unskilled workers widened in the U.S.A. but narrowed in Britain and

Germany. The suggestion is that these processes are linked to broader patterns of technological learning and industrial production. In particular, America integrated skill formation into the 'American system of manufactures' (based on the extensive use of interchangeable parts) and into new forms of corporate organization; this produced a major wage premium for 'operational' skills in such sectors as mechanical engineering. In the longer run, different modes of integration of creative engineering skills into management had profound effects on technological and economic trajectories for these economies.[48]

The essays in this volume place the political economy of technological change at the forefront of explanations for the emergence and progress of Europe's technological revolutions in the eighteenth and nineteenth centuries. They also point the way to a new social and cultural turn in the study of technologies, one not to be side-stepped by economic historians, as in the sociological explanations of the past, but to be faced and absorbed. Public spheres of scientific and technological discourse, speech communities among craftsmen and technicians, and public policies shaped the learning economies and knowledge bases which developed in their different ways across Europe, and set the terms for receptivity to new technologies. Technology is, after all, not a thing but a culture. So, too, explanations for those concentrated phases of invention appearing between the seventeenth century and the present have their own cultural histories; these have moved from providential revelation in the eighteenth century to individual heroism in the nineteenth century, and on to systemic evolution in our own time.[49]

NOTES

1. We wish to thank Peter Mathias for his written comments summing up the conference on which this book is based.
2. J. Mokyr (1992), 'Technological Inertia in Economic History', *Journal of Economic History*, **52**, 325–38; J. Mokyr (1990), *The Lever of Riches*, Oxford: Oxford University Press.
3. G.N. von Tunzelmann (1995), *Technology and Industrial Progress. The Foundations of Economic Growth*, Cheltenham, U.K. and Brookfield, U.S.: Edward Elgar, p. 2.
4. C. MacLeod (1995), 'The Springs of Invention and British Industrialization', *ReFresh*, **20**, 3.
5. P. Mathias (1972), 'Who Unbound Prometheus?', in P. Mathias (ed.), *Science and Society 1600–1900*, Cambridge: Cambridge University Press. In comparison see M.C. Jacob (1997), *Science and the Making of the Industrial West*, New York: Oxford University Press.
6. On the historical school see A. Kadish (1989), *Historians, Economists and Economic History*, London: Routledge; and A.W. Coats (1954), 'The Historicist Reaction in English Political Economy 1870–1890', *Economica*, **21**, 143–53. On the new institutional economic history see D.C. North (1990), *Institutions, Institutional Change and Economic Performance*, Cambridge: Cambridge University Press; C. Poni and R. Scazzieri (1994), 'Production Networks: Market Rules and Social Norms', unpublished C. Session Papers, Eleventh International Economic History Congress, Bologna.
7. See N. Rosenberg (1982), *Inside the Black Box: Technology and Economics*, Cambridge:

Cambridge University Press; G.N. von Tunzelmann (1993), 'Technological and Organizational Change in Industry during the Early Industrial Revolution', in P.K. O'Brien and R. Quinault (eds), *The Industrial Revolution and British Society*, Cambridge: Cambridge University Press, pp. 254–82.

8. von Tunzelmann, *Technology and Industrial Progress*, p. 62.
9. D.S. Landes (1969), *The Unbound Prometheus. Technological Change and Industrial Development in Western Europe from 1750 to the Present*, Cambridge: Cambridge University Press, pp. 84–114.
10. J. Schmookler (1966), *Invention and Economic Growth*, Cambridge, MA and London: Harvard University Press and Oxford University Press; J. Schmookler (1972), *Patents, Invention, and Economic Change: Data and Selected Essays* (ed. Z. Griliches and L. Hurwicz), Cambridge, MA: Harvard University Press.
11. N. Rosenberg (1976), 'The Direction of Technological Change: Inducement Mechanisms and Focusing Devices', in Rosenberg, *Perspectives on Technology*, London &c.: Cambridge University Press, pp. 108–25.
12. See T. Griffiths, P. Hunt and P.K. O'Brien, 'The Curious History and Imminent Demise of the Challenge and Response Model', Chapter 7 of this volume.
13. Mokyr, *The Lever of Riches*, p. 104.
14. P.K. O'Brien (1991), 'The Mainsprings of Technological Progress in Europe, 1750–1850', in P. Mathias and J.A. Davis (eds), *Innovation and Technology in Europe from the Eighteenth Century to the Present Day*, Oxford: Blackwell, pp. 6–17, p. 13.
15. See R.R. Nelson and S.G. Winter (1982), *An Evolutionary Theory of Economic Change*, Cambridge, MA: Harvard University Press, pp. 9, 17, 349; G. Dosi (1988), 'Sources, Procedures and Microeconomic Effects of Innovation', *Journal of Economic Literature*, **26**, 1131; Mokyr, *The Lever of Riches*, pp. 494–9.
16. J. Mokyr (1994), 'Technological Change, 1799–1830', in R. Floud and D.N. McCloskey (eds), *The Economic History of Britain since 1700*, second edition, Vol. 1, Cambridge: Cambridge University Press, pp. 12–43, p. 42.
17. D.N. McCloskey (1994), '1780–1860: a Survey', in Floud and McCloskey, *The Economic History of Britain*, pp. 242–70, p. 268.
18. K.L. Sokoloff and B.Z. Khan (1990), 'The Democratization of Invention during Early Industrialization: Evidence from the United States 1790–1846', *Journal of Economic History*, **50**, 363–78.
19. D.N. McCloskey (1994), *Knowledge and Persuasion in Economics*, Cambridge: Cambridge University Press.
20. McCloskey, '1780–1860: a Survey', p. 269.
21. McCloskey, ibidem, p. 270.
22. D.S. Landes, 'East is East and West is West', Chapter 2 of this volume.
23. See C. MacLeod, 'James Watt, Heroic Invention and the Idea of the Industrial Revolution', Chapter 6 of this volume.
24. M.C. Jacob, 'The Cultural Foundations of Early Industrialization: A Summary of the Project and its Conclusions', Chapter 4 of this volume. See also Mokyr, 'Technological Change, 1700–1830', p. 37.
25. P.K. O'Brien, T. Griffiths and P. Hunt (1996), 'Technological Change During the First Industrial Revolution: the Paradigm Case of Textiles, 1688–1851' in R. Fox (ed.), *Technological Change: Trends and Themes in the History of Technology*, London: Harwood, pp. 155–76; Mokyr, 'Technological Change, 1700–1830', pp. 36, 38; von Tunzelmann, *Technology and Industrial Progress*, p. 121.
26. See R. Fox, 'Science, Practice, and Innovation in the Age of Natural Dyes, 1750–1860', Chapter 5 of this volume.
27. J.R. Harris (1971), 'Industry and Technology in the Eighteenth Century: Britain and France' (Inaugural Lecture), Birmingham: University of Birmingham Press; J.R. Harris (1992), *Essays in Industry and Technology in the Eighteenth Century*, Aldershot: Variorum; P. Mathias (1979), 'Skills and the Diffusion of Innovations from Britain in the Eighteenth Century', in P. Mathias (ed.), *The Transformation of England: Essays in the Economic and Social History of England in the Eighteenth Century*, London: Methuen; Mokyr, 'Technological Change, 1700–1830', p. 41.
28. B.-A. Lundvall and B. Johnson (1994), 'The Learning Economy', *Journal of Industry Studies*, **1** (2), 23–42, pp. 38–40; K. Bruland, 'Skills, Learning and the International Diffusion of Technology: a Perspective on Scandinavian Industrialization', this volume.

29. O'Brien refers to 'illiterate craftsmen' or 'humble and unsung craftsmen who designed and constructed optical and mechanical instruments' ('The Mainsprings of Technological Progress', pp. 15–16). Mokyr cites Mathias on the mechanical skill which came in handy during an industrial revolution ('Technological Change, 1700–1830', p. 40). But see also von Tunzelmann, *Technology and Industrial Progress*, p. 121, on the long-term negative consequences of the failure of educational improvement in Britain.
30. See Evans and Rydén, 'Kinship and the Transmission of Skills: Bar Iron Production in Britain and Sweden, 1500–1860', Chapter 10 of this volume.
31. R. Schofield (1973), 'Dimensions of Illiteracy, 1750–1850', *Explorations in Economic History*, **10**, 437–54, pp. 442–6, 450; P. Earle (1989), 'The Female Labour Market in London in the Late Seventeenth and Early Eighteenth Centuries', *Economic History Review*, **42**, 328–53; M. Berg (1994), *The Age of Manufactures, 1700–1820: Industry, Innovation and Work in Britain*, second edition, London and New York: Routledge, p. 276.
32. G. Tortella (1994), 'Patterns of Economic Retardation and Recovery in South-Western Europe in the Nineteenth and Twentieth Centuries', *Economic History Review*, **47**, 1–21, pp. 11, 13. See also L. Sandberg (1982), 'Ignorance, Poverty and Economic Backwardness in the Early Stages of European Industrialization', *Journal of European Economic History*, **11**, 675–97.
33. North, *Institutions*, pp. 110–12, 135–6, 99, 114.
34. P. David (1996), 'Technology and Economic Progress: Two Paths and the Question of Path Dependence', paper given to the conference on Technological Revolutions in Europe, Oslo. See also P. David (1994), 'Why Are Institutions "the Carriers of History": Path Dependence and the Evolution of Conventions, Organizations and Institutions', *Economic Dynamics and Structural Change*, **5**, 205–30. J. Mokyr, 'The Political Economy of Technological Change: Resistance and Innovation in Economic History', Chapter 3 of this volume.
35. See A. Johansson, 'Why Change to Steam Power? Institutional Constraints, Risk-Aversion and Path-Dependence in the Swedish Sawmill Industry, 1850–1900', F. Sejersted, 'An Old Production Method Mobilizes for Self-Defence', and F. Caron, 'The Birth of a Network Technology: the First French Railway System', Chapters 11, 12 and 14 of this volume.
36. David, 'Technology and Economic Progress'; see also P. David (1985), 'Clio and the Economics of QWERTY', *American Economics Review*, **75**, 332–7.
37. O'Brien, Griffiths and Hunt, 'Technological Change During the First Industrial Revolution', p. 175.
38. See Griffiths, Hunt and O'Brien, 'The Curious History', and M. Berg, 'Product Innovation in Core Consumer Industries in Eighteenth-Century Britain', Chapters 7 and 8 of this volume.
39. See B.Z. Khan and K.L. Sokoloff, 'Patent Institutions, Industrial Organization, and Early Technological Change: Britain and the United States, 1790–1850', Chapter 15 of this volume.
40. S. Pollard (1981), *Peaceful Conquest. The Industrialization of Europe 1760–1970*, Oxford: Oxford University Press.
41. M. Berg (1991), 'Markets, Trade and European Manufacture', in M. Berg (ed.), *Markets and Manufacture in Early Industrial Europe*, London: Routledge, pp. 3–29, p. 17.
42. J.K.J. Thomson (1991), 'The Catalan Calico-Printing Industry', in Berg, *Markets and Manufacture*, pp. 57–89, pp. 61, 75; J.K.J. Thomson (1992), *A Distinctive Industrialization: Cotton in Barcelona 1728–1832*, Cambridge: Cambridge University Press.
43. J.A. Davis (1991), 'Innovation in an Industrial Late-Comer: Italy in the Nineteenth Century', in Mathias and Davis, *Innovation and Technology in Europe*, pp. 83–106, pp. 88–90.
44. von Tunzelmann, *Technology and Industrial Progress*, p. 142.
45. Davis, 'Innovation in an Industrial Late-Comer', p. 97.
46. K. Bruland (1991), *British Technology and European Industrialisation. The Norwegian Textile Industry in the Mid-Nineteenth Century*, Cambridge: Cambridge University Press.
47. von Tunzelmann, *Technology and Industrial Progress*, p. 164.
48. See G.N. von Tunzelmann and E. Anderson, 'Technology, Trade and Wage Inequalities in the Late Nineteenth Century', Chapter 13 of this volume.
49. MacLeod, 'James Watt', Chapter 6 of this volume; S. Lindqvist (1993), 'Social and Cultural Factors in Technology Transfer', in K. Bruland (ed.), *Technology Transfer and Scandinavian Industrialisation*, Oxford and New York: Berg Publishers, pp. 15–36; Mokyr, *The Lever of Riches*, pp. 273–99.

PART 1

Technology, Cultures and Political Economy

2. East is East and West is West

David S. Landes

The world history of technology is the story of a long, protracted inversion. As late as the millennium, the civilizations of Asia were well ahead of Europe in wealth and knowledge. The Europe of what we call the Middle Ages (say, the tenth century) had regressed from the power and pomp of Greece and Rome, had lost much of the science it had once possessed, had seen its economy retreat into generalized autarky. It traded little with other societies, for it had little surplus to sell, and it paid for foreign goods largely with human beings. Nothing testifies better to deep poverty than the export of slaves or the persistent exodus of job-hungry migrants.

Five hundred years later, the tables had turned. I like to sum up the change in one tell-tale event: the Portuguese penetration into the Indian Ocean. This was an extraordinary achievement. Some scholars hold that it was some kind of accident; that it could just as easily have been Muslim sailors, or Indian, or Chinese, to make the connection from the other direction. Did not the Chinese send a series of large fleets sailing west as far as the East African coast in the early fifteenth century – bigger, better and earlier than anything the Portuguese had to show?

This kind of optative history (history not as it was, but as one would have it) is untrue and misleading. These affirmations of Asian priority are especially prominent and urgent nowadays because a new grand inversion is bringing Asia to the fore, and history (as a discipline) runs to justify and sanctify events *ex post*. A new 'multicultural' world history finds it hard to live with a Eurocentric story of achievement and transformation. So a new would-be (politically correct) orthodoxy would have us believe that a sequence of contingent events (gains by Portugal and then others in the Indian Ocean; conquests by Spain and then others in the New World) gave Europe what began as a small edge and was then worked up into centuries of dominion and exploitation. A gloss on this myth contends that a number of non-European societies were themselves on the edge of a technological and scientific breakthrough; that in effect European tyranny 'froze the genial current of the [Asian] soul'.

This is not the place to enter fully into the revisionist debate. Let it suffice to say here:

(1) The Portuguese success was the result of decades of rational exploration and extension of navigational possibilities in an ocean that was hostile to traditional coasting techniques. This technological enhancement rested in turn on a systematic utilization of astronomical observations and calculations, taken from the Muslims and transmitted largely by Jewish intermediaries.
(2) The Chinese abandonment of the programme of oceanic voyages was partly the result of contingent political events; but at bottom it reflected the values and structures of Chinese society and civilization.
(3) European exploitation of the breakthrough rested on a disparity of power technology (better powder and better guns) as well as on navigational superiority.

The extension of European power into other parts of the world was an expression of these and other disparities. Why Europe should have been pulling ahead is an important historical question, for one learns almost as much from failure as from success. It is not possible in brief compass, of course, to pose this question for every non-European society or civilization; but three do deserve serious reflection: Islam, India, and China.

ISLAM

The first is Islam, which inherited and developed the knowledge and ways of predecessor civilizations (Greek, Persian) in the area of its extension, which by the period we are talking about (roughly A.D. 1000 to 1500) went from the western end of the Mediterranean to the Indies. Islamic science and technology far surpassed those of Europe, which needed to recover its heritage and did so to some extent through contacts with the Muslim world in such frontier areas as Spain. Islam was Europe's teacher.

And then something went wrong. Islamic science, even at its peak, was denounced as heresy by religious zealots. Under theological pressures towards spiritual conformity, thinkers and searchers bowed to curbs on intellectual speculation and the dissemination of ideas. Very understandable: discretion could be a matter of life and death. The point of the defenders of the faith was that the truth had already been revealed; what led to and promoted the truth was useful and permissible; all the rest was deceit and error.[1]

As for technology, Islam knew areas of change and advance: one thinks of the rapid adoption of paper; or the introduction and diffusion of new crops such as coffee and sugar; or the Ottoman Turkish readiness to learn the use of cannon. But most of this was learned from outside and continued to depend on outside support. The springs of autonomous invention, such as they were, seem to have dried up. Even in the golden age (A.D. 750 to 1100), there seems to have been a disconnection between thought and application: 'For nearly five hundred years the world's greatest scientists wrote in Arabic, yet a flourishing science contributed nothing to the slow advance of technology in Islam.'[2]

Even the Ottoman Empire, at its peak in the sixteenth century, revealed great limitations in its military technology, both in artillery and in battleships. And this was not all. Over time the trade relations between Europe and the Levant were reversed. Where once Eastern craftsmen had supplied the Europeans with fine cloth, carpets, tapestries, faience and the like in exchange for metal (copper and tin), slaves, and specie, from the sixteenth century onwards it was Europe that sold the manufactures in exchange for dried fruit, spices, cotton and cereals. The same for silk: in the Middle Ages Europe had bought Byzantine silk fabrics; now it imported raw silk, and local producers in Turkey found it hard to compete with European buyers. And paper: this writing material was eagerly adopted in the Middle East (8th century) from the Chinese example; because forage was short in China, hides were scarce and so was parchment. The technology was slower to take root in Europe, where parchment was relatively abundant; but once European makers learned to produce paper, they far surpassed their Levantine predecessors and were soon selling large quantities in the East.[3] Even substances (coffee and sugar) that had originally come to Europe from the East were now going the other way – in the case of sugar, after refining and processing.[4]

The most serious mistake was the Muslim refusal of the printing press, which was seen as a potential instrument of sacrilege and heresy. Nothing did more to cut them off from the mainstream of knowledge. As a result of this intellectual segregation, technical lag, and industrial dependency, the balance of economic forces shifted inexorably against the Ottomans, while the experience of military defeat undermined their assumptions of superiority and paralysed their ability to respond.

In these circumstances, the continued transformation of European technology, in particular the Industrial Revolution, sounded the death knell for what remained of Ottoman industry. Except for some local specialities, nothing could compete with cheap factory-made cottons and silks. The nineteenth century saw Britain stand up for the Ottoman empire and protect it from the territorial ambitions of its adversaries, whilst blithely killing off its manufactures. But from the British point of view, that was as it should be: British goods were cheaper. The Ottomans did not know enough, they did not have the capital, and they could not count on political stability. Let them grow grapes and turn them into raisins; let them export nuts and opium. The British would supply them with industrial products.

INDIA

What about India? Why no Industrial Revolution?

India had the world's premier cotton industry in the seventeenth and eighteenth centuries, unbeatable for quality, variety and cost. This was an industry that not only satisfied the large domestic demand but exported roughly half its output throughout the Indian Ocean and indirectly to southeast Asia and China. Then, beginning in the seventeenth century, there came the stimulus of European demand – a huge shot in the arm that inevitably aggravated old supply problems and created new ones. Why, then, was there no interest in easing these difficulties by substituting capital (machines) for labour?

Indian historians have tended to overlook or reject this issue. Some, especially Indian nationalists, feel that they know the answer without seeking: it's all the fault of the Europeans, and most particularly the British. This was a prosperous, resourceful, developing economy until these intruders burst on the scene, began to mix into Indian politics, and fomented rebellion and conflict destructive of capital and industry.

One way to approach the problem of non-development, which is worth taking seriously, is to ask the critical question, *cui bono* – who would have benefitted? Who would have gained from mechanization and transformation? Three groups of interests were involved: the workers (spinners and weavers); the middlemen, who typically advanced capital to the weavers against the promise of delivery; and the European traders and chartered companies, who wanted to buy both for the country (intra-Asian) trade and for their European clientèle.

It would be unreasonable to expect capital-using technological innovations from the first group. They had an obvious interest in improving the availability of materials (cotton fibre for spinners, yarn for weavers), but here they were simply dependent on merchant intermediaries. They had neither means nor the habit of command. If there was to be any initiative in the direction of technological change, then, it would have to come from the Indian middlemen, who had both interest and (some of them) means; or from the European chartered companies. Yet neither moved.

Why not? The explanations that have been advanced have been based on an implicit law of conservation of energy. The supply of labour was elastic, so it was easier and more economical to hire additional workers, from among untouchables and poor women for spinning, and from agricultural labourers for weaving, than to look to change the technology; and that may well be the whole of the story.[5] Since demand was segmented and different markets wanted different goods, it was also possible, within limits, to shift goods among markets, from domestic to foreign and from one foreign to another.

It was even possible, though very difficult, to assemble large numbers of workers 'under one roof' (in one place), to toil under supervision. This was the sort of thing the foreign trading companies tried to do, by way of ensuring prompt completion of tasks. In some instances, such concentration yielded economies of scale and materials – in fuel-using branches, for example, or in assembly work such as shipbuilding. Technological change, then, in the form of organizational innovation, was not unknown. Such enterprises, however, remained the exception; the 'small-scale family-based unit [retained] its position of primacy'.[6]

Hardware – instruments, equipment, machines – was another matter. This is what it took to make an industrial revolution, and India was not ready. 'In India it is seldom that an attempt is made to accomplish anything by machinery that can be performed by human labour.'[7] One reason for this 'general indifference' was limited division of labour and specialization. Indifference, moreover, was promoted by functional segmentation: it was not the merchant's job to find, assemble and deliver the raw materials. He advanced capital, and it was up to the weaver and spinner to do the rest. This was significantly different from putting-out as practised in Europe, where the merchant was directly linked to the preparation and processing (preparing and finishing) of the materials. In India, then, the final buyer was cut off from the means of remedy. The worker did what he had always done, and so did the merchant. In the Coromandel, for example, the raw cotton was moved from the interior to the spinning and weaving villages on and near the coast, by huge trains of bullocks numbering in the thousands and tens of thousands, the whole shapeless mass feeding while shambling along at the rate of a couple of miles a day. Since the trek covered some 300 miles or more, it took about half a year to deliver the goods.[8]

Meanwhile the European companies' own rhythm of purchases and shipments was spasmodic because of the constraints of navigation and availability of capital, to say nothing of fluctuations in supply. Besides, it was not obvious to these companies – certainly not to the East India Company – that direct assistance to the Indian industry was politically wise. British manufacturing interests would have seen that as an act of treason.

Finally, where were the ideas and concepts of mechanization to come from? It was not that Indian society was unfamiliar with technological change: the most important change in the textile manufacture came with the substitution of the wheel for the distaff (though not for the finest muslin yarn). But these changes took place within the conventional manual context, and a big conceptual and social difference separates machines from hand tools, even improved tools. One must distinguish further between all-purpose tools and specialized tools: Indian artisans, however skilled, had scarcely started on the path to instrumentation. Worse yet, Indian craftsmen avoided using iron, and iron (and steel) is indispensable to precision work. This was not a ferruginous society.[9]

Under the circumstances, the move to machinery in India was not to be envisaged. Such a leap would have entailed a shift away from hand skills nurtured from childhood, in some instances linked to and reinforced by caste identity and sex and age. It would also have required an effort of imagination that lay outside the Indian cultural and intellectual experience. As Chaudhuri puts it: 'In eighteenth-century India the empirical basis for an Industrial Revolution was conspicuously lacking. There had been no marked progress in scientific knowledge for many centuries, and the intellectual apparatus for a diffusion and systematic recording of the inherited skills was seriously defective.'[10]

And so it was still in the nineteenth century: the British engineers who built the Indian railways understood that Indian labour, cheap as it was, would move earth and rock by hand; but they also took it for granted that the Indians would use wheelbarrows. Not at all: the Indians were used to moving heavy burdens in baskets on their heads, and refused to change.

On the other hand, if Indian cotton producers of the seventeenth and eighteenth centuries were content to work as before, the Europeans were not. The influx of Indian cottons was a shock to producers of other textiles, who saw these imports as a threat to profits and pleaded the need to defend employment. The result in England was the so-called 'Calico Acts', designed not only to keep cottons out but also to ban the wearing of some of these un-British fabrics. But whereas the aim of these Acts was to protect existing industrial interests, that is, wool and silk, the effect was to promote the development of an indigenous cotton industry – unanticipated consequences.

So it was that when attempts to mechanize the spinning of wool failed, English inventors turned to the vegetable fibre, more uniform and manipulable by the insensate motions of a machine. The Indians could not and did not; the British did, and changed history.

CHINA

Now England is paying homage.
My Ancestors' merit and virtue must have reached their distant shores.
Though their tribute is commonplace, my heart approves sincerely.
Curios and the boasted ingenuity of their devices I prize not.
Though what they bring is meagre, yet,
In my kindness to men from afar I make generous return,
Wanting to preserve my good health and power.

(Poem by the Qienlong Emperor on the occasion of the Macartney embassy, 1793.)

The Empire of China is an old, crazy, first rate man-of-war, which a fortunate succession of able and vigilant officers has contrived to keep afloat these one hundred and fifty years past, and to overawe their neighbours because of her bulk and appearance, but whenever an insufficient man happens to have the command upon deck, adieu to the discipline and safety of the ship. She may perhaps not sink outright; she may drift some time as a wreck, and will then be dashed to pieces on the shore; but she can never be rebuilt on the old bottom.

(George, Lord Macartney, to his journal.)[11]

The one civilization that was in a position to match and even to anticipate the European achievement was China. China had two chances: first, to generate a continuing, self-sustaining process of scientific and technological advance on the basis of its indigenous traditions and achievements; and, second, to learn from European science and technology once the foreign 'barbarians' entered the Chinese domain in the sixteenth century. China failed on both counts.

The First Chance

The first failure has elicited much scholarly inquiry and analysis. And yet it remains a mystery – in spite of a monumental effort by the late Joseph Needham and others to collect the facts and clarify the issues. The China specialists tell us, for example, that in a number of areas of industrial technique, China long anticipated Europe: in textiles, where the Chinese had a power-driven spinning machine in the thirteenth century, some 500 years before the England of the Industrial Revolution knew water frames and mules; or in iron manufacture, where the Chinese early learned to use coal and probably coke (as against charcoal) in blast furnaces for smelting iron, and were turning out perhaps as many as 125,000 tons of pig iron by the later eleventh century – a figure not achieved by Britain until 700 years later.[12] In general, one can establish a long list of instances of Chinese priority: the wheelbarrow, the stirrup, the rigid horse collar (to prevent choking), the compass, paper, printing, gunpowder, porcelain.

The mystery lies in the failure of China to realize the potential of some of the most important of these inventions. One generally assumes that knowledge and know-how are cumulative and that a superior technique, once known, will dominate older methods and remain in use. But Chinese industrial history offers a number of examples of technological regression and oblivion. The machine to spin hemp was never adapted to the manufacture of cotton, and cotton spinning was never mechanized; and coal/coke smelting was allowed to fall into disuse, along with the iron industry. Why? asks Mark Elvin:

> It would seem that none of the conventional explanations tells us in convincing fashion why technical progress was absent in the Chinese economy during a period that was, on the whole, one of prosperity and expansion. Almost every element usually regarded by historians as a major contributory cause to the industrial revolution in north-western Europe was also present in China. There had even been a revolution in the relations between social classes, at least in the countryside; but this had had no important effect on the techniques of production. Only Galilean-Newtonian science was missing; but in the short run this was not important. Had the Chinese possessed, or developed, the seventeenth-century European mania for tinkering and improving, they could easily have made an efficient spinning machine out of the primitive model described by Wang Chen ... A steam engine would have been more difficult; but it should not have posed insuperable difficulties to a people who had been building double-acting piston flame-throwers in the Sung dynasty. The crucial point is that nobody tried. In most fields, agriculture being the chief exception, Chinese technology stopped progressing well before the point at which a lack of scientific knowledge had become a serious obstacle.[13]

Why indeed? Sinologists have put forward several partial explanations. The two that I find the most persuasive are the following:

1. China had no free market and no institutionalized property rights. The Chinese state was always stepping in to interfere with private enterprise – to take over certain activities, to prohibit and inhibit others, to manipulate prices, to exact bribes. At various times the government was motivated by a desire to reserve labour to agriculture, or to control important resources (salt and iron, for example); by greed (the story of the goose that laid the golden eggs is a leitmotif of Chinese history); by fear and disapproval of self-enrichment, except by officials, giving rise in turn to abundant corruption and rent-seeking; and by a distaste for maritime trade, which the Heavenly Kingdom saw as a diversion from imperial concerns, as a divisive force and a source of income inequality in the ecumenical empire, and, worse yet, as an invitation to leave. Matters reached a climax under the Ming dynasty (1368–1644), when the state attempted to prohibit all trade

overseas.[14] Such interdictions led of course to evasion and smuggling, with concomitant corruption (protection money), searches for contraband, confiscations and punishment. All of this necessarily acted to strangle initiative, to increase risk and the cost of transactions, and to divert talent from commerce and industry.

2. Another explanation may lie in the larger values of the society. The great Hungarian-German-French sinologist, Etienne Balazs, sees China's abortive technology as part of a larger pattern of totalitarian control. He does not explain this by hydraulic centralism, but he does recognize the absence of freedom and the weight of custom and consensus, of what passed for higher wisdom. His analysis is worth repeating:

> The ingenuity and inventiveness of the Chinese, which have given so much to mankind – silk, porcelain, paper, printing, and more – would no doubt have enriched China further and probably brought it to the threshold of modern industry, had it not been for this stifling state control. It is the State that kills technological progress in China. Not only in the sense that it nips in the bud anything that goes against or seems to go against its interests, but also by the customs implanted inexorably by the *raison d'Etat*. The atmosphere of routine, of traditionalism, and of immobility, which makes suspect any innovation, any initiative that is not commanded and sanctioned in advance, is unfavourable to the spirit of free inquiry.[15]

In short, to go back to Elvin, no one was trying. Why try?

In all this, the contrast with Europe was marked. Where fragmentation and national rivalries compelled European rulers to pay heed to their subjects, to recognize their rights and to cultivate the sources of wealth, the rulers of China had a free hand.[16]

Whatever the mix of factors, the result seems to have been a curious pattern of isolated initiatives and Sisyphean discontinuities – up, up, up, and then down again – almost as though the society were constrained by a homeostatic braking mechanism or held down by a silk ceiling. The result, if not the aim, was a kind of change-in-immobility, or maybe immobility-in-change. Innovation was allowed to go (was able to go) so far and no farther.[17]

The Europeans knew much less of such interferences. Instead, they entered during these centuries into an exciting world of innovation and emulation that challenged and tempted vested interests and kept the forces of conservatism scrambling. Changes were cumulative, news of novelty spread fast, and a new sense of progress and achievement replaced an older, effete reverence for authority. This intoxicating sense of freedom touched (infected) all domains. These were years of heresies in the church, of popular initiatives that, we can see now, anticipated the rupture of the Reformation; of new forms of expression and collective action that challenged the older organization of society and posed a threat to other polities; of new ways of doing and making things that made newness a virtue and a source of delight.

Important in all this was the rôle of the Christian church as custodian of knowledge and school for technicians. One might have expected otherwise: that an organized spirituality, with its emphasis on prayer and contemplation, would have had little interest in technology; and that, with its view of labour as penalty for original sin, it would have had no concern to save labour. And yet everything seems to have worked in the opposite direction. The desire to free clerics from time-consuming earthly tasks led to the introduction and diffusion of power machinery and, beginning with the Cistercians, to the hiring of lay brothers (*conversi*) to do the dirty work, which led in turn to an awareness of and attention to time and productivity. All of this gave rise, on monastic estates, to remarkable assemblages of powered machinery – complex sequences designed to make the most of the water power available and to distribute it through a series of industrial operations. A description of the abbey of Clairvaux in the mid-twelfth century exults in this versatility: 'coquendis, cribrandis, vertendis, terendis, rigandis, lavandis, molendis, molliendis, suum sine contradictione praestans obsequium' (submitting without demur to cooking, sieving, turning, grinding, watering, washing, milling, and moistening). The author, clearly proud of these achievements, further tells his readers that he will take the liberty of joking (the medieval clerical equivalent of 'if you'll pardon the expression'): the fulling hammers, he says, seem to have spared the fullers the penalty for their sins; and he thanks God that such devices can mitigate the oppressive labour of men and spare the backs of their horses.[18]

Why this peculiarly European *joie de trouver*, this pleasure in new and better, this cultivation of invention – or what some have called 'the invention of invention'? Different scholars have suggested a variety of reasons, typically related to religious values:

1. The Judaeo-Christian tradition respected manual labour, and this respect is summed up in a number of Biblical injunctions. One example will suffice: when God warns Noah of the coming flood and tells him that he will be saved, it is not God who saves him: 'Build thee an ark of gopher wood,' He says, and Noah builds an ark to the divine specifications.
2. The Judaeo-Christian tradition subordinates nature to man. This is a sharp departure from widespread animistic beliefs and practices that saw something of the divine in every tree and stream (hence the dryads and naiads). Ecologists today would say that these animistic beliefs were preferable to what was put in their place, but no one was listening to pagan nature-worshippers in Christian Europe.

3. The Judaeo-Christian tradition features a sense of linear time. Other societies thought of time as cyclical, returning to earlier stages and starting over again. Linear time can be thought of as progressive or as regressive, as moving on to better things or as declining from some earlier, happier state. For Europeans in our period, it was the progressive view that prevailed.
4. In the last analysis, however, I would stress the rôle of the market: the fact that enterprise was free in Europe, that innovation worked and paid, that rulers and vested interests were narrowly constrained in what they could do to prevent or discourage innovation. Success bred imitation and emulation, and also a sense of power that would in the long run raise men almost to the level of gods. The old legends remained – the expulsion from the Garden, Icarus who flew too high, Prometheus in chains – to warn against hubris. The very notion of hubris – cosmic insolence – is testimony to some men's pretensions, and the efforts of others to curb them.

   But the doers were not paying attention.

The Second Chance

At the time when the first Europeans arrived in the Indian Ocean and made their way to China, the Celestial Empire (as it was called) was, at least in its own eyes, the premier political entity in the world – first in size and population, first in age and experience, untouchable in its cultural achievement, apparently imperturbable in its sense of moral and spiritual superiority.[19] The Chinese lived, as they thought, at the centre of the universe; around them, lesser breeds basked in their glow, reached out to them for light, gained stature by doing obeisance to them and offering tribute. Their emperor was the 'Son of Heaven', the unique, godlike representative of celestial power. Those few who entered his presence showed their awe by kowtowing – kneeling and touching their heads nine times to the ground. Others kowtowed to anything emanating from him – a letter, a single handwritten ideograph. The paper he wrote on, the clothes he wore, everything he touched partook of his divine essence.

Those who represented the emperor and administered for him were chosen on the basis of competitive examinations in Confucian letters and morals. These mandarin officials were in effect the embodiment of the higher Chinese culture, invested with its prestige, imbued with its wholeness and its sublime superiority. Their self-esteem and haughtiness had ample room for expression and exercise on their inferiors, and were matched only by their 'stunned submissiveness' and self-abasement to superiors.[20]

Such cultural triumphalism combined with petty tyranny made China a singularly bad learner. What was there to learn? This rejection of the strange and foreign was the more anxious for the very force of the arrogance that justified it. For that is the paradox of the superiority complex: it is an expression of insecurity. It is intrinsically brittle; those who nourish it need it, depend on it, and fear nothing so much as contradiction. The French today are so persuaded of the superiority of their language that they dither and tremble at the prospect of a borrowed word, especially from English. Ming China was so convinced of its ascendancy that it quaked before the challenge of Western technology, which was there for the learning.

The irony is that those first Portuguese visitors and Catholic missionaries used the wonders of Western technology to charm their way into China. The mechanical clock was the key that unlocked the gates. This was a European mega-invention of the late thirteenth century, crucial not only for its contribution to temporal discipline and productivity, but also for its susceptibility of improvement and its rôle at the frontier of instrumentation and mechanical technique. The water clock was a dunce by comparison.

For the Chinese in the sixteenth century, the mechanical clock came as a wondrous machine capable not only of keeping time, but also of amusing and entertaining. Some clocks played music; others were automata with figurines that moved rhythmically at intervals. Clocks, then, were the sort of thing that the emperor would want to see, that had to be shown him if only to earn his favour, that a zealous courtier had to show him before someone else did. But that was not so easy. This magical device had to be accompanied. Where all Chinese instincts and practice dictated that foreigners should be kept at a distance, confined to some peripheral point like Macao and allowed to proceed to the centre only by exception, the clock, in its sixteenth-century avatar, needed its attendant clockmaker.

There is no question that the Chinese loved clocks and watches. They were less happy, though, with their European attendants. The problem here was the Chinese sense of the wholeness of culture, the link between things, people and the divine. The Catholic priests who first brought them these wonderful machines were salesmen of a special kind. They sought to convert the Chinese to the one true God, the trinitarian God of the Roman church, and the clocks were not only an entry ticket but also an argument for the superiority of the Christian religion. Were not those who could make these things, who possessed all kinds of special astronomical and geographical knowledge into the bargain, were they not superior in the largest moral sense? Was not their faith truer, wiser? The Jesuits were prepared to make such an argument, stretching the while the rules and rites of the Church to fit the premises and win the sympathy of an understandably sceptical Chinese élite. (The Chinese ideographs for ancestor worship, for example, became the signifiers of the Christian mass.) But European laymen used it as well. Here is Gottfried Wilhelm von Leibniz (1646–1716), mathematician (co-inventor of the calculus) and philosopher:

> What will these peoples [Persians, Chinese] say, when they see this marvellous machine that you have made, which represents the true state of the heavens at any given time? I believe that they will recognize that the mind of man has something of the divine, and that this divinity communicates itself especially to Christians. The secret of the heavens, the greatness of the earth, and time measurement are the sort of thing I mean.[21]

This argument, whether explicit or implicit, did carry occasionally. The Catholic missionaries had some small success, although they had trouble persuading their open-minded 'converts' to be good exclusivists (no other faith but the 'true' faith) in the European tradition. But most Chinese saw these pretensions for what they were: an attack on Chinese claims to moral superiority, an assault on China's self-esteem.

The response, then, had to be a repudiation or depreciation of Western science and technology. Here is the K'ang Hsi emperor, the most open-minded and curious of men in his pursuit of Western ways, the most zealous in teaching them:

> even though some of the Western methods are different from our own, and may even be an improvement, there is little about them that is new. The principles of mathematics all derive from the *Book of Changes*, and the Western methods are Chinese in origin...[22]

That was the heart-warming myth. So the Chinese, who were not prepared to give up clocks, who wanted clocks but who recognized their Western origin, trivialized them as toys, which for many they were; or as non-functional symbols of status, unaffordable by or inaccessible to most. Pre-modern imperial China did not think of time knowledge as a personal right. The hour was sounded by the authorities, and the right to own a timepiece was a rare privilege. As a result, although the imperial court set up workshops to make clocks and got their Jesuit clockmakers to train some native talent, these Chinese makers never arrived at the level of Western horologists for want of the best teachers and lack of commercial competition and emulation. Nor did imperial China ever develop a clockmaking trade comparable to that found in European countries.

The same sin of pride (or indifference) shaped the Chinese response to European armaments. These were anything but toys. Cannon and muskets were instruments of death, hence of power, and the Chinese had every reason to interest themselves in them, the more so as the seventeenth century saw the progressive dissolution of the Ming dynasty and the conquest of China by a Tartar people from the north. These were decades of war, and the balance of power might well be tilted by access to these artefacts.

And yet the Chinese never learned to make guns – not modern guns. Worse yet, they had known and used cannon as early as the thirteenth century, but had forgotten much of what they had once known. Their city walls and gates had emplacements for cannon, but no cannon. Who needed them? The enemies of China did not have them. Yet China did have enemies, without and within, and no European nation would have been deterred from armament by enemy weakness; when it came to death, as in so many other things, the Europeans were maximizers. European technology was also monotonic-increasing: each gain was the basis for further gain. The Chinese record of advance *cum* regression, step-forward, step-back, signalled an entirely different process. (I shall return to this below.) The Chinese, we are told, had a proverb: he who does not go forward will go backward.[23] It must have been more observation than prescription.[24]

So it was that in the seventeenth century, when the Portuguese in Macao offered three cannon to the emperor by way of gaining favour, they had to send three cannoneers along with them. Similarly, the Chinese on occasion hired Portuguese musketeers to do some fighting for them, and they got their Jesuit theologian-mechanicians to make them cannon. These seem to have been among the best they had, so good compared to the run-of-the-foundry product that some were still in use in the nineteenth century, some 250 years later. If most Chinese guns did not last that long, it was because they were notoriously unreliable, more dangerous to the men who fired them than to the enemy. We even have one report of the use of clumps of dried mud as cannonballs. These at least had the merit of allowing the force of the explosion to escape by the mouth of the tube. In general, the authorities frowned on the use of firearms, perhaps because they doubted the loyalty of their subjects. In view of the inefficacy of these pieces, one wonders what they had to fear – presumably the improvement that comes with use.[25]

All of this may seem irrational to a means-and-ends-oriented person, but it was not quite that; the ends were different. The European may have thought that the purpose of war was to kill the enemy and win; the Chinese, strong in space and numbers, thought otherwise. Here is Mu Fu-sheng (a pseudonym) on the imperial viewpoint:

> military defeat was the technical reason why Western knowledge should be acquired, but it was also the psychological reason why it should not be. Instinctively the Chinese preferred admitting military defeat, which could be reversed, to entering a psychological crisis; people could stand humiliation but not self-debasement ... The mandarins sensed the threat to Chinese civilization irrespective of the economic and political issues and they tried to resist this threat without regard to the economic and political dangers. In the past the Chinese had never had to give up their cultural pride: the foreign rulers always adopted the Chinese civilization. Hence there was nothing in their history to guide them through their modern crisis.[26]

Along with indifference to technology went imperviousness to European science. The same conditions applied. The Jesuits and other Christian clerics brought in not only clocks but also knowledge (sometimes obsolete) and ideas. Some of this was of interest to the court: in particular, astronomy and techniques of celestial observation were extremely valuable to a ruler who claimed a monopoly of the calendar and used his mastery of time to impose on the society as a whole. The Jesuits, moreover, trained gifted Chinese students who went on to do their own work: mathematicians who learned to use logarithms and trigonometry; astronomers who prepared new star tables.

Little of this got beyond Peking, however, and the pride which some took in the new learning was soon countered by a nativist reaction that reached back to long-forgotten work of earlier periods. One leader of this return to the sources (Wen-Ting, 1635–1721) examined the texts of mathematicians who had worked under the Song dynasty (tenth to thirteenth centuries) and proclaimed that the Jesuits had not brought much in the way of innovations. Later on, his manuscripts were published by his grandson under the title of 'Pearls Recovered from the Red River'.[27] The title was more eloquent than was intended: by this time much of Chinese scientific 'inquiry' took the form of raking alluvial sediment.

Meanwhile European science marched ahead, and successive churchmen brought to China better knowledge than their predecessors had (though still well behind the frontier). Here, however, they were thwarted by the constraints of their mission. The Christian missionaries had laid so much stress on the link between scientific knowledge and religious truth that any revision of the former implied a repudiation of the latter. When in 1710 a Jesuit astronomer sought to use new planetary tables based on the Copernican system, his superior would not permit it, for fear of 'giving the impression of a censure on what our predecessors had so much trouble to establish and occasioning new accusations against [the Christian] religion.'[28]

To recall these many instances of intellectual xenophobia is not to imply that all Chinese were hostile to European ideas and innovations. We know that a few far-sighted officials and at least one emperor understood that the empire had much to gain by learning these new ways.[29] They were thwarted, however, not only by the studied complacency of an insecure superiority – and by a sense of completeness[30] – but also by the intrigue of a palace milieu where innovations were judged by their consequences for the balance of power and influence. No proposal that did not incite resistance; no novelty that did not frighten vested interests. At all levels, moreover, the fear of reprimand (or worse) outweighed the prospect of reward. A good idea brought credit to one's superior; a mistake was invariably the fault of subordinates.

One consequence was a prudent, almost instinctive, resistance to change. Another was a plague of lies and misinformation: officials wrote and told their superiors what they wanted to hear; or what the subordinate thought the superior would want to hear.[31]

The smothering of incentive and the cultivation of mendacity are characteristic weaknesses of large bureaucracies, whether public or private (business corporations). These are composed of nominal colleagues, who are supposedly pulling together but in fact are adversarial players; what is more, they compete within the organization, not in a free market of ideas, but in a closed world of guile and manoeuvre. Here the advantage lies with those in place. Reformers and subversives, beware.

The rejection of foreign technology was the more serious because China itself had long slipped into a régime of technological and scientific inertia, coasting along on the strength of previous gains and slowly losing speed as a result of the inevitable frictions of vested interest and of the diversion of talent and wealth into the comfort and gratification of gentility.

It has been argued that such retirements from the fray should not deter ambitious newcomers; on the contrary, the prospect of such a happy exit should encourage entry, and the departures should make room for others. But in most aristocratic societies, the availability of more esteemed careers seems to divert talent from commerce and industry by offering short-cuts to high status, while the withdrawal of successful merchants into land and office is seen not as a logical promotion but rather as an escape. In such circumstances, the presence of groups precluded from access to office and honours by birth (such as merchants in Tokugawa Japan) or by belief (Protestant dissenters in England) – the existence, in other words, of a reserved pool of talent – may, paradoxically, be a powerful contribution to otherwise inhibited economic development.

In any event, one of the great mysteries of Chinese history is why China did not produce from within the kind of scientific and industrial revolutions that gave Europe world dominion. A thousand years ago, the Chinese were well ahead of anyone else and certainly of Europe. Some would argue that this superiority held for centuries thereafter. Why, then, did China 'fail'?

Some China scholars would mitigate the pain by euphemism: 'Chinese society, though stable, was far from static and unchanging ... the pace was slower ... the degree of change less ...'.[32] (True, but the issue remains.) Others would dismiss the question as unanswerable or illegitimate: unanswerable because it is said to be impossible to explain a negative (this is certainly not true in logic; the explanation of large-scale failure and success is inevitably complicated, but that is what history is all about); and illegitimate because where is the failure? The very use of the word imposes non-Chinese standards and expectations on China. (But why not? Why should one not expect China to be interested in economic growth and development – to be curious about nature and to want to understand it – to want to do more work with less labour?) The earlier successes of China in these respects make these questions the more pertinent and acute.[33]

What about the relations between science and technology? Did the one matter to the other? After all, science was not initially a major contributor to the European industrial revolution, which was built largely on empirical advances by practitioners. What difference would it have made to Chinese technology if science had slowed to a crawl by the seventeenth century?

The answer, I think, is that in both China and Europe, science and technology were (and are) two sides of the same coin, two manifestations of a common approach to problems and experience. The response to new knowledge of either kind is of a piece, and the society that closes its eyes to novelty from one source has already been closing them to novelty from the other.

In addition, China lacked the institutions that made for a cumulative process of finding and learning – the schools, the academies, the learned societies, the challenges and competitions. The sense of give-and-take, of standing on the shoulders of giants, of collective as well as individual achievement, of an inherited but ever imperfect treasure, of progress – all of these were weak or absent. And this is another paradox. On the one hand, the Chinese formally worshipped their intellectual ancestors; in 1734 an Imperial decree required court physicians to make ritual sacrifices to their departed predecessors.[34] On the other hand, they showed a deplorable tendency to let the findings of each new generation slip into oblivion, to be recovered perhaps at a later date by antiquarian and archaeological research.[35]

The history of Chinese advances, then, is one of points of light, separated in space and time, unlinked by replication and testing, obfuscated by metaphor and pseudo-profundity, limited in diffusion (since China had nothing comparable to European printing) – in effect, a succession of ephemera. Much of the vocabulary was invented for the occasion and fell as swiftly into disuse, so that scholars today spend much of their effort trying to decipher these otherwise familiar ideograms. Much thought remained mired in metaphysical scepticism and speculation. Here Confucianism, with its easy disdain for scientific research, which it disparaged as 'interventionist' and superficial, contributed its discouraging word. 'With the microscope you see the surface of things ... But do not suppose you are seeing the things in themselves.'[36]

The effect was discredit or indifference, the greater for the want of mutual verification and support. It is this want of continuing intellectual exchange and reinforcement, this subjectivity, that more than anything explains the uncertainty of gains and the easy loss of impetus. Chinese savants had no way of knowing when they were right; it is subsequent research, mostly Western, that has discovered and awarded palms of achievement to the more inspired.

Small wonder that China reacted so unfavourably to European imports. European knowledge was not only strange and implicitly belittling. In its ebullience and excitement, its urgency and competitiveness, its brutal commitment to truth and efficacy (Jesuits excepted), it went against the Chinese genius.

So the years passed, and the decades, and the centuries. China saw Europe leave it far behind. At first it was unbelieving and contemptuous. Later it became increasingly anxious and frustrated. From asking and begging, the Westerners became insistent and impatient. The British saw two embassies dismissed. The third time, in 1839, they blew the door down. Other Western nations followed suit, and then, after the Meiji restoration, the Japanese, with their own pretensions to dominion, moved to secure their place alongside Britain, France, Germany and Russia.

NOTES

1. In A.D. 885, all professional copyists in Baghdad were required to swear an oath not to copy books of philosophy. And although the Arabs were quick enough to use paper, they refrained from printing for almost three centuries after its introduction in Europe. On the conflicts of Muslim science and Islamic doctrine, see Pervez A. Hoodbhoy (1991), *Islam and Science: Religious Orthodoxy and the Battle for Rationality*, London and Atlantic Highlands, NJ: Zed, especially chapters 9 and 10.
2. Lynn White, jr. (1978), *Medieval Religion and Technology*, Berkeley: University of California Press, p. 227.
3. Eric L. Jones (1987), *The European Miracle: Environments, Economies and Geopolitics in the History of Europe and Asia*, Cambridge: Cambridge University Press, 2nd edition, p. 177; Bernard Lewis (1982), *The Muslim Discovery of Europe*, New York: W.W. Norton, p. 199.
4. Lewis, *The Muslim Discovery*, pp. 195–6.
5. Cf. Tapan Raychaudhuri (1982), 'Non-Agricultural Production: Mughal India', in T. Raychaudhuri and Irfan Habib (eds), *Cambridge Economic History of India, vol. I: c. 1200–c. 1750*, Cambridge: Cambridge University Press, p. 295: 'If necessity is the mother of invention, its pressure in the Indian case was not insistent.'
6. Ibidem, pp. 286–7.
7. F. Buchanan (1807), *A Journey from Madras ...*, cited in Raychaudhuri, 'Non-Agricultural Production', p. 291. Raychaudhuri agrees.
8. On these caravans of the Banjara (a nomadic caste), see Irfan Habib (1963), *The Agrarian System of Mughal India (1556–1707)*, London: Asia Publishing, p. 62, and Joseph J. Brennig (1990), 'Textile Producers and Production in Late Seventeenth Century Coromandel', in Sanjay Subrahmanyam (ed.), *Merchants, Markets and the State in Early Modern India*, Delhi: Oxford University Press, pp. 68–9. As Brennig puts it, 'time was of little importance'.
9. This ferruginous temper was an old story in Europe. Jean Gimpel (1976), *The Medieval Machine: The Industrial Revolution of the Middle Ages*, Harmondsworth: Penguin Books, pp. 65–6, cites the number and variety of nails kept in store: half a million in Calais in 1390; tens of thousands in a dozen different sizes (listed with their prices) at York Castle

in 1327. The specialization of nails by use is indicative of the sophistication of this technology.

10. K.N. Chaudhuri (1978), *The Trading World of Asia and the English East India Company 1660–1760*, Cambridge: Cambridge University Press, pp. 273–4. Cf. François Bernier ([1670–72], 1981), *Voyage dans les états du Grand Mogol*, Paris: Fayard, p. 168.

11. Cited in Frank Welsh (1993), *A Borrowed Place: the History of Hong Kong*, New York: Kodansha International, p. 33.

12. Mark Elvin (1973), *The Pattern of the Chinese Past. A Social and Economic Interpretation*, Stanford: Stanford University Press, p. 85. Elvin gives the figure as 'between 35,000 to 40,000 tons and 125,000 tons', but prefers the higher estimate. He relies here on Yoshida Mitsukuni, a Japanese specialist writing in 1967. Subsequent work by Robert Hartwell, 'Markets, Technology, and the Structure of Enterprise', p. 34, also advances the higher figure. In John A. Hall (1986), *Powers and Liberties: the Causes and Consequences of the Rise of the West*, Harmondsworth: Penguin and Oxford: Basil Blackwell, p. 46, this becomes 'at least 125,000 tons'. In this regard, Elvin, p. 285, quotes a description by Yen Ju-yu of iron works on the Hupei/Shensi/Szechwan borders with blast furnaces 18 feet high, using charcoal and hand-operated bellows (more than ten persons working in relays) and working continuously. The iron was apparently used for castings, and there is no indication of further refining either as wrought iron or as steel.

13. Elvin, *The Pattern of the Chinese Past*, pp. 297–8.

14. The imperial authorities vacillated in their attitude to foreign trade, now favouring it, now clamping down; and these tergiversations were in themselves a deterrent to stable enterprise and capital accumulation. In addition, even when the state relented, it did so in circumstances that pushed the traders into illicit operations. Thus the early Mongol (Yuan) dynasty (1280–1368) allowed freedom of enterprise, but then succumbed to the temptation of instituting a licensing system. This enabled officials to play the rôle of capitalist, financing venturers and dividing profits 70–30: 70 for the official, 30 for the working trader. That was greedy; compare the European 50–50 split. The traders presumably sought to conceal gains, but in the long run trade had to suffer.

15. Etienne Balazs (1968), *La bureaucratie céleste: recherches sur l'économie et la société de la Chine traditionelle*, Paris: Gallimard, pp. 22–3, translated by the present author.

16. Elvin, *The Pattern of the Chinese Past*, pp. 224–5. On the economic significance of nationalism, cf. Liah Greenfeld (1995), 'The Worth of Nations: Some Economic Indications of Nationalism', *Critical Review*, **9** (4), 555–84.

17. Max Weber argues that the administrative bureaucracy was undermanned, so that the government came to know and to respond to changes only after they had got under way. Hence a pattern of 'intermittent and jerky' homeostatic interventions. Max Weber (1951), *The Religion of China: Confucianism and Taoism*, Glencoe, Ill.: Free Press, p. 134, cited Hall, *Powers and Liberties*, p. 41.

18. Anonymous, cited in White, *Medieval Religion and Technology*, pp. 245–6.

19. These Portuguese sailors of the sixteenth century were of course not the first Europeans to make their way to China. The best known of the earlier visitors is Marco Polo, who came in the thirteenth century from Venice, then the richest city in Europe, yet thought it a small town by comparison with what he saw in Cathay. See Marco Polo (1965), *The Travels*, transl. by Ronald E. Latham, Harmondsworth: Penguin.

20. From Welsh, *A Borrowed Place*, p. 16, who quotes without reference.

21. From a letter to the French minister Colbert, undated but of 1675, my translation. David S. Landes (1983), *Revolution in Time: Clocks and the Making of the Modern World*, Cambridge, MA: Harvard University Press, p. 45.

22. Jonathan Spence (1974), *Emperor of China: Self-Portrait of K'ang-Hsi*, New York: Knopf, p. 74. On this story, see Carlo M. Cipolla (1967), *Clocks and Culture, 1300–1700*, New York: Walker; also Landes, *Revolution in Time*, chapter 2.

23. Alain Peyrefitte (1992), *The Immobile Empire*, New York: Knopf, p. 157.

24. Students of the history of Chinese technology and science, most notably Joseph Needham and his team, have made much of Chinese priority in discovery and invention, pushing the origins of important techniques and devices far back, well before their appearance in Europe. They see this, quite properly, as a sign of exceptional creativity and precocity, but they would do better then to ask why the subsequent retreat and loss.

25. On all of this the best source is still Carlo M. Cipolla (1965), *Guns, Sails, and Empires: Technological Innovation and the Early Phases of European Expansion 1400–1700*, New

York: Pantheon, especially pp. 116–9. Cipolla is not a sinologist and had to rely exclusively on European sources, including the testimony of Christian missionaries and travellers; but his 'global vision' gives him crucial insights that are missing in the specialist literature.
26. Mu Fu-sheng (1962), *The Wilting of the Hundred Flowers: Free Thought in China Today*, New York: Heinemann, pp. 76–7, cited in Cipolla, *Guns, Sails, and Empires*, p. 120.
27. René Taton (ed.) (1964), *A General History of the Sciences, II: The Beginnings of Modern Science from 1450 to 1800*, London: Thames and Hudson, p. 592.
28. Ibidem, p. 590.
29. The curse of foreignness, however, remained. In a letter of November 1640, the Jesuit von Bell wrote: 'The word *hsi* [Western] is very unpopular, and the Emperor in his edicts never uses any word than *hsin* [new]; in fact the former word is used only by those who want to belittle us.' Ibidem, p. 589 n. 1.
30. Cf. Patricia Crone (1989), *Pre-Industrial Societies*, Oxford: Blackwell, pp. 172–3: 'China is a star example of a successful civilization ... China reached the pinnacle of economic development possible under pre-industrial conditions and stopped: no forces pushing it in a different direction are in evidence ...'.
31. This is one of the major contributions of Peyrefitte's book, *The Immobile Empire*. Because he gained access to the Chinese archives, including papers read and annotated by the emperor, Peyrefitte is able to show the inner workings of bureaucratic equivocation.
32. John King Fairbank and Edwin Oldfather Reischauer (1965), *East Asia: the Modern Transformation*, London: Allen and Unwin, p. 291, cited in Harry T. Oshima (1987), *Economic Growth in Monsoon Asia: A Comparative Survey*, Tokyo: University of Tokyo Press, p. 34. Fairbank and Reischauer suggest that the reason for Chinese 'stability' was 'the very perfection that Chinese culture and social organization had achieved by the thirteenth century'. The contrast with Europe, roiling with imperfection, could not be sharper.
33. Cf. Eric L. Jones (1990), 'The Real Question about China: Why was the Song Economic Achievement not Repeated', *Australian Economic History Review*, **30**, 5–22, pp. 8–9, who is equally nonplussed by these *a priori* objections to this line of inquiry: 'I cannot see why; [the sinologists] are not being blamed and the question does not seem tendentious with respect to a society that had achieved so much and then passed so many centuries without achieving it again.'
34. Taton, *General History of the Sciences, II*, p. 590.
35. This is true in spite of considerable effort to collect knowledge and present it in encyclopaedias. One such encyclopaedia, really a kind of anthology, may well have been the biggest project of its kind ever attempted: it ran to 800,000 pages (Jonathan Spence (1990), *The Search for Modern China*, New York: W.W. Norton, p. 86). But a plethora of encyclopaedias is a bad sign: like still photographs, they are an effort to fix knowledge at a point of time. They are useful as reference works, especially for historians, but they can impede free inquiry.
36. From a poem, early nineteenth century, by the son of the then Prime Minister – himself a high state dignitary – quoted in Taton, *General History of the Sciences, II*, p. 593. Of course, when the time came, one could find support in Confucianism for other positions. That is the nature of sacred writ: one can quote it to one's purpose. This does not prevent people from using it to bad purpose.

3. The Political Economy of Technological Change: Resistance and Innovation in Economic History

Joel Mokyr

INTRODUCTION

If technology is one of the prime movers of economic growth, how exactly are technological decisions made? A technological decision is taken whenever a new technique is proffered, and firms have to decide whether to adopt it or not. It might seem that in the vast majority of cases this decision is trivial: if the new technique increases efficiency and profits it will be adopted, otherwise it will not. But few economies have ever left these decisions entirely to the decentralized decision-making processes of competitive firms. There usually is, at some level, a non-market institution that has to approve, license, or provide some other *imprimatur* without which firms cannot change their production method. The market test by itself is not always enough. In the past, it almost never was.

Much as economists might deplore the fact, therefore, the acceptance of innovation is more than an economic phenomenon, and certainly far more than a pure advance in productive knowledge. The concept of competition remains central here, but it is not the neo-classical concept of price competition of firms in the marketplace. Rather it is Schumpeter's concept of competition between different techniques struggling to be adopted by existing firms, or between different final products 'slugging it out' over the consumer's preferences. At times individual techniques may be identified with a firm, but often techniques struggle for adoption within a *single* organization. How are these decisions made? Could it be that even when a new and superior technology is made available at zero marginal costs, the economy to which it is proposed may choose to reject it?

New technologies have failed and opportunities have been missed despite their ostensible economic superiority. The idea that seemingly superior inventions are spurned or rejected is hardly new. In 1679, William Petty wrote that

> Although the inventor often times drunk with the opinion of his own merit, thinks all the world will invade and incroach upon him, yet I have observed that the generality of men will scarce be hired to make use of new practices, which themselves have not been thoroughly tried ... for as when a new invention is first propounded, in the beginning every man objects, and the poor inventor runs the gantloop of all petulent wits ... not one [inventor] of a hundred outlives this torture ... and moreover, this commonly is so long a doing that the poor inventor is either dead or disabled by the debts contracted to pursue his design.[1]

Petty speaks of scepticism and doubt. When radically novel technological ideas are first proposed, a normal reaction is that it will not work because otherwise we would have thought of this ourselves. Yet throughout history technological progress has run into another, more powerful, foe: the purposeful, self-interested resistance to new technology. Without an understanding of the political economy of technological change, the historical development of economic growth will remain a mystery to us. In the very long run, technological progress in its widest sense remains indispensable to sustainable economic growth. Of course, the failure to adopt a new technology can have many reasons: new technology is often embodied in expensive capital goods; it often requires scarce complementary factors such as infrastructural capital or a highly-skilled labour force. Yet outright resistance is a widely observed historical phenomenon.[2] Precisely because such resistance must work outside the market and the normal economic process, artificial distinctions between the 'economic sphere' and the 'political sphere' for this class of problems are doomed. The adoption of a wholly new technology is often the target of long debates and public discourse, unlike many other technical and economic choices. The role of *persuasion* and rhetoric in these decisions is something economists have paid scant attention to, and hence they have not had much success in understanding why, for example, some economies have adopted nuclear power or why some have allowed experimental drugs to be sold and others have not. Furthermore, not all resistance is purely social. There are instances in which the technological 'system' resists a novel and improved component because it does not fit the operation of the whole.

Technological inertia in many societies has often been ascribed to irrationality, technophobia, a blind adherence to traditional but outmoded values and customs. Yet as Timur Kuran has shown, conservatism and rationality are not always mutually exclusive.[3] In what follows, I hope to establish two basic propositions. One is that inertia is usually a characteristic widely observed in complex systems that follow an evolutionary dynamic. The second is that technological inertia is usually the outcome of rational behaviour by utility-maximizing individuals, and that we do not have to fall back on differences in preferences to explain why some societies are more amenable to technological change than others.

INERTIA AND EVOLUTION

Many scholars have recognized that new techniques emerge in a manner that is in some ways analogous to the emergence of new species and variations on existing ones in the evolution of living beings.[4] The choice of techniques is akin to the process of natural selection; natural selection is really a metaphor for an impersonal process in which no concrete entity actually does the selecting. New technologies are similarly selected (although here, at least in some cases, the selecting is done by conscious individuals making deliberate choices). The market is of course one arena in which this selection takes place; the political sphere is another. 'Fitness' in this context does not mean just technical superiority but also political power and social preferences, based on beliefs, rhetoric, prejudice, marketing, propaganda and bargaining.

With reference to the natural world, the concept of fitness is well understood. Yet the selection of new life-forms of superior fitness is the exception rather than the rule. Despite the seemingly unbelievable diversity of life-forms, actual phenotypical change is quite unusual and runs into many barriers. That natural selection is inherently a *conservative* process was first understood by Alfred Russel Wallace, who likened natural selection to a governor on a steam engine, essentially a device to correct deviations automatically.[5] System externalities have an equivalent in biology known as 'structural constraints'. Genetic material is transmitted in 'packages' and thus sticks together. At the same time, evolution tends to be localized and cannot change too much at once. As François Jacob put it in a famous paper, evolution does not so much create as tinker: it works with what is available, odds and ends, and much of it involves therefore minor variations on existing structures.[6]

Furthermore, the emergence of new species (speciation), analogous to the emergence of new techniques, is both rare and poorly understood. Although the resistance to change in natural systems is of an entirely different nature from that in technological systems, it too implies a cohesive force that limits the amount and rate of change. Stability in the systems of living beings is maintained by what biologists term genetic cohesion. While not wholly understood, this stability is the key to the dynamics of evolutionary systems. Those that are too conservative will end up in complete stasis; too much receptivity to change will result in chaos.[7]

In the economic history of technology, too, change occurs despite the formidable forces pitted against it. Radically new technological ideas, from antibiotics to nuclear power to telegraphy, have emerged time and again despite the systemic inertia from which the technological domain seems to be suffering. The dynamic may be similar: a complex system which struggles to change against built-in inertia is more likely to change in sudden bursts than in slow, continuous fashion. The idea of 'punctuated equilibria' in evolutionary change can be projected to historical processes to cast light on the question why so much of historical change occurs in concentrated spurts of intense technological activity such as the British Industrial Revolution.[8] Most recent research in modern evolutionary biology suggests that the dynamic of evolution, too, proceeded in intensive spurts separated by long periods of stasis rather than in linear progressions.[9] At times, it seems, more resistance to technological change might have reduced costly experimentation, needless multiplication of standards, and continuous model changes. An example is the frenetic rate of change in the US aeroplane industry in the years after World War I, in which a feverish drive for continuous progress probably slowed down the emergence of a mass-produced standard plane which would have reduced the cost of flying.[10] Clearly, then, not *all* resistance to technological change is necessarily socially detrimental.

The analogy with evolutionary biology underlines the rather unlikely nature of continuous technological progress. Technology may be said to be subject to *technological cohesion*, meaning that on the whole technological systems will be stable and inert. It could be the case, of course, that the agents of change, whether they are mutations in DNA or new ideas occurring to people, are themselves highly non-linear in their frequency. It is more plausible, however, to assume that changes in 'mutagens' are relatively rare and that mutations occur at more or less uniform rates but are constrained by the inertia and resistance to change within the system. The likelihood of change taking place depends on the outcome of the struggle between novelty, thirsting for a chance to take its place, and the old, fearful of any threat to the status quo. As Wesson has pointed out, 'the most important competition is not among individuals and their lineages, but between new forms and old. The old must nearly always win, but the few newcomers that score an upset victory carry away the prize of the future.'[11] This paragraph, written as a comment on Darwinian evolution, mirrors the one written decades earlier by Schumpeter: 'In capitalist reality, as distinguished from its textbook picture, it is not [price] competition which counts but the competition from the new commodity, the new technology ... which strikes not at the margins of the profits of the existing firms but at their ... very lives.'[12] Schumpeter believed that pure competitive capitalism ensured that cases in which a superior technology would be rejected would be rare, but he also understood the fragility of capitalism in a democratic society.

In the context of such a struggle between the *status quo* and novelty, the non-linear dynamic of historical evolution becomes more plausible. The technological *status quo* will create barriers that make it more difficult for new ideas to catch on, and at times may succeed in rigging the decision-making process so that novelty becomes almost impossible. Once these dams are broken, however, the torrent of innovation may be unstoppable, at least for a while. When the political arguments are cast not in terms of the perceived costs and benefits of the new technology itself, but rather in terms of the rules that are to be followed in making these decisions, such non-linearities become understandable.

The story becomes considerably more involved but also richer if we regard not only technology but also institutions as subject to evolutionary forces. Douglass C. North has stressed the idea that institutions evolve in that their dynamic can be described by stochastic shocks subject to selective filters, even if not all the implications of this approach have been fully explored.[13] What we have, then, is two evolutionary systems, one epistemological (technology) and one political and social (formal institutions, customs, and other informal rules of behaviour) that co-evolve over time.[14] An example is the emergence of American industrial capitalism after the Civil War, in which the technology of interchangeable parts and mass-production assembly lines implied an enormous growth in the optimal scale of much of the manufacturing. This technology co-evolved with changes in the structure of business institutions, including the emergence of the modern hierarchical business corporation and of labour unions, and the growth in efficiency and scope of capital and labour markets.[15] Such a continuous, interactive co-evolution means that if a foreign technology was transplanted into a society where the adapted institutions had not evolved jointly, serious incongruities and disruptions could be the result. The consequent resistance to technological change can, in this fashion, be reinterpreted in a wider context as a defence mechanism against the possibly chaotic consequences of disruptions in the fabric of a traditional society.

SYSTEMIC RESISTANCE

A different reason why society might resist innovation has to do with cross-technique spillover effects. As we have seen, *all* evolutionary systems have some source of resistance to change, or else they might collapse into the indeterminacy which Kauffman calls his 'supracritical region'.[16] Yet the technological choices offer some sources of inertia that are not found in nature. Unlike biology, production technology can mould its own selection environment by the development of rules of behaviour which evolve spontaneously but the purpose of which is presumably to preserve the *status quo* and to protect existing firms. Nelson points out that such action may be central in determining what design or system becomes dominant.[17]

Technology, too, occurs in 'systems' – that is, components that are changed will have effects on other parts with which they interact. This implies that a change in technique from T_0 to T_1 is likely to affect costs through unintended consequences to other components. Many of these occur through a variety of externalities or network effects: electrical equipment, trains, software, telephones, farming in open field agriculture, and all mechanical devices using interchangeable parts, all shared the problem of interrelatedness. In order to work, they require a uniformity which we call standardization, and thus single members cannot change a component without adhering to the standards. Yet here, too, the analogy can be pressed too far: in technology – but not in nature – we can invent 'gateway' technologies in which the incompatibilities are overcome, including for instance electrical adaptors from 115V to 220V or railroad cars with adjustable axes that travelled on different gauges. Positive feedback traps can occur in technological systems, but tend to be rare in open economies because of competitive pressures from outside. Yet they do occur: American colour television has been 'stuck' now for decades at a low-quality screen (of a low definition). For a long time, IBM-based computers struggled with the often paralysing constraint of 640 K RAM in 'conventional memory', the undoing of computer games and of many multi-media applications. In both cases it has turned out to be costly and tricky but not impossible to devise a 'gateway' solution.

The complementarities involved (broadcast–reception in the case of television; software–hardware in the case of computers) are characteristics of one of the most widely occurring sources of technological inertia in history: frequency dependence.[18] Frequency dependence occurs when a new technique cannot be successful until it is already adopted by a sufficiently large number of users. Similarly, in natural selection, new species cannot propagate unless their members can mate with sufficiently similar creatures. This kind of model sounds almost discouraging, since in its strictest sense it means that only success succeeds, a blueprint for total stasis. As noted, in *some* cases such hurdles could be and have been overcome, but in normal situations new technological ideas that might on the face of them work well do not 'catch on' and eventually vanish without a trace. IBM's OS/2 operating system, much superior to MS/DOS, was rejected because it was not sufficiently 'compatible', as were DAT tape players and Beta-system VCRs.[19] A special case of frequency dependence is learning by doing, where average costs decline with cumulative output. It is not always possible to know exactly how important these learning effects would have been in products that never made it into mass production. They are the subject of an experiment never performed. Would airships have become safe and fast (in addition to being quiet and fuel-efficient) had the world of aviation not switched to fixed-wing aircraft in the inter-war period? Would small mass-produced 'flivver' personal planes have dominated the civilian air travel market if their production had been pursued vigorously? If Volkswagen and Toyota had tried to implement a steam engine in their mass-produced models, would steam cars have been perfected to the point where they could have put up as good competition to the four-stroke internal combustion engine as the Diesel engine? Could the same be said for two-stroke engines, Wankel engines, and so on?

RULES AND RESISTANCE

I now turn to a brief analysis of the political economy of technological change. Any change in technology leads almost inevitably to an improvement in the welfare of some and a deterioration in that of others. To be sure, it is *possible* to think of changes in production technology that are Pareto superior, but in practice such occurrences are extremely rare. Unless all individuals accept the 'verdict' of the market outcome, the decision whether to adopt an innovation is likely to be resisted by losers through non-market mechanism and political activism.[20] Two recent books dealing with social response to technology, while totally different in tone and background, implore social scientists to pay more attention to the question of resistance to the seemingly inexorable march of new technology.[21] One important distinction should be made between the introduction of a totally new invention in the economy in which it originates, and the transfer of existing technology into new places after it has already been practised and tried elsewhere. In both cases resistance may emerge, but its nature may differ substantially between the two. Either way, however, markets judge techniques by profitability and thus, as a first approximation, by economic efficiency. How, then, does conflict occur?

To simplify matters, let me define the adoption of a new technique as a binary process: either it is adopted or it is not. Each individual has a set of idiosyncratic exogenous variables (preferences, age, endowments, education, wealth, and so on) which lead him or her either to 'support' or to 'object to' the innovation. To reach this decision, society follows what I shall call an aggregation rule, which maps a vector of *n* individual preferences into a <0,1> decision. This aggregation rule may be a market process (as would be the case in a pure private economy) but this would be a very special case. The pure market outcome is equivalent to an aggregator which weights preferences by their income. The optimality of the outcome will vary with the income distribution even for the market aggregator.

I should like to suggest that the argument about a new technique is conducted at two levels. One is an argument about the nature of the aggregators that make the decision. Should there be licensing of new techniques; how is the patent office to judge novelty; to what extent can production be codified in official rules? A second level of discourse occurs once the institutions exist. Only in a pure market aggregator is there no room for politics to enter the decision-making process. To start with, different groups in the economy favour different aggregation rules. In the terminology of the new Historical Institutional Analysis, an aggregator is an *institution*, that is, a non-technologically determined constraint on economic behaviour.[22] If the market outcome rules in favour of one group, another might find it in its interest to circumvent the market process. If the supporters and opponents of the new technique could form separate societies, the optimal outcome would be to separate them. Because they cannot, and one of them has to live with the undesirable outcome, the struggle consists of the attempt of the members of each group to set up an aggregation rule (for example, the market) that is most consistent with its interests.[23]

An example would be when the majority of poor people resist an innovation attractive to wealthy consumers. In that case there will be a difference between the market, in which 'votes' are weighted by purchasing power, and the democratic process, where each person has one vote. In decisions about technology, at least, there could be a serious conflict between democracy and continuous innovation.[24] In other words, contrary to the optimism of free market advocates in the Friedman tradition, it may well be that democratic decision processes do not maximize the long-term economic welfare of economies. This dilemma is faced by democratic countries which wish to undergo rapid development, as has long been recognized.[25] The technological decision-making in democratic societies is clearly inefficient, but the experience of the twentieth century, at least, is that totalitarian societies by and large do even worse.[26] In so far as technological decision-making is made in the political market, there is no reason to believe that the decisions will be in any definable sense efficient – we are strictly in worlds of second- and third-best.

A major reason why people tend to remove the market as the sole arbiter of technological decisions and to delegate part of the decision-making process to political bodies is that markets effectively truncate preferences over technology at zero. If one supports a new technique, one can vote 'yes' by buying the new product or switching to the new technique. By not buying the product or refusing to switch, one can express indifference or dislike, but individuals have no control over what others do even if they feel it might affect them. In markets it is difficult to express a 'no' vote.

Another reason is that so much technology is part of the public sector: transport, public health, infrastructure, and the military require political approval of changes simply because these are sectors in which some form of prior market failure has been observed. In his classic article 'Gunfire at Sea', Elting Morison has described the resistance put up by the navy to the introduction of continuous-aim firing in the U.S. Navy in the first decade of our century. In this case the resistance was overcome by the officer in question appealing directly to President Roosevelt over the head of his immediate superiors and the officers in charge of the Bureau of Navy Ordnance.[27] Market failure occurs when network- or other forms of externalities are often observed, and hence the resistance to technology is often most pronounced when there is already another reason to take decision-making out of the hands of private enterprise.

Above all, consumers seem to distrust the free market as an arbiter of new technology just because it is new. Whereas in a technologically static economy there may be no reason to distrust the invisible hand, the informational asymmetries and irreversibilities associated with the generation and adoption of new techniques seem to demand a cool and unbiased arbiter. It is feared that greedy entrepreneurs will sell asbestos-type products to the public and then abscond. Thalidomide-type disasters, however minor when compared with the benefits of advances in medical technology, produce a constant demand for government assurances that new products and techniques are safe. At the same time it needs stressing that not all resistance to technological progress is necessarily conservative and in defence of some technological *status quo*. Many cases of social resistance to a new technique occur because there are two alternatives to T_0, T_1 and T_2. If the decision is left to the market, T_1 will be chosen; if some interest group wishes to use non-market mechanisms to bring about some alternative T_2, it is the *nature* of technological change they wish to influence, not its very existence.[28]

Formally, we may distinguish between the following decision rules. G_M, which is the pure market aggregator, means that the new technology will be adopted by profit-maximizing firms following exclusively the dictates of the market. G_D is a decision rule that designates an *authorized subset* such as a representative parliament or a panel of technical experts, a violent mob, a court, or a single dictator, to decide whether to permit and/or support the new technology. G_V is a voting rule, say one-person-one-vote, in which a new technology is voted in or out by some kind of referendum. In most realistic situations the actual decision rule or aggregator which maps individual preferences to the decision space $<0,1>$ is

$$G = \alpha G_M + \beta G_D + (1-\alpha-\beta)G_V \text{ where } \alpha+\beta \leq 1.$$

The pure market outcome occurs only when $\alpha = 1$. The social decision process may thus be viewed as consisting of two stages. First, society determines the political rules of the game, that is, it sets α and β. Then, depending on the aggregator chosen, it determines whether the new technique will be adopted or not. An obvious elaboration of the simple model is that one decision-maker may delegate decisions to another: the authorized subset can decide to hand things over to a referendum or leave it up to the market. An election, on the other hand, can appoint delegates who can either make technological decisions themselves or decide not to act at all, effectively leaving the decision to the market. The interpretation of α and β as probabilities or proportions of the 'cases' that are decided in one arena or another thus lends some intuitive meaning to G.

A great deal of political and social struggle involves not only the implementation of new technology itself, but the decision rules themselves, as it is reasonably believed that some decision rules favour one interest group more than another. Economists, in particular, are concerned by the size of α, that is, how much of the decision is left to the market and how much will be decided on by other aggregators. In part, the aggregator will be determined by the nature of the product: technological change in public goods and other areas of market failure will obviously be largely outside the market decision process; but there is a huge grey area of private goods where there is room for political action. It may be thought that societies will be more creative and technologically successful the larger α is, but this is by no means certain. It may well be that the free market, for reasons of its own, foregoes technological opportunities. For instance, the new technology may require unusually large capital spending or a co-ordination between existing firms that cannot be materialized without direct intervention. In that case, the government may step in to make up for the market failure. Pre-Revolution-ary France, especially, saw a great deal of government involvement in trying to encourage French entrepreneurs to accept British techniques.

When the aggregator has been decided upon, as long as $\alpha < 1$, so that – as is often the case – some non-market decision is necessary to approve the new technology, opposition occurs within given political structures, such as a courtroom or a Parliamentary Committee. Of course, many new technologies are too trifling to be the matter of public debate; one hears little of a public outcry over the switch, say, from spark-plugs to fuel injection or from dot-matrix to ink-jet printers. In those cases the decision will normally be delegated to the market. But when there are major technical choices to be made that involve public expenditures, complementary or substitute relations with other technologies, or other types of spillover effects, they will end up being judged by non-market criteria.[29] Similarly, uncertainty of any nature regarding possible externalities, especially when these concern public health and safety, almost invariably lead to a reduction of the market component in the aggregator. In those cases, political lobbying about the new technology is natural. The usual rules of political economy and collective decision-making by interest groups apply, with the additional complication that the introduction of a new technology is by definition a highly uncertain event, involving known and unknown dangers that play no rôle in, say, political decisions about tariff policy or public work procurements. Moreover, the technical and scientific issues are often highly complex and even a phrasing of the correct *questions* (let alone the answers) is often beyond the intellectual capability of decision-makers. Precisely for that reason, there is more reliance on the opinion of 'experts' but also, paradoxically, frequent appeal to emotions, fears, and religious and nationalist sentiments. As litigation becomes increasingly important, technological decisions are relegated to courts, and rhetorical imagery and other persuasive tools, from television advertisements to neighbourhood rallies, become a means by which technological decisions are made. Reliance on technical expertise, a long-standing practice in the West, is weakened by disagreements among experts and even disagreements as to *who* is an expert to begin with.[30]

An anti-technological and conservative bias can be built into a culture, so that the decision-making body becomes technologically reactionary. In this fashion, the technological *status quo* does not have to fight battles against hopeful innovations over and over again. This cultural bias can be introduced through an education system that fosters conformist values in which traditions are held in respect and deviancy and rebellion are made highly risky.[31] Morris lists the sources of technological reaction in traditional India: there was 'no organization for the propagation or dissemination of knowledge, and an unbridgeable social barrier between theorists and craftsmen'.[32] Eric L. Jones has argued that the Indian caste system was a deeply conservative and rigidified institution, in which ascriptiveness is pervasive and personal achievement 'was excluded in principle'. Jones realizes that a caste system, too, could never be an *absolute* constraint on economic growth, it 'may constitute an infuriating brake, yet it will not be able to switch off a motor located somewhere else in society'.[33] The argument made here is exactly about such brakes; societies with such brakes would develop much slower than those without. Perhaps that is as much as we will ever be able to say about economic growth.

The political battles over technology have profound implications for economic history. One is, as I have emphasized elsewhere, that technological progress in a given society is by and large a temporary and vulnerable process, with many powerful enemies whose vested interest in the *status quo* or aversion to change of any kind continuously threaten it.[34] The net result is that changes in technology, the mainspring of economic progress, have been rare and that *stasis* or change at very slow rates has been the rule rather than the exception. It is our own age, and especially the rapid technological change in the Western World, that is the historical aberration. Another implication is that most underdeveloped countries cannot take technology transfer for granted. Even when capital is available and when complementary inputs such as skilled labour and infrastructure are present, attempts to transplant technology from one society to another are likely to run into social barriers that economists may find difficult to understand. Before we can delve into the economic and social causes of resistance, I need to place its importance in a theoretical framework.

MARKETS OR POLITICS?

Although the terminology here is different, the concept of heterogeneous aggregators is closest to the concepts enunciated by Mancur Olson.[35] Consider for simplicity an economy that has to make a binary choice whether to adopt T_1 or not. While in a market economy such decisions are, of course, made by individuals, in most societies discontinuous and discrete changes in the main technique in use involve to some extent public decision-making. Patents have to be issued, environmental impact statements are filed, and often outright licences and support from some public authority are required.[36]

When, then, will opposition to the market as the arbiter of innovations emerge? One issue is that technology may appear *directly* in people's utility function. Such a concept may appear bizarre to economists, but not to sociologists or psychologists.[37] For economists, moreover, it has been deemed traditionally uninteresting to ascribe differences in behaviour to different utility functions. Historically, however, cultural and religious elements may have had a big influence on technological decision-making.[38] Technology is something profoundly unnatural, as Freud observed in his *Civilization and its Discontents*[39] when he compared it to an artificial limb.

While many societies encourage the search for new inventions, one author sighs, 'we remain, in part, appalled by the consequences of our ingenuity and ... try to find security through the shoring up of ancient and irrelevant conventions'.[40] Technology is regarded by many writers as something uncontrollable and incomprehensible and thus somehow evil in itself. The literature on this issue is rather large and I cannot do justice to it here.[41]

The distinction I drew above between indigenous technologies and imported technologies was of considerable historical importance. When a technology has never been tried before and is genuinely novel, there is a serious fear of the unknown, resulting from risk aversion or deeper fears of 'devils we do not know'. The resistance to a new technique has two separate roots: heterogeneous preferences and heterogeneous expectations. If there is a probability that a technique may mal- or dysfunction, people with high risk-aversion will resist it. Secondly, precisely because it is new and there are no exact precedents, people could disagree about the magnitude of the probability of a failure, so that even people with the same rate of risk-aversion would have different attitudes. In societies that adopt known and tried technological changes from other countries, such fears of the unknown are secondary, and the resistance is more likely to come simply from having observed the negative effects of a new technology elsewhere. But such 'learning' effects are relatively rare.[42] More likely are what we could call 'correlation effects': that is, technology is viewed and depicted as 'packaged' in a cultural–political deal that is undesirable even if the new technology in and of itself is desirable. This kind of ambiguity flavours much of the political argument in non-Western nations and is often coupled with a cultural suspicion of foreigners. There is a sense that 'the magical identity is development = modernization = Westernization'. Especially when new technology takes the form of new products, it is often regarded as correlated with undesirable cultural and social side-effects.

Self-interest, of course, counts too. Economists have used the term 'rent-seeking' for the replacement of market decisions by government control or some other form of collective decision-making that benefits a small group or individual. Here I expand the standard definition of rent seeking to include 'loss-avoidance'. Historically, most of the resistance to new technological change had economic reasons: potential losers set up obstacles to obstruct innovation. At this stage, I shall first assume identical utility functions and attribute the differences in opinion to observable parameters such as differences in information, economic costs, and endowments. Proposed technological changes are expected to benefit one segment of society and harm another; the market may determine one outcome, which could be circumvented by another aggregator. To start with, assume for the sake of argument simply that all utility functions contain only income as an argument, and that the only effect of the transition to the new technique is to increase total income so that the gains of the winners exceed the losses of the losers. This means that the invention is socially preferable, but the potential for conflict is only resolved if the gainers use part of their augmented incomes to compensate the losers. Compensation would seem at first glance a reasonable way to resolve the problem but in fact rarely occurs directly because of the formidable problems of identifying the losers, measuring the dimensions of their loss and overcoming the problems of moral hazard among losers as well as collective action among gainers. All the same, compensation does occur. The welfare and farm support systems in modern Western economies could be interpreted, at least in part, as mechanisms designed to compensate and placate groups that ended up with the short end of the stick in rapid industrialization and subsequent de-industrialization. If compensation does not occur, the losers will have an interest to band together to try to change the social decision rule from G_M to a rule that is more favourable to them. The way for them to do this is to circumvent the market, in our terms by reducing α, and then to try to affect the aggregators G_D and/or G_V by political action. It is in this fashion that persuasion and rhetoric enter the story; in a 'pure' market system, they need not be part of the debate. The main question is why technological change is income-reducing for some individuals. Below I provide a typology of some of the more obvious sources of purely rational resistance to innovation.

a. Unemployment

One obvious reason, widely believed since Ricardo's famous chapter on 'Machinery',[43] is that labour-saving technological change reduces the demand for undifferentiated labour, thus leading to unemployment and a possible decline in wages. As economists have long understood, this statement cannot be accepted in and of itself without working through the general equilibrium properties of an exogenous change in the production function. An invention that replaces workers by machines will have effects on *all* product and factor markets. An increase in the efficiency of production which reduces the price of one good will increase real income and thus increase demand for other goods; the replaced workers may find employment in other industries, and their real wage may go up or down. In an abstract general equilibrium world, without adjustment costs, in which all workers and productive assets can be costlessly converted from one usage to another, there is no *a priori* expectation that changes in production technology will necessarily reduce labour income and employment. In the real world, of course, temporary disequilibria can cause hardship to large subgroups of the population. Yet in some of the most widely studied instances, the feared patterns of technological unemployment did not materialize. Notwithstanding the long and intricate national debate about the 'Machinery Question' raised by Ricardo, nineteenth-century Britain did not suffer from the secular increase in structural employment feared by Ricardo and the Luddites alike.[44] In a very different environment, it was widely feared that the mechanization of agriculture in Asia in the 1970s would lead to widespread rural unemployment; this did not occur.[45] Recent studies by labour economists find that the introduction of new technology is on balance associated with positive job *growth*. One such study flatly declares that 'job growth and the introduction of new technology appear to be complements rather than substitutes. The Luddites were wrong'.[46] The danger here is one of overaggregation: it is likely that compensating fluctuations in labour demand in different sectors will spawn substantial resentment even if the total demand for labour is unchanged. The cost of making the transition is often non-negligible, and workers are likely to observe the decline in their own sector before they perceive better opportunities elsewhere.

b. Capital Losses

A different problem occurs when physical capital is of a 'putty-clay' variety; once shaped, it is difficult to convert to another use. This can be seen in a simple vintage model in which one product is produced by machines of different efficiency. The lowest-ranked machine earns a rent of zero; all other machines earn a rent that is proportional to the difference between the production cost of the least efficient machine in use and their own. The value of the asset can thus be determined by the present discounted value formula, in which the value of the asset is a function of this difference and *expected* future technological depreciation. A rise in the rate of technological change will reduce the market value of existing machines of older vintage and thus it might be expected that the owners of such machines will find a way to avert it if they can.

Yet in practice this rarely happens. The cases in which the owners of physical capital have fought against the introduction of new techniques are comparatively few. The reason must be that while the physical qualities of machines can only rarely be altered, capital goods – including ownership in patents – can be bought and sold.[47] Thus the owner of a set of machines that become obsolete will take a loss on those machines, but he can always buy into the new technology by buying the new machines that yield the higher profits through lower costs. This explains, for instance, the relatively weak resistance to the introduction of steam engines despite the huge locational rents that were being secured by the owners of water mill sites. Industrialists using water power might have been losing when their mills fell into disuse, but they could make up for those losses by buying into steam technology themselves, which is precisely what happened in Lancashire during the British Industrial Revolution. In those cases in which capital markets favoured some existing producers over others, however, this principle is violated and in such cases resistance is to be expected.[48]

c. Non-Pecuniary Losses

Another source of resistance to technological change is that it changes not just the *level* of average costs, but the overall *shape* of the cost function. While new technology thus reduces overall costs and increases efficiency, it may also change the minimum efficient size of the firm and the entry conditions to the industry. Thus, when the minimum efficient size of firms in the textile industry was hugely increased during the first Industrial Revolution, artisans and small domestic producers were effectively driven out of the industry. In a world without transaction and information costs and hence 'perfect' capital markets, the costs of these changes would be mitigated by small producers combining into large firms and exploiting some of the economies of scale. This did occur to a larger extent than is usually appreciated.[49] The so-called 'workshop system' in which workers hired space and a piece of equipment in a large building and worked on their own account without hierarchy and discipline was prominent in many industries until deep into the nineteenth century. All the same, during the British Industrial Revolution, even before the famous Luddite and Captain Swing disturbances, there were some riots by artisans and self-employed producers threatened by factories.[50]

Workers, moreover, care about non-pecuniary characteristics of the workplace, from safety and noise on the shop floor to job satisfaction and decision-making authority. If a new technology affects these characteristics negatively, workers will resist it unless they can be bought off by employers through fully-compensating wage increases or unless they can find new jobs similar to their old ones at zero cost to themselves. During the Industrial Revolution, a particular bone of contention was the attempt by employers to standardize products and reduce the leeway that artisans and domestic workers had in setting the parameters of the product. When the advantages of product standardization led to lower tolerance boundaries on the characteristics of output, from cotton cloth to musket balls, repeated attempts to enforce such standards ran into determined opposition.[51] Beyond that, technological change affects the regional distribution of production and employment, thus forcing workers to move from one region to another, or from a rural to an urban area. New technology is often felt to destroy traditional communities. For some members of those communities that counts for little, whereas others care about it a great deal; thus any kind of aggregator will lead almost inevitably to some subset of the population being dissatisfied.

d. Human Capital

The opportunities for conflict are much wider when we consider human capital.[52] Skills and experience are acquired over a lifetime, but the ability to learn new skills declines over the life cycle.[53] Workers beyond the student or apprentice stage can be expected to resist new techniques in so far as innovation makes their skills obsolete and thus irreversibly reduces their expected life-time earnings. The new technology may be inaccessible to them for more than one reason; factories require a willingness to submit to discipline and hierarchy that the independent artisan is too proud to submit to. It is no consolation to the older generation that their children may have no difficulty in adjusting to the new regime, mastering the new technique and thus improving their material standard of living. Again, the example of the British Industrial Revolution illustrates this point vividly. As the old domestic industries came increasingly under pressure from the more efficient factories, the older artisans by and large refrained from seeking employment in them; the reliance of factories on child and teenage labour was motivated by the ability of youths to learn the skills and adopt the docility required for the factory environment.[54] Some new technology was in fact deliberately designed to exclude males and favour women and children, as was the case in the early factories of the Industrial Revolution.[55]

The protection of skills and specific human capital is often combined with other forms of rent-seeking through the creation of barriers to entry and the control of output. This is clearly a widespread interpretation of the European craft-guild system which ruled urban artisans in many areas for many centuries. In pre-modern urban Europe these guilds enforced and eventually froze the technological *status quo*.[56] Similar phenomena, *mutatis mutandis*, occurred in China.[57] It is important to stress that many of these guilds were originally set up to fulfil different functions, acting as clearing houses for information, organizational devices to set up training, mutual insurance support organizations, and sincere attempts to prevent opportunism and free riding on others' reputations. Yet over time many of them degenerated into technologically conservative bodies.[58]

In most of Europe, then, craft guilds eventually became responsible for a level of regulation that stifled competition and innovation. They did this by laying down meticulous rules about three elements of production that we might term 'the three Ps': prices, procedures, and participation. As guilds gained in political power, they tried to weaken market forces as aggregators and tended increasingly to freeze technology in its tracks. The regulation of *prices* was inimical to technological progress because process innovation by definition reduces costs, and the inventor makes his profits by underselling his competitors. Regulating prices may still have allowed some technological progress because innovators could have realized increased profits by lowering costs even if they could not undersell their competitors. To prevent this, *procedures* stipulated precisely how a product was supposed to be made, and such technical codes, while originally designed to deal with legitimate concerns such as reputation for quality, eventually caused production methods to ossify altogether. Enforcing these procedures, however, was far more difficult than enforcing pre-set prices. Finally, and in the long run perhaps the most effective brake on innovation, there was *participation*: by limiting and controlling the number of entrants into crafts, and by forcing them to spend many years in apprenticeship and journeymanship, guild members imbued them with the conventions of the technological *status quo* and essentially cut off the flow of fresh ideas and the cross-fertilization between branches of knowledge that so often is the taproot of technological change.[59] A particularly pernicious custom was the rigid division of labour between craft guilds so that each guild was confined to its designed occupation, a practice that required royal intervention from time to time to prevent egregious abuses.[60] The exclusion of innovators by guilds did not end with the Middle Ages or even with the Industrial Revolution. In 1855, the Viennese guild of cabinetmakers filed a suit against Michael Thonet, who had invented a revolutionary process for making bentwood furniture. The *Tischlermeister* claimed that the inventor was not a registered cabinetmaker. The suit was dismissed when the court made his workshop an 'Imperial privileged factory'.[61] The rôle of the guilds can go some way in explaining the series of technological successes which we usually call the British Industrial Revolution, and why it occurred in Britain as opposed to in Europe, although clearly this was only one of many variables at work.[62] The one industry in which Britain fell behind technologically, even during its era of industrial triumph in the first half of the nineteenth century, was in watchmaking, where both labour and entrepreneurs were resistant to innovation.[63] Resistance was not confined to manufacturing; when large department stores were introduced into Germany in the later nineteenth century, following the French model of retailing technology, small shopkeepers banded together and were able to convince the major states in Germany to pass a special tax on large stores to protect the small merchants from the threat of modernization.[64]

Perhaps the arena in which the largest number of technological battles have been fought since the Industrial Revolution has been in free trade. Protection for domestic industries was often identical to protection for obsolete technology. While the battles against free trade and technological progress by no means coincide, their overlap is considerable, and free trade and an open economy are by far the best guarantees that an economy will use best-practice technology. This idea goes back at least as far as David Hume, who pointed out in 1742 that 'nothing is more favourable to the rise of politeness and learning than a number of neighbouring and independent states, linked together by commerce and policy. The emulation which naturally arises among those neighbouring states is an obvious source of improvement. But what I would chiefly insist on is the stop [i.e. constraint] which such limited territories give both to power and authority.'[65] At the same time, free trade was hardly a *necessary* condition: Britain remained protectionist until the 1840s, and the United States followed highly protectionist policies in the last third of the nineteenth century, yet both were highly open to innovation.[66]

In the twentieth century, resistance to new production technology has come in part from labour unions. There is no compelling reason why labour unions must always resist technological change: after all, as 'encompassing organizations', they ought also to be aware of the undeniable benefits that new technology brings to their members *qua* consumers.[67] The growth of the labour movement's power in Britain is often held responsible for the declining technological dynamism of post-Victorian Britain. Resistance from organized labour slowed down technological progress in mining, shipbuilding and cotton weaving.[68] Such resistance was not a hundred per cent effective, but Coleman and MacLeod may well be right when they judge that it 'reinforced the increasingly apathetic attitude of employers toward technological change'.[69] In printing, London's notorious Fleet Street earned a reputation for stormy industrial relations, where management's major preoccupation was with avoiding disruptions to production, even at the expense of high unit labour costs and restrictions on technological innovation.[70] The crisis in the Bombay cotton industry in the 1920s and 1930s, when Bombay lost much of its market share to other areas, is attributed to the militancy with which Bombay trade unions fought against a technical and administrative rationalization of cotton mill practices.[71] In a recent paper, Susan Wolcott documents in detail how Indian workers blocked the implementation of larger spindles in the cotton spinning industry, not only in Bombay but in Ahmedabad and Sholapur as well.[72]

In our own time, labour unions have been held responsible for impeding technological progress in many industries. In the European and American motor car industry, for instance, they have resisted the closing of outdated plants and the introduction of the flexible work practices and reduced job classifications that have increased the efficiency of Japanese car manufacturers.[73] Needless to say, not *all* unions have taken a consistently conservative stance against new technology: in post-1945 Sweden and

Germany, for example, unions were induced to join coalitions aimed at increasing productivity. These unions were large and encompassing groups, and as the Olsonian theory suggests, their membership benefitted enough from technological progress for the benefits to outweigh the costs.

e. Externalities

The non-pecuniary aspects of new technology raise particular concerns when there are 'external effects', that is, when new technology affects common resources. Much of the resistance by the environmental movement to super-fast railroads, nuclear power, and advanced pesticides, for instance, deals precisely with the non-income effects of technological change. Again, such non-pecuniary effects are valued differently by different individuals, and thus the outcome that political aggregators determine will differ from the market outcome. In the standard case of externalities, common resources are not priced at their marginal social cost. In a static economy, arrangements will often emerge that minimize such discrepancies. New technology compounds the transaction costs with information problems. Thus it is difficult enough to limit the use of *known* atmospheric pollutants, but far harder to enforce agreements when the damage is unknown or in dispute. Unknown effects on shared resources thus aggravate disagreement and political resistance to technological progress.

To conclude, then, there are good reasons for subgroups within an economy to try to dethrone the free market as the sole aggregator, that is, to prevent the competitive price mechanism by itself from determining which technologies will be adopted and which will not. This effort has been undeniably successful; almost everywhere some kind of non-marketing control and licensing system has been introduced that has some agency or group of experts approve new technology *before* it is brought to the market.

CONCLUDING REMARKS

One of the main re-discoveries of the new growth theory and recent thinking about economic development is the possibility of the poverty trap or multiple equilibria. Whereas the new Growth Theory has emphasized the importance of technological progress and its relation to capital formation, both human and physical, it has failed, thus far, to build a bridge to the 'New Institutional Economics' in looking at the kind of institutional set-up that is most conducive to technological progress. A plausible way to do this is to understand that resistance to progress can be based entirely on rational and selfish reasons and that there are therefore causes, deeply rooted in the logic of collective action, that impose an element of inertia and stasis on the technology used by any economy. Another way of thinking about the issues discussed in this paper is that it is possible for an economy to be 'stuck' at a low level of income because the institutions it has are somehow inappropriate for technological progress. Usually the literature has thought of institutions as affecting the allocation of resources or the formation of capital. What I have suggested here is that the suitability of such institutions to the successful adoption of new ideas is an important question. Simple economic models may be difficult to construct here, but by a combination of political economy with the lessons of economic history, some inferences about the causes and consequences of resistance and opposition to technological change can be drawn.[74] The deeper question is whether sustained economic growth is the exception and stagnation the default, or whether, as argued especially by E.L. Jones, economic growth is a natural condition for most economies, but more often than not political and cultural impediments drag an inherently dynamic economy into stagnation and poverty.[75] This debate may seem to some a bit like an argument whether a zebra is black with white stripes or the other way around. In either case, the political economy of technological progress must occupy its rightful place at centre stage.

NOTES

1. William Petty (1679), *A Treatise of Taxes and Contributions*, London: Obadiah Blagrave, p. 53. I am indebted to Patrick O'Brien for bringing this text to my attention.
2. For some historical detail, see Joel Mokyr (1994), 'Progress and Inertia in Technological Change', in John James and Mark Thomas (eds), *Capitalism in Context: Essays in Honor of R.M. Hartwell*, Chicago: University of Chicago Press, pp. 230–54.
3. Timur Kuran (1988), 'The Tenacious Past: Theories of Personal and Collective Conservatism', *Journal of Economic Behaviour and Organization*, **10**, 143–71.
4. For a recent summary see Joel Mokyr (1996), 'Evolution and Technological Change: a New Metaphor for Economic History?' in Robert Fox (ed.), *Technological Change*, London: Harwood Publishers, pp. 63–83.
5. Gregory Bateson (1979), *Mind and Nature: a Necessary Unity*, New York: Dutton, pp. 175–76.
6. François Jacob (1977), 'Evolution and Tinkering', *Science*, **196**, no. 4295 (June 1977), 1161–6.
7. Ernst Mayr (1991), *One Long Argument: Charles Darwin and the Genesis of Modern Evolutionary Thought*, Cambridge, MA: Harvard University Press, pp. 160–61. For a detailed argument along these lines, see Stuart Kauffman (1995), *At Home in the Universe: The Search for the Laws of Self-Organization and Complexity*, New York: Oxford University Press, p. 73.
8. Joel Mokyr (1990), 'Punctuated Equilibria and Technological Progress', *American Economic Review*, **80**, 350–54; Joel Mokyr (1991), 'Was There a British Industrial Evolution?', in J. Mokyr (ed.), *The Vital One: Essays Presented to Jonathan R.T. Hughes*, Greenwich, CT: JAI Press, pp. 253–86.
9. For an accessible summary, see Niles Eldredge (1995), *Reinventing Darwin: The Great Evolutionary Debate*, London: Weidenfeld and Nicholson.

10. C. Derek Johnson, 'Markets in the Air: the Development of American Aviation Culture, 1918–1934', unpublished paper, Northwestern University, 1997.
11. Robert Wesson (1991), *Beyond Natural Selection*, Cambridge, MA: MIT Press, p. 149.
12. Joseph A. Schumpeter (1950), *Capitalism, Socialism and Democracy*, 3rd edition, New York: Harper and Row, p. 84.
13. See Douglass C. North (1990), *Institutions, Institutional Change and Economic Performance*, Cambridge: Cambridge University Press, p. 87.
14. For a further discussion of biological co-evolution, see Geerat Vermeij (1994), 'The Evolutionary Interaction Among Species', *Annual Reviews of Ecology and Systematics*, 1994, 219–36. Whereas Vermeij's analysis deals primarily with interaction between two evolving species, there is no reason why it cannot be extended to larger groups.
15. Richard R. Nelson (1995), 'Recent Evolutionary Theorizing About Economic Change', *Journal of Economic Literature*, **33**, p. 64.
16. Kauffman, *At Home in the Universe*, p. 294. Kauffman conjectures that 'the enhanced diversity of goods and services can lead to a further explosion of the technological frontier ... if the social planner deems them useful to the king'.
17. Nelson, 'Recent Evolutionary Theorizing', p. 77.
18. For a recent survey of this literature, see Brian Arthur (1994), *Increasing Returns and Path Dependence in the Economy*, Ann Arbor: University of Michigan Press. See also Paul A. David (1992), 'Path Dependence in Economic Processes: Implications for Policy Analysis in Dynamical System Contexts', CEPR working papers, April 1992.
19. A famous but controversial example is the DVORAK keyboard, thought to be superior to the standard QWERTY system. See Paul A. David (1986), 'Understanding the Economics of QWERTY: the Necessity of History', in William N. Parker (ed.), *Economic History and the Modern Economist*, Oxford: Basil Blackwell, pp. 30–49; and S.J. Liebowitz and Stephen E. Margolis (1990), 'The Fable of the Keys', *Journal of Law and Economics*, **33**, 1–25.
20. See Allan C. Mazur (1993), 'Controlling Technology', in Albert Teich (ed.), *Technology and the Future*, New York: St. Martin's Press, p. 217.
21. Martin Bauer (ed.) (1995), *Resistance to New Technology*, Cambridge: Cambridge University Press; Kirkpatrick Sale (1995), *Rebels Against the Future: the Luddites and their War on the Industrial Revolution*, Reading, MA: Addison Wesley.
22. The terminology is borrowed from Avner Greif, 'Micro Theory and the Study of Economic Institutions Through Economic History', paper prepared for a symposium on Economic history at the 7th World Congress of the Econometric Society, Tokyo 1995.
23. Joel Mokyr (1992), 'Technological Inertia in Economic History', *Journal of Economic History*, **52**, 325–38; Mokyr, 'Progress and Inertia'.
24. The notion that democracy endangers technological creativity was particularly embraced by nineteenth-century reactionary writers opposed to the extension of the franchise, such as Sir Henry Maine, who argued that universal suffrage would have prevented most of the major technological breakthroughs of the Industrial Revolution. See Albert Hirschman (1991), *The Rhetoric of Reaction*, Cambridge: Harvard University Press, pp. 97–100, who adds that the argument was palpably absurd and was immediately proven to be so. Yet it is not impossible that democracy could under certain circumstances be *less* hospitable than other political régimes to technological progress.
25. For an interesting discussion which concludes firmly that 'democracy entrenches economic freedoms, and in doing so underpins growth', see 'Why Voting is Good for You', *The Economist*, 27 August 1994, pp. 15–17.
26. See Barbara Ward (1964), *India and the West*, New York: W.W. Norton, pp. 150–52.
27. Elting E. Morison (1966), 'Gunfire at Sea; A Case Study of Innovation' in E.E. Morison (ed.), *Men, Machines, and Modern Times*, Cambridge, MA: MIT Press, pp. 17–44.
28. This is what sets aside the literature advocating 'alternative' or 'soft' technology from shriller and more technophobe positions. For the former, see especially Amory Lovins (1977), *Soft Energy Paths: Toward a Durable Peace*, Harmondsworth & New York: Penguin. For the latter, see for example Chellis Glendinning (1990), *When Technology Wounds*, New York: William Morrow, or Ivan Illich (1977), *Towards a History of Needs*, New York: Bantam Books.
29. The adoption of fluoridation of drinking water in the United States, the use of insecticide in mosquito abatement, and all matters pertaining to military technology are prime examples of such public technical choices.

30. See Dorothy Nelkin (1992), 'Science, Technology and Political Conflict', in D. Nelkin (ed.), *Controversy: Politics of Technical Decisions*, London: Sage Publications, p. xx.
31. Bernard Lewis has pointed out that in the Islamic tradition the term *bidaa* (innovation) eventually acquired a seriously negative connotation, much like 'heresy' in the West and that such subtle cultural changes account for much of the technological slow-down of the Islamic Middle East after 1400. Cf. Bernard Lewis (1982), *The Muslim Discovery of Europe*, New York: W.W. Norton, pp. 229–30. This is not to argue that *any* religion is inherently anti-technological, even in a relative sense. Yet there are many subtle ways in which an entrenched élite can manipulate institutions and culture in order to make any contemplated challenge to their dominance more difficult.
32. Morris D. Morris (1983), 'The Growth of Large-Scale Industry till 1947', in Dharma Kumar (ed.), *The Cambridge Economic History of India*, Vol. 2, Cambridge: Cambridge University Press, p. 563.
33. Eric L. Jones (1988), *Growth Recurring*, Oxford: Oxford University Press, pp. 103–06.
34. Joel Mokyr (1994), 'Cardwell's Law and the Political Economy of Technological Progress', *Research Policy*, **23**, 561–74.
35. Mancur Olson (1965), *The Logic of Collective Action*, Cambridge, MA: Harvard University Press; Mancur Olson (1982), *The Rise and Decline of Nations*, New Haven: Yale University Press.
36. An example would be the adoption of railroads after the 1830s, which involved varying combinations of private and public decision-making in different countries. In Britain, the decision to build railways was largely a private decision made in the context of the free market, though requiring Parliamentary consent; in other countries the government played a direct entrepreneurial rôle.
37. In psychology there is a great deal of emphasis on seemingly 'irrational' phenomena such as fear of new technology. Psychological 'diagnosis' of 'cyberphobia', 'technophobia' and even 'neophobia' is common. For a thoughtful debunking of this literature, see Martin Bauer (1995), '"Technophobia": a Misleading Conception of Resistance to New Technology', in M. Bauer (ed.), *Resistance to New Technology*, pp. 87–122.
38. In most of humanity's history, *all* technology has been inextricably mixed up with religion, and not just medical and biological research as in our time. For some introductory notes, see Joel Mokyr (1990), *The Lever of Riches: Technological Creativity and Economic Progress*, New York: Oxford University Press, pp. 170–73, 200–206.
39. Sigmund Freud (1961), *Civilization and its Discontents*, New York: W.W. Norton (originally published 1930), p. 39.
40. Morison, 'Gunfire at Sea', p. 43.
41. For a historiographical introduction, see Langdon Winner (1977), *Autonomous Technology*, Cambridge, MA: MIT Press, pp. 107–34.
42. The most obvious example is the prohibition of firearms in Tokugawa Japan, where the government succeeded in eliminating the production and use of muskets in its attempt to retain a monopoly on violence.
43. David Ricardo (1971), *Principles of Political Economy*, 3rd edition, ed. by R.M. Hartwell, Harmondsworth: Penguin Books, pp. 378–89.
44. As Berg has noted, Ricardo did not imply that technological unemployment was inevitable. It did not occur because machines were substituted for labour, but only because they reduced the stock of 'circulating capital'. It would thus only occur when a country's capital stock was very small and where the construction of machinery demanded a 'strong switch to fixed capital' – hardly a description of nineteenth-century Britain. See Maxine Berg (1980), *The Machinery Question and the Making of Political Economy, 1815–1848*, Cambridge: Cambridge University Press, p. 67. Cf. John Hicks (1969), *A Theory of Economic History*, Oxford: Oxford University Press, pp. 148–54, 168–71. There are several theoretical demonstrations that in certain unlikely configurations some (temporary) unemployment can be caused by the introduction of 'machinery'; but none of these is tantamount to a demonstration that such technological unemployment did in fact occur on a large scale. It is telling that working-class leaders, in Berg's view, resisted the machine because of the economic distress it caused, such as 'technological unemployment, long hours of alienated factory labour, and the smoking blight of rapidly expanding industrial towns' (Berg, *Machinery Question*, p. 17) – the former clearly being contradicted by the latter two.

45. M.J. Campbell (1990), 'Technology and Rural Development: The Social Impact', in M.J. Campbell (ed.), *New Technology and Rural Development: The Social Impact*, London: Routledge, p. 26.
46. David G. Blanchflower and Simon M. Burgess (1995), 'New Technology and Jobs: Comparative Evidence from a Two-Country Study', paper presented to the National Academy of Sciences Conference on Technology, Firm Performance and Employment, Washington DC, May 1995 (version cited dated December 1995), p. 18.
47. It is critical for this argument that patents do not categorically exclude existing producers from licensing patents or from having them assigned to them. When this happens, it is of course quite likely that existing producers will not be able to jump onto the new bandwagon. For a survey of how common patent licensing and assignment already was in nineteenth-century America, see Naomi Lamoureaux and Kenneth Sokoloff (1995), 'Long-Term Change in the Organization of Inventive Activity', paper presented to the National Academy of Sciences Colloquium on Science, Technology, and the Economy, Irvine, CA, 20–22 October, 1995.
48. A recent example is provided by Kristine Bruland (1995), 'Patterns of Resistance to New Technologies in Scandinavia: an Historical Perspective', in M. Bauer (ed.), *Resistance to New Technology*, p. 131. Norwegian fishermen in the eighteenth century resisted a new technique of multiple lines, which enhanced productivity but whose use was 'confined to relatively well-off fishermen who could afford to invest in extra equipment and suitable boats'.
49. Gregory Clark (1994), 'Factory Discipline', *Journal of Economic History*, **54**, 132–35. In other societies, too, such workshops occurred early on in the industrialization process. In India, in industries such as cotton ginning, rice polishing, and flour milling, entrepreneurs often just provided the machines and their maintenance, and charged a fee for processing from the workers. See Morris, 'The Growth of Large-Scale Industries', p. 675.
50. Adrian Randall (1991), *Before the Luddites*, Cambridge: Cambridge University Press.
51. Ken Alder (1996), *Engineering the Revolution: Arms, Enlightenment, and the Making of Modern France*, Princeton: Princeton University Press, chapters 4 and 5.
52. In a formal analysis of the emergence of resistance among skilled workers, Krusell and Ríos-Rull ingeniously capture an example of this kind of problem: Per Krusell and Jose-Victor Ríos-Rull (1992), 'Vested Interests in a Positive Theory of Stagnation and Growth', unpublished manuscript. They model an economy in which all capital is technology-specific human capital, and show that older workers who have invested in a skill that is specific to a technology threatened by obsolescence can be modelled as a 'vested interest' for whom it is optimal to try to block the new technology. For an analysis along similar lines and dealing with the important constraint on the effectiveness of such resistance by the openness of the economy, see Thomas J. Holmes and James A. Schmitz (1995), 'Resistance to New Technology and Trade Between Areas', working paper, Federal Reserve Bank of Minneapolis.
53. As *The Economist* put it recently, 'What grown-up who spent years of childhood learning to tie shoes, to count to ten, to parse Greek or to find triple integrals does not now sigh at having to lipread the baffling instructions for a video recorder or for Windows 95? Almost every generation gets overtaken in some department of knowledge as new discoveries and unfamiliar technologies replace yesterday's learning.' See 'Cranks and Proud of It', *The Economist*, 20 January 1996, pp. 86–87.
54. The classic text on this is still Sidney Pollard (1965), *The Genesis of Modern Management*. Harmondsworth: Penguin Books, pp. 213–25. See also Arthur Redford ([1926], 1964), *Labour Migration in England, 1800–1850*, Manchester: Manchester University Press; for a recent restatement see John S. Lyons (1989), 'Family Response to Economic Decline: Handloom Weavers in Early Nineteenth-Century Lancashire', *Research in Economic History*, **12**, 45–91.
55. Maxine Berg (1994), *The Age of Manufactures, 1700–1820*, second edition, London: Routledge, pp. 144–55. Carolyn Tuttle (1996), 'Children Hard at Work During the Industrial Revolution', unpublished manuscript, Lake Forest College, pp. 156-175.
56. Kellenbenz, for example, states that 'guilds defended the interests of their members against outsiders, and these included the inventors who, with their new equipment and techniques, threatened to disturb their members' economic status. They were just against progress.' Herman Kellenbenz (1974), 'Technology in the Age of the Scientific Revolution, 1500–1700', in Carlo Cipolla (ed.), *The Fontana Economic History of Europe*, London: Fontana, Vol. 2, p. 243. Much earlier, Pirenne pointed out that 'the essential aim [of the

craft guild] was to protect the artisan, not only from external competition, but also from the competition of his fellow-members.' The consequence was 'the destruction of all initiative. No one was permitted to harm others by methods which enabled him to produce more quickly and more cheaply than they. Technical progress took on the appearance of disloyalty.' (See Henri Pirenne (1936), *Economic and Social History of Medieval Europe*, New York: Harcourt Brace & World, pp. 185–6.) For a similar description of the Italian guilds, see Carlo Cipolla (1968), 'The Economic Decline of Italy', in Brian Pullan (ed.), *Crisis and Change in the Venetian Economy in the Sixteenth and Seventeenth Centuries*, London: Methuen.

57. See Olson, *The Rise and Decline of Nations*, p. 150, and Mokyr, *The Lever of Riches*, pp. 232–33.
58. In a recent paper, S.R. Epstein has defended the technological rôle of craft guilds, pointing out that they played an important part in the dissemination and intergenerational transmission of technical information. There is no contradiction between such a rôle and the inherently conservative rôle played by craft guilds. More controversial is his claim that guilds provided a cloak of secrecy which worked as a protection of property rights for inventors. Even if such a system could be demonstrated to have existed, most authorities are in agreement that eventually much of the guild system was overtaken by technologically reactionary forces which, instead of protecting innovators, threatened them. See S.R. Epstein (1995), 'Craft Guilds, Apprenticeship, and Technological change in Pre-Modern Europe', unpublished mimeograph, London School of Economics. An extreme example is the printers' guild, one of the most powerful and conservative guilds in Europe, which steadfastly resisted any innovation and as late as 1772 legally restrained one of its members from building an improved press. Cf. Maurice Audin (1979), 'Printing', in Maurice Daumas (ed.), *A History of Technology and Invention*, Vol. 3, *The Expansion of Mechanization, 1725–1860*, New York: Crown, p. 658.
59. Particularly restrictive was the custom confining the inter-generational transmission of skills to kinship. In some industries, particularly in ironmaking, skills were the traditional realm of dynasties in which technological knowledge was kept as much as possible within the family. See Chris Evans and Göran Rydén, Chapter 10 of this volume.
60. Thus in the 1560s, three Parisian coppersmiths invented improved *morions* (military helmets), but were prevented from producing them because the armourers held the exclusive rights to defensive weapons. In this case they were overruled by King Charles IX. Cf. Henry Heller (1996), *Labour, Science, and Technology in France, 1500–1620*, Cambridge: Cambridge University Press, pp. 95–96.
61. Ekaterini Kyriazidou and Martin Pesendorfer (1996), 'Viennese Chairs', unpublished manuscript, University of Chicago and Yale University, June 1996, p. 4.
62. In pre-Revolutionary France the network of craft guilds and small producers, often supported by local authorities, was adamantly opposed to all technical innovation. See Pierre Deyon and Philippe Guignet (1980), 'The Royal Manufactures and Economic and Technological Progress in France before the Industrial Revolution', *Journal of European Economic History*, **9**, 611–32.
63. David Landes (1983), *Revolution in Time*, Cambridge, MA: Harvard University Press, chapters 17 and 18, esp. pp. 300–301.
64. E. Andrew Lohmeier (1995), 'Consumer Demand and Market Responses in the German Empire, 1879–1914', unpublished Ph.D. dissertation, Northwestern University, chapter 2.
65. David Hume ([1742], 1985), 'On the Rise and Progress of the Arts and Sciences', in David Hume, *Essays: Moral, Political and Literary*, ed. Eugene F. Miller, Indianapolis: Liberty Fund.
66. The strong connection between openness and economic growth was recently demonstrated by Jeffrey Sachs and Andrew Warner (1995), 'Economic Reform and the Process of Global Integration', *Brookings Paper on Economic Activity*, **1**, 1–95. Oddly enough, the technological implications of the open economy are entirely neglected by Sachs and Warner in their list of links between openness and more rapid economic growth.
67. See Alan Booth et al. (1997), 'Institutions and Economic Growth', *Journal of Economic History*, **57**, forthcoming.
68. For the cotton industry, see especially William Lazonick (1990), *Competitive Advantage on the Shop Floor*, Cambridge, MA: Harvard University Press, pp. 78–114. In shipbuilding, for example, the boilermaker union limited the ability of employers to introduce pneumatic machinery after 1900. See Edward H. Lorenz (1991), *Economic Decline in Britain: The Shipbuilding Industry*, Oxford: The Clarendon Press, pp. 58–59.

69. Donald Coleman and Christine MacLeod (1986), 'Attitudes to new Techniques: British Businessmen, 1800–1950', *Economic History Review*, **39**, 606.
70. Roderick Martin (1995), 'New Technology in Fleet Street, 1975–1980', in M. Bauer (ed.), *Resistance to New Technology*, p. 194.
71. Morris, 'Growth of Large-Scale Industry', pp. 622–3.
72. Susan Wolcott (1994), 'The Perils of Lifetime Employment Systems: Productivity Advance in the Indian and Japanese Textile Industries, 1920–1938', *Journal of Economic History*, **54**, 307–324.
73. Holmes and Schmitz, 'Resistance to New Technology', p. 29. See also Martin Kenney and Richard Florida (1993), *Beyond Mass Production*, New York: Oxford University Press, p. 315.
74. See Joel Mokyr (1993), 'Institutions, Technological Creativity, and Economic History', *Innovazione e Materie Prime*, **1**, 27–42 (in Italian); reprinted in A. Quadrio Curzio, M. Fortis and R. Zoboli (eds) (1994), *Innovation, Resources and Economic Growth*, Berlin: Springer Verlag, pp. 39–59 (in English).
75. Eric L. Jones (1988), *Growth Recurring*, Oxford: Oxford University Press.

PART 2

Cultures of Invention

4. The Cultural Foundations of Early Industrialization: A Project

Margaret C. Jacob

My project on culture and industrialization was started in 1990. At this time there was a clear divide in social scientific thinking between culture and economic history. Economic historians and historians of technology claimed that culture played little part in developments tacitly defined as purely economic or purely technological. Historians in the 1960s and 1970s and before, such as Landes, Nef, Cardwell, Musson and Robinson, and Crouzet, had written about the values and attitudes of early entrepreneurs and engineers, and the social settings in which they operated. Their work was, however, based on the assumption of the autonomy of the 'cultural'. My aim was to build on the work of these historians, but to use distinctly different, more complex understandings of culture.

The epistemology commonplace in economic history was precisely what I sought to avoid when invoking the cultural. Culture can simply be conceived as being present in any setting – in no sense sufficient or determining. Cultural factors may undergird, or retard, the deployment of technology and technical knowledge; they need not be sufficient to explain failure or success. The first results of the project were recently published as *Scientific Culture and the Making of the Industrial West* (New York: Oxford University Press, 1997). There is still work to be done in France and the Low Countries. For example, one discovery, that of the earliest women's scientific society, which met from 1785 in Middelburg, Zeeland, still awaits a monographic study. Here I shall summarize the results of the project, documented in greater detail in *Scientific Culture*, and present new material from Zeeland in the Dutch Republic.

Culture is now back on the historical agenda in the way that twenty years ago 'the social' occupied centre stage. Conceived as a nexus of belief-systems, attitudes, gestures and symbolic systems – or more broadly as what Bourdieu would call the *habitus*, the locus of social interaction – culture, I believe, can be best seen through a comparative lens. As employed in our research project, comparisons were intended to reveal the culturally distinctive elements in a setting. Thus we developed research sites in four countries, Britain, France, Belgium (formerly the Austrian Netherlands) and the anomalous case of the Netherlands. I say anomalous because despite an enormous commercial expansion during the seventeenth century, the Dutch Republic experienced little mechanized industrial activity in the period from 1750 to 1830. Our comparative effort relied upon a theoretical approach that properly credits both the power and importance of people *and* the social and cultural settings wherein they operate. None of these theoretical issues about how the settings and the agents interact can be imagined as being free from controversy, but that should only testify to their importance. Yet the matter of theory requires further, brief elaboration.

Since the mid-nineteenth century, historical approaches to work, industry, and the deployment of technology have been torn between two theoretical approaches: one best described as 'instrumental', that is, de-cultured free and free-willed agents naturally pursue their interests, and the alternative approach privileging the cultural, that is, the images, symbols, concepts, beliefs and practices offering explanations for the social and economic behaviour of agents. A hundred years ago Weber was the most important representative of the latter. The best practitioners of the former included Lord Acton and J.B. Bury, and the worst were the hagiographers who made the entrepreneurs into Horatio Algers.[1] In the post-war era, John U. Nef, and, to a more sophisticated extent, David Landes, belong essentially to the instrumental camp.

Among historians of thought and culture in my generation, few may be said to be instrumentalists, yet none of the leading practitioners of cultural history had tried their hand at the early industrialization and culture problem.[2] In the division of academic labour, industrialization was presumed to belong to the economic and social historians. As a result, the problem of the relationship between culture and the turn toward the mechanization of industry became a hard historiographical problem, both practically in terms of its being a relatively impoverished genre, and theoretically because of the tension imagined between the freedom and rationality of economic agents and the constraint suggested by a recourse to cultural explanations. By contrast, I began with the assumption that culturally informed agents do also pursue their interests in ways we may identify as 'rational'. Yet agents, with early entrepreneurs among their most aggressive exemplars, act in pursuit of their economic interests within a cultural framework. They are both voluntary and involuntary participants in its permissions and limitations, just as they are knowledgeable, and yet at moments unwitting, about market forces. They embody and manipulate culture just as they are inscribed by it.[3] When they succeed through the deployment of technology, they acquire wealth and influence. They thus also inscribe their cultural values into the economic and social order, creating models – companies, workshops, sites of engines and canals, educational curricula – ready to be imitated. The gradual triumph of one cultural matrix over another less successful one obscures the era of contest; success also makes the major, by then famous, agents of change seem so atypical as to be irrelevant to the mass of their less creative or original imitators. Going back to the earliest inscribers permits us to witness industrial culture in formation, to unearth its undergirding values, assumptions, attitudes, and symbolic guideposts.

One of the focal points of the British side of our study became the enormous collection of manuscripts from the families and firm of Boulton and Watt. Midway in our project, we also got lucky. The Welsh branch of the Watt family finally put its manuscript collection, dating back to the 1680s, on the market. The Birmingham City Library suddenly acquired an immense archive, nearly doubling the size of its already substantial Boulton–Watt holdings. The archive owned by Lord Gibson-Watt had long been central to early industrialization and almost impossible to see. It had much to tell us about the values and personal development of the Watt family, and the collection became critical for the British side of the story we wanted to tell.

Before James Watt became Watt of separate condenser fame, he was from an informally educated Dissenting family where technical knowledge, self-discipline and eventually republican politics abounded. His extraordinary artisanal skill was encouraged by a cultural universe that also included secular and scientific values. From his days as a young apprentice in London Watt participated in what we came to see as the distinctively British culture of applied science.[4] He had a mathematics tutor; he attended scientific lectures and he read scientific textbooks, some possibly inherited from his uncles. He went from owning vices, anvils, quadrants and compasses to studying the construction of mathematical instruments from slide rules to levels, astrolabes, sectors and telescopes.[5] By 1783 his scientific self-confidence was such that he could imagine that Lavoisier might have borrowed some of his ideas.[6] The early aspirations of Watt and his family were shared by thousands of other families, and its values were enhanced in the minds of still more people by his success. The culture of applied science at work in the deployment of technology, as seen among entrepreneurs, engineers, and agents of the state, lay at the heart of what our research effort sought to uncover.

The Watts and their circle gave us a paradigm that replaced the myth of the uneducated tinkerer with a template of the self-educated and tutored Watt who knew Newtonian mechanics and mathematics for surveying, as well as the skills of business and commerce. He and his wife, Annie, in turn made science a key element in the education of their children, who then developed their own scientific and industrial interests.[7] One other element in the culture of applied science that we saw in Britain was the relatively high literacy in basic science available to some women. By contrast to what could be obtained in the school system in a town like Birmingham, for example, French education for girls was woeful in matters scientific and numerate.[8] By 1730 it was also possible for men to learn more applied mechanics at a lecture series in a London coffee house than was the case in most French colleges, where Cartesian physics still predominated.

Any study that searches for cultural patterns must include key agents, those major entrepreneurs and engineers who have left a significant body of personal records. We went first to family papers like those of the Watts because, in them, values, beliefs and assumptions are more readily revealed. Found in four countries, family conversations revealed the cultural universe most immediately influencing entrepreneurs and engineers. When looking across cultures and across national borders, it was also possible to see variations and limitations of which the agents themselves (and the historians who have studied them locally or simply as individuals) appear to be unaware. The study of letters between husbands and wives, and the relatives of both, in entrepreneurial families (such as the Watts in England, the Lièven Bauwens in Belgium,[9] the Overkampfs in France) revealed remarkably similar attitudes and values present among people from different cultures, writing in different languages.

Yet because they wrote as families engaged in very similar economic activities, they present a pattern to the cultural historian. The letters reveal a distinctive set of ideas about family life: the agents saw the family as an intrinsic part of economic life, not a refuge from it; in some areas of both private life and public business, husbands and wives were interchangeable, both possessed technical and technological knowledge, and so on. Yet the exact contemporaries of the Watts, the French Overkampfs of cotton manufacturing fame, although Protestant and eager to apply machinery to production, saw no need to impart technical, scientific knowledge to their children.[10] They knew about spinning jennies, but steam was outside the range of their vision. In 1792 their factory in Rouen employed 1200 workers at 150 tables, made finer colours than could be dyed in England – or so James Watt junior said – but their records evince no interest in further mechanization.[11]

To complicate the comparative project, I should also note, the Continental evidence revealed a different, more convoluted relationship between industrial actors and the state. On the Continent, to a greater degree than one sees in England, entrepreneurs 'looked over their shoulders', and were constantly concerned about political powers and opponents perceived to be stronger than themselves, even dangerous to their economic livelihood and survival. They perceived the state and its agents as active competitors, but also occasionally as powers to be enlisted, that gave access to markets, technology, subsidies and skills. This ever-watchful stance was exacerbated in the revolutionary 1790s, for obvious reasons, but it is present before and after. The letters undermine the constantly repeated, older historical assessment that the Continental *ancien régime* states simply inhibited industry by over-regulation. Sometimes the French and the Austrians (who controlled the southern Netherlands) did inhibit industry, but their impact was more complex and interesting than that. The Ghent textile manufacturing of Lièven Bauwens benefitted enormously from the contract, awarded late in the 1790s, to manufacture French military uniforms.

The framework of early industrial culture was far more complicated than the old 'British freedom/French restraint' model suggested. The new discoveries showed that in the early stages of industrialization the active intervention of the absolutist state was sometimes quite effective. Its effectiveness, however, depended on how its representatives negotiated within a cultural setting that could foster or inhibit the deployment of technical, mechanical knowledge. Contrary to the theory of the instrumental model, technical knowledge was never simply found when needed and deployed accordingly. Cultural settings in large measure determined who possessed technical knowledge, how the possessors saw their chances of success, and how they were able to use their knowledge. In brief, the civil engineer emerged in Britain by 1750; his French counterpart was a military man, and thereby hangs an important comparative and cultural story. Standing aloof from the entrepreneur, French engineers shared their technical knowledge only among themselves or with other academicians. National settings, as revealed in local case studies, illustrate the differences between Britain and its Continental rivals.

In an earlier study that set the agenda for this research project, I presented a case study of efforts made by Bristol's élites from the 1770s onward to address the inadequacy of their harbour for the greater volume of new and larger ships. The Bristol study revealed the extraordinary depth of technical knowledge possessed by the urban gentry who hired and worked with the engineers.[12] Other research also found peers in the House of Lords who assessed the technical merits of canal bills, and mining entrepreneurs in the north country who, being knowledgeable about technology, bought steam engines to their own specifications. All illustrated the significant technical competence possessed by literate Britons. The relationship of entrepreneurs with civil engineers was close, even one of superiority, because the entrepreneurs, just like the peers and the mine owners, possessed real technical sophistication. Engineers complained that their expertise had to compete with what their employers or the Parliamentary Committees knew about hydrostatics or mechanics.[13] Smeaton or Jessop could not imagine how across the Channel their state-employed counterparts were objects of deference, seen as the 'experts'.

If one crosses the Channel to Nantes, to a French Atlantic port also in need of refurbishing by the 1770s, there one observes a very different relationship between local élites and engineers. In Nantes, as in Bristol, the *cultural* setting affected the deployment of technical knowledge. In the massive effort to improve Nantes' harbour, the engineers were not partners of the local merchants, and certainly not their hired and sometimes frustrated assistants, but in the words of the merchants and magistrates, 'experts' (a term rarely used in Britain even to describe John Smeaton). In an elaborate and complex controversy between the central government and the Nantes merchants, the engineers became identified as academicians, classified by their status as graduates of the prestigious state-sponsored engineering school. Although honoured, knowledgeable and powerful, the engineers became pawns in the political games of *ancien régime* France, in the interaction between the absolutist centre and the self-interested periphery.[14] The engineers and their knowledge remained separate players, and both were perceived, perhaps incorrectly, as more allied with the state that paid their salary than with the merchants who needed their services. Sometimes these perceptions, as well as the state's actual rôle, could conspire to produce technological development such as occurred in the French glass industry. But throughout the eighteenth century the political culture wherein the new mechanical technology was deployed projected its instabilities onto moments of invention and innovation.

The outcome for the harbour at Nantes was interrupted by the French Revolution, but the records of these pre-1789 encounters add a piece to a comparative puzzle that can situate the French deployment of technology and technical knowledge within a distinctive cultural setting. In pre-Revolutionary France the culture of deference limited access to technical knowledge, or it restricted the interest that entrepreneurs had even in acquiring that knowledge. The setting also determined the amount of trial and error that could occur. Testing of devices became a formal, often once-only procedure, reflecting the procedures used by the official academies to assess inventions. In effect, rather than culture being incidental to the economic process, the French setting might be seen as an example of cultural factors overdetermining responses to market forces. Individuals *and* structures need to be seen in dynamic relation, one to the other.

Culture does not create society or politics, but it organizes perceptions of both, possesses a constitutive logic, and marks the possible avenues of intervention in the world. The autonomy of agents permits mediation within and around the thicket of those perceptions, but the field for action is not infinite.[15] The old model – give free agents a wide berth and industrial development will occur – amounts to a gloss on a much more complex set of historical circumstances. A comparison of Troyes and Manchester cotton manufacturers in the period from 1770 to 1800 once again revealed a hesitancy about mechanical power, and a lack of understanding on the French side that made steam power less attractive. The cost of fuel remained prohibitive in France, but that does not explain why, when Watt's engine finally arrived in the late 1780s, few if any modifications were wrought on it for at least two decades. By 1810 its application to manufacturing remained negligible.

The French system of privileges, monetary grants or rewards for new technological enterprises meant that entrepreneurs and engineers, or foreigners drawn there in search of profit, either kept their technical knowledge a secret (as did John Badger who brought the Calandre Anglaise to the silk industry in Lyon and then to Nîmes) or saw their knowledge as mental labour to be 'purchased' through a system of privileges or special funds established by bureaux of the state for that purpose.[16] When being secretive, agents refused to avail themselves of the technical knowledge possessed by state academicians or engineers. Yet when seeking to sell their skills, agents of technological change found the French state mediating the value of invention, with those same academicians and military engineers playing key rôles as technical assessors.[17] Again, in neither case were entrepreneurs themselves encouraged to enhance their technological proficiency, and the distance between mechanical knowledge and its commercial users increased by comparison with what we have discovered in the British or Belgian cases. In the silk industry at Nîmes, secrecy also exacerbated the endemic quarrelling between the absolutist state and local Protestant merchants, a fraught political culture very similar to what can be seen in Nantes.

If the state's representatives failed as suppliers of expertise, or, as in the Austrian–Belgian case, just never sought that rôle in the first place, then entrepreneurs had to acquire their own technical knowledge. As became increasingly clear to us, the interaction of entrepreneurs, the state, and old local élites was different in the Belgian and the French cases. In the southern Netherlands the Austrians, like all good absolutists, also operated a system of privileges awarded to industry, whether in mining or manufacturing. But given the power of the old élites in what was a colony, the state's representatives had less room to manoeuvre than did their French counterparts. In this elaborate chess game, engineers and technical knowledge became allies of the state, and to its representatives even the market seemed less capricious than the ancient privileges over land and mineral resources inherited from time out of mind. In this semi-feudal setting, technical knowledge became a political weapon; only belatedly did it become an economic one. In addition, the Austrians kept a small cadre of military engineers in the colony solely for military purposes, and there was no system of state-employed technical experts who could be sent to sites in need of development. Entrepreneurs were basically on their own when it came to finding technical support or knowledge.[18] In the major coal-mining district of Liège, technical education began in earnest only after the expulsion of the Jesuits in 1773. In the Belgian case, entrepreneurs, generally without sufficient technical knowledge, either had to hire engineers and let them get on with it – as occurred with the installation of a steam engine in Mons in the 1770s – or they had to reform the educational system and start training the next generation in knowledge useful to industry. With the French invasion in 1795, that process of educational reform began in earnest.

Where the schools and academies failed to provide technical knowledge, often because of clerical control in many parts of northern and western Europe prior to 1800, leading scientists, engineers and scientific societies, in touch with an international body of knowledge and practices, became key players. Throughout this research project, attention increasingly focused on local scientific societies, their experimental life, ideological perspectives, and so on. Knowing the mindset of individual key players in local societies – such as Jean Chaptal, chemist, entrepreneur in Montpellier, and then minister of the interior in Napoleonic France – provided access to policies and the transmission of values which after 1800 became commonplace. Chaptal brought the international culture of applied science to his ministry, and its efforts at curriculum reform, and the establishment of festivals of industry, prizes and expositions, broke the mould inherited from the old order.[19] The state remained a key player, but it promoted the cultural values of the Chaptals and the Overkampfs, of the inventors and industrialists. Sometimes agents who are atypical and the harbingers of change, because of the impact they can have in the aftermath of a revolution, merit the special attention of historians.

Such reformers particularly fretted over the quality of technical education, and by 1800 it had became an issue in France and the Low Countries. Relevant educational records before and after the revolutions in all three countries became part of the history of the cultural foundations upon which Western industrialization developed. The shift toward applied technical knowledge in the schools of the 1790s was the work of pro-revolutionary republican reformers operating in the corridor from Paris to Amsterdam after 1795. In 1808 the science faculty of the University of Brussels was completely reorganized and applied mathematics was introduced into the curriculum. The move was made so quickly that staff had to be taken from the local *lycée*. Libraries in both Belgium and the Netherlands were examined by the French occupying ministers and the quality of their science collection was commented upon.[20] In the first half of the nineteenth century the applied and technical education put in place by the French revolutionaries became a model for European-wide emulation.[21]

Having examined the important rôle played by technical knowledge in the hands of entrepreneurs, and the differences among the cases from Britain, France and the Austrian Netherlands, we considered the Dutch Republic the obvious next place to look. Predictably, we sought a harbour project comparable with the Bristol and Nantes efforts. The question we wanted to answer concerned the rôle of technical knowledge in the problem of Dutch retardation. This led us to Middelburg in Zeeland. There, too, by the second half of the eighteenth century the harbour needed massive work (partly because of silting).[22] We examined how the Dutch engineers were treated by the merchants who hired them. They were members of the most important mercantile élite in the Republic.[23] Other records established what knowledge the mercantile directors of the Indies companies could bring to bear on the project. Were they similar to their Bristol counterparts, or did these Dutch merchants stand aloof from engineers? In addition, how did the highly localized system of political authority affect knowledge deployment, given the inevitability that political interactions must accompany large expenditures of public monies whether in a monarchy or a republic?

Leading Orangists such as Van Goens defined the wealth of the nation in trade and agriculture and saw science as 'pure', not the possession of mere engineers. More than a fair share of the Middelburg directors were Orangist.[24] Attitudes toward the various branches of eighteenth-century science in Patriot circles – their political opponents – were markedly different and much more 'applied'. Such differences in the scientific culture embraced by the literate had never before been figured into the problem of Dutch retardation.[25] From the 1790s we found, in the papers of the Middelburg *burgermeester*, a rudimentary, handwritten text on hydrostatics. It was quite elementary in that it simply communicated the notion of the relative force of moving water dependent upon its volume and speed.[26] The tract coincided precisely with the moment when the water problems of the harbour had become so urgent that it had to be closed and the canal leading into it deepened, at the eventual cost of 400,000 guilders.[27] Clearly the struggling *burgermeester* was trying to make himself at least minimally literate in what the engineers were saying.

By 1802, deliberations about water in Middelburg had become more complex; not only the harbour but also the off-shore islands were afflicted with serious water problems. Also by that year it was clear at least to some leaders that steam machines were more efficient than wind-driven mills in moving large amounts of water with regularity. By the 1780s the Dutch Republic had become a leading importer of steam engines, largely for the draining of polders. The key trader had been Watt's good friend and member of the *patriotten*, J.D.H. van Leinder. Yet by 1800 the reality was that steam engines remained foreign products and that neither engines nor their parts could be procured in the domestic market. In war-time amid French occupation money was short and the absence of an indigenous metal industry of sufficient skill to make the parts – never mind the engines – had become a real handicap. When foreign experts in water engineering were discussed for what light they might shed on Middelburg's problems, they were invariably Italian, French or German and not British.[28] By contrast, over twenty years earlier, artisanal and mechanical expertise in steam had come to exist in the southern Netherlands, for example at coal mines in Hainaut and near Mons. Late in the 1770s a local firm installed a steam engine there and by 1807 Watt's engine had also come to the site.[29] No one in Middelburg seemed to be conversant with the new engines, or connected with their British exporters.

Sometimes technological retardation becomes cumulative. As late as the 1830s the main scientific society in the Netherlands was putting out prize questions that asked competitors to address the problem of how to improve steam technology.[30] Predictably, answers ranged across the spectrum of available explanations, from the cost of steam to the hostility to new ideas commonplace in the Republic. On the whole Dutch engineers were well-trained, but there was a deep reluctance – exacerbated by the regional and national debts – to use their services. The library of the Scheepvaart Museum in Amsterdam contains an excellent collection of tracts that permit a reconstruction of Dutch water engineering practices from the 1720s to the early 1800s, when A. Schraver acted in the capacity of civil engineer in Zeeland both before and during the Napoleonic occupation.[31] Yet the Middelburg magistrates collected for their reading and lending library far more books on physico-theology than on advanced mechanics.[32] A similar bibliographical backwardness has also been found in other late eighteenth-century Dutch institutions of higher learning. After 1750 Dutch literacy in matters mechanical had come to lag behind what could routinely be seen across the Channel.[33]

No country on the Continent was more aware than France of what was happening industrially in Britain in the late eighteenth century.[34] French ministers of state began to watch British technological activity as early as the 1740s, when the ministry of the interior was gathering information about British-made textile machines in use at Mons.[35] By the 1770s there were full-time secret agents in London whose purpose was to gather information and recruit trained mechanists.[36] We may be good cultural relativists and dismiss the determinism implicit in asking 'why was Britain first', but two generations of old régime French ministers, as well as numerous Dutch scientists of the 1770s onwards, were asking that very question. Laying emphasis upon protoindustrialization, a system basically dependent upon hand-labour, simply provides background to an historical phenomenon where mechanical technology was still essential. It does not explain the 'why'.

In the late 1780s French cotton manufacturers were convinced that the British machines were destroying their businesses. Theorists pointed to the superiority of English technology and also worried about the growing prosperity of Flanders.[37] From 1798 onward, annual public expositions organized by the French government, and intended to inculcate industrial values, showed models of the Newtonian universe, as well as 'instruments de physique', samples of machine-made cotton from many parts of Europe, and just about every conceivable mechanism for manufacturing, printing, and farming.[38] The purpose of these expositions was to bridge the gap between 'les savans et les fabricans' by favouring 'le mécanicien', that is, go-betweens, 'artists who transmit to them the results of their research'.[39] Year by year the expositions increasingly emphasized industrial innovation. In 1795 the Committee on Public Safety decreed that a factory or workshop should be set up in the centre of Paris where all could receive 'a true idea of mechanics and have at the same time a double rapport of theory with practice'.[40] In this revolutionary decade 'Newtonian atheists' turned their attention to the schools, bypassing the universities with their entrenched faculties, and sought to use a new science curriculum 'to change the world'.[41] In 1807 Napoleonic ministers frantically surveyed the state of French industry and reported on 'the absolute necessity for the creation and support of it'. They commented on the widespread stagnation caused by war and lack of capital, and on the need for more canals and machines; not least, they lamented bad manufacturing techniques. As for England, in its workshops lay 'le secret de ses mécaniques. Dans ce secret réside sa puissance industrielle.' A police report of 1807 about impoverished Parisian spinners of cotton who worked the new machines described them as 'les véritables machines'. At the same time, the 332 technologically skilled machinists in Paris were among the best paid workers in the city.[42] Clearly the Industrial Revolution had come to Paris.

The move to inculcate industrial values began with the new men brought into power by the French Revolution. A few years before it, self-taught engineers like Coulomb had turned to explicating publicly the principles of the new steam engines; after 1789 such non-noble engineers came into their own within technical and applied science.[43] As early as 1793 the revolutionary government instituted a reformed *bureau des arts et métiers*, led by men skilled in 'the happy mixture of theory and practice which is absolutely necessary for making good judgments in the productive arts'.[44] In 1795 the army as it moved north was instructed to send any foreign machines that were found back to Paris. These could be used in the model factory.[45] A few years later the French minister in charge of encouraging industry brought a British engineer to Paris, set him up with steam engines and had people come to observe him.[46] The minister, J. Chaptal, had an entire theory to account for English superiority, one which combined cultural and economic explanations and which when published in 1819 also offered one of the first histories of industrialization.[47] He regarded the fact that France held second place to the English among the manufacturers of Europe as an 'inversion of the natural order'.[48]

If there had been no Industrial Revolution in the period before 1830 – as some economic historians would have us believe – this would have been news to the Watts and the Chaptals who lived through the period. Few of them would have agreed with the 'new' economic historians. Neither would contemporaries have endorsed our reluctance to see the cultural dimension, which they struggled to understand precisely so that they might be able to imitate British successes. By 1816 the French ministers of state were anxiously counting every steam engine that could be found in the country; they came up with the figure of 48 in that year.[49]

Suppose, however, that we grant Britain its Industrial Revolution after 1780. Then need any other country at that time have also experienced it? Surely there is no template, either economic or cultural, that should be fitted over the terrain bounded by Edinburgh, London, Amsterdam, Cologne and Lyon, and against which success or failure need be measured and adjudicated. Why then engage in a comparative approach, to compare Britain, France, Belgium and the Netherlands, when it may be argued that industrial phenomena occurred randomly, at different times, generated only by a small minority of eccentric, atypical mechanists and entrepreneurs whose genius could be admired by others but never replicated? The Watts, Chaptals, Periers may have possessed heroic dimensions – especially after the nineteenth-century hagiographers were finished with them – but since they are not typical, why study them in search of a cultural template that could not have existed even if the pioneers would have liked it to have been there?

Unfortunately, the questions of typicality or randomness are not quite that simple. By the 1780s it was clear to a wide variety of reformers, scientific practitioners, engineers and machine makers, even to government ministers throughout Western Europe, that what the British were doing technologically in manufacturing and transportation was distinctive and worthy of imitation. In Britain dozens of efforts were made to break or copy Watt's patent on the engine. He may have been a mechanical genius, but there were many men around eager to try their mechanical skills at achieving the prosperity and fame that Watt had acquired by 1790. On the Continent as early as the 1740s, just mentioning that a mechanist had learned his skills in Britain – from Desaguliers, for example – or, later, that a technician had experience at British industrial sites, gave a cachet that none could surpass. By the 1740s in Britain, the technological complexity that contemporaries reported was real: 'we daily see the same number of men or of horse, or the same stream of water, perform ten times [more] with one machine than with another.'[50] In addition, British engineers were all over the Low Countries and France by the 1750s, looking for work or for a government grant to practise their new techniques. With their techniques they spread values and cultural styles which could be imitated or at least admired.

Take also the larger cultural picture. The literatures of Enlightenment, republicanism, reform in education and child-raising practices, even novels and reformist theatre, preached in many languages a remarkably common set of values pertinent to industrial development: merit as superior to birth or blood; practical, sense-based knowledge as superior to doctrines and dogmas; thinking men, and not old élites, as the true political agents; scientific theory as a necessary accompaniment to practice; the disciplined, interior self as the measure or judge of achievement. Arguably these cultural values were absorbed and internalized selectively among literate élites on both sides of the Channel, earlier in some places, notably Britain, and there too among some families before all others. But these values soon – by the early 1800s – became commonplace (with important variations) among the educated on both sides of the Channel. They were more commonplace, earlier, among the mechanically knowledgeable and entrepreneuring. Might not their prosperity have privileged their values, suggesting that imitation would bring comparable success? British entrepreneurs abroad consistently remarked on how difficult it was to work with foreign workers in mechanical ways. Gradually after 1800 those complaints lessened, but there remained culturally different styles of mechanical work encoded in the manufacturing processes well into the second half of the nineteenth century (if not also in contemporary manufacturing processes – think of recent studies of American and Japanese automobile manufacturing).[51]

The thrust of the democratic revolutions of the late eighteenth century was to transform French, Belgian and Dutch society in ways that facilitated the emergence of industrialists, entrepreneurs, and civil engineers. There may have been no cultural or economic template, but by the 1840s the agents of industrialization were apparent in almost every European country. National and cultural differences would remain and continue to this day. Culture constantly changes, like all else in human history. But it seems reasonable to look for the earliest similarities, as well as the differences, that existed in the first societies anywhere in the world to break through the limitations imposed by human strength. There is such a thing as Western culture. Its components differ from country to country at any given moment, yet by the mid-eighteenth century there was one common element that would prove immensely germane to the mechanization of the manufacturing and transportation systems: the belief that science and technology could control nature and that creativity in both was desirable, a sign of Western industrial superiority.[52]

NOTES

1. For the context in which Acton et al. wrote, see Christopher Parker (1983), 'English Historians and the Opposition to Positivism', *History and Theory*, **22**, 120–45.
2. A start was made by William Reddy (1987), *Money and Liberty in Modern Europe*, Cambridge: Cambridge University Press; see also Patrick Joyce (ed.) (1987), *The Historical Meanings of Work*, Cambridge: Cambridge University Press. I am indebted to a work that should change the way historical sociologists approach culture: Richard Biernacke (1994), *The Fabrication of Labor in Germany and Britain, 1640–1914*, Berkeley: University of California Press. A controversial book by Martin J. Wiener (1981), *English Culture and the Decline of the Capitalist Spirit 1850–1980*, Cambridge: Cambridge University Press, helped to put culture on our agenda. Note the response in Sidney Pollard (1990), 'Reflections on Entrepreneurship and Culture in European Societies', *Trans. of the Royal Historical Society*, **40**, 153–74.
3. See Pierre Bourdieu (1984), *Distinction. A Social Critique of the Judgment of Taste*, Cambridge, MA: Harvard University Press. For culture as simply a set of rules in the rational choice mould, see James S. Coleman (1990), *Foundations of Social Theory*, Cambridge, MA: Harvard University Press.
4. For another local example of the culture to which I refer, see David L. Bates (1993–94), 'All Manner of Natural Knowledge: the Northampton Philosophical Society', *Northamptonshire Past and Present*, **8**, 363–77. Here leaders included the Dissenting minister, Philip Doddridge, and the engineer, Thomas Yeoman. I am grateful to Dr. Bates for these references. See also A.E. Musson and E. Robinson (1969), *Science and Technology in the Industrial Revolution*, Manchester: Manchester University Press; reissued by Gordon and Breach, New York, 1990. For essential background as to how this culture developed, see Keith Thomas (1987), 'Numeracy in Early Modern England', *Trans. of the Royal Historical Society*, **37**, 103–32.
5. Birmingham City Library, James Watt Papers, 6/46: a list dated 1756 sent to his father back in Scotland that indicates his efforts to master Bion's *Construction and Principal Uses of Mathematical Instruments* in the more complex English version done by Edmund Stone.
6. Birmingham City Library, James Watt Papers: Watt to Mr. De Luc, Birmingham, 30 November [1783].
7. See for example letters to Gregory Watt from William Creighton, dated 1804: James Watt Papers, Birmingham City Library, 6/10.
8. When Diderot – hardly a wealthy man – sought scientific education for his daughter, he had to hire a tutor: see Monique Ruffet (1992), 'La physique pour débutants: Angélique Diderot et les leçons de l'abbé Nollet', *Recherches sur Diderot et sur l'Encyclopédie*, **13**, 57–78.
9. See Kristien Depoorten et al. (1982), *Inventaris van het Fonds Napoleon de Pauw*, Ghent, vols. 1–3; and as a further guide Fernand Leleux (1969), *A l'Aube du Capitalisme et de la Revolution Lièven Bauwens Industriel Gantois*, Paris; S.E.V.P.E.N.
10. Archives nationales, Paris, MS 93.M.I; MS 44.a.Q.1.
11. Birmingham City Library, M/7/1 no.14, 3 June 1792; and on 16 June 1792, 'New improvements I have seen none, nor any process which we have not in England, on the contrary they are considerably behind hand here in the Manufactory, but yet not so much

as I expected ...' James Watt junior notes the high price of fuel as decreed by the National Assembly (they cannot burn wood and must import their coal from England).

12. The Bristol story is first told in Margaret C. Jacob (1988), *The Cultural Meaning of the Scientific Revolution*, New York: Knopf/McGraw-Hill. The kind of evidence for this knowledge – aside from its being employed – could be found in, for example, John R. Millburn (1983), 'The London Evening Courses of Benjamin Martin and James Ferguson, Eighteenth-Century Lecturers on Experimental Philosophy', *Annals of Science*, **40**, 452n. Ferguson had over eighty subscribers for his course in 1774 in Bristol. His lectures were almost entirely on the application of mechanics, on pumps, engines, and so on. Cf. G. Timmons (1983), 'Education and Technology in the Industrial Revolution', *History of Technology*, **8**, 135–49.
13. Jacob, *The Cultural Meaning of the Scientific Revolution.*
14. For the pattern into which this status can be fitted, see Terry Shinn (1994), 'Science, Tocqueville, and the State: The Organization of Knowledge in Modern France', in Margaret C. Jacob (ed.), *The Politics of Western Science, 1640–1990*, Atlantic Highlands, NJ: Humanities Press. For the harbour, see Archives départementales, Loire-Atlantique, Nantes C677, carton 19, and C630 carton 1 côtes 1–4. The complex story of why Watt's engine never got to France before 1789 can be further augmented by letters in the Gemeente Archief, Rotterdam, 'Brieven aan Huichelos van Liender, 1770–80' (typed copies of originals lost in a bombing raid in 1940).
15. For some theoretical assistance, see Mary Douglas (1975), *Implicit Meanings*, London: Routledge & Kegan Paul; and Marshall Sahlins (1976), *Culture and Practical Reason*, Chicago: University of Chicago Press.
16. See letter of 23 October 1753: John Kay tries to talk sense to Badger, explaining that he has got to reveal his 'secret'; Archives nationales, F12 1442. Badger wants the subsidy up-front before he will reveal anything.
17. For an excellent essay with an eye for culture, see Edward A. Allen (1990), 'Business Mentality and Technology Transfer in Eighteenth-Century France: the Calandre Anglaise at Nîmes, 1752–92', *History and Technology*, **8**, 9–23. For the sheer complexity of the governmental system, see Liliane Hilaire-Pérez (1991), 'Invention and the State in 18th-Century France', *Technology and Culture*, **32**, 911–31. This article cites other secondary sources where it is claimed that French administrators 'did not feel that English industry was much more advanced than their own, and other historians have said much the same thing'. None of this research, however, has been comparative, and in addition there is a wealth of primary source material that contradicts the assessment. The recent work of Phillippe Minard also undercuts the Hilaire–Pérez assessment.
18. F. Macours (1952), 'L'enseignement technique à Liège au xviiie siècle', *Bulletin de l'Institut archéologique liègeois*, **69**, 131–85.
19. Jeff Horn and Margaret Jacob, article on Chaptal, forthcoming, in *Technology and Culture*.
20. Archives nationales, Paris, F17 1098. See also Simon Schama (1970), 'Schools and Politics in the Netherlands, 1796–1814', *Historical Journal*, **13**, 589–610.
21. See the entire issue of *History and Technology*, **12** (1995), with articles devoted to Belgium, Germany, Russia, etc.
22. See inv. 170 and inv. 201, Rijkswaterstaat, Zeeland.
23. For an overview of Dutch engineering practices at this time, see P. van Schaik (1984), *Christiaan Brunings. Waterstaat in opkomst*, Zutphen: De Walburg Press, esp. pp. 78–87.
24. Margaret C. Jacob (1991), *Living the Enlightenment. Freemasonry and Politics in Eighteenth-Century Europe*, New York: Oxford University Press, p. 177. Cf. J.J. Kloek and W.W. Mijnhardt (1988), *Leescultuur in Middelburg aan het begin van de negentiende eeuw*, Middelburg: Van Benthem & Jutting.
25. H.W. Lintsen (ed.) (1992), *Geschiedenis van de Techniek in Nederland. De wording van een moderne samenleving 1800–1890*, vol. I, *Techniek en modernisering, Landbouw en voeding*, Zutphen: De Walburg Press. From the brochure: 'technologies ... are embedded in much larger technological, economic and social environments. This is what in recent historiography of technology is referred to as the systems-approach.' The word 'culture' never appears.
26. See the records of Willem A. Van Citters, inv. 17, Rijkswaterstaat, Zeeland. The engineering reports were quite technical and obsessed with the issue of cost. The Raden paid the engineers directly.

27. See P.W. Sijnke (1976), 'Rondom de Middelburgse haven in de 18e eeuw', *Bulletin van de werkgroep histoire en archeologie*, **25**, 3–11. In 1809 the contract for improving the harbour was given to the engineers Schraver and Dingemans, and it was ready to receive a frigate from New Orleans in 1817.
28. See MS 236, 'Verhandelingen, ingekomen bij het Zeeusch Genootschap ter Wetenschappen, 1801–02', found in the archives of the water authority: Archieven der besturen van het Eiland Walcheren, 1511–1870, Waterschap Walcheren, Groenmarkt, Middelburg. See also J. Tann and M. Breckin (1978), 'The International Diffusion of the Watt Engine, 1785–1825', *Economic History Review*, **30**, 541–64. According to Tann and Breckin, the highest number of engines exported was to the Netherlands, 21. This is almost certainly an underestimation. Atmospheric engines were supposedly being made in the Liège area by the 1730s.
29. Archives d'état, Mons, Charbonnages, Bois du Luc, Régistre no. 1. See also Archives nationales, Paris, 109/Mi, on Anzin mines and engines in place by 1771–72. For the educational side of these developments, see Hervé Hasquin (ed.) (1983), *La Vie Culturelle dans nos Provinces au xviiie Siècle*, Brussels: Crédit Communal de Belgique, pp. 132–3.
30. J.G. De Bruijn (ed.) (1977), *Inventaris van de prijsvragen uitgeschreven door de Hollandsche Mattschappij der Wetenschappen, 1753–1917*, Haarlem: Hollandsche Mattschappij der Wetenschappen, p. 202. Answers have been preserved in the Society's archives in Haarlem. By the mid-nineteenth century there were approximately 300 steam engines at work in the Netherlands and over 2000 in Belgium; see Harry Lintsen and Rik Steenaard (1991), 'Steam and Polders. Belgium and The Netherlands, 1790–1850', *Tractrix. Yearbook for History of Science, Medicine...*, **3**, 122. They claim that in 1800 there were fewer than six engines in the Dutch Republic and 67 in Belgium, but that figure is contradicted by Tann and Breckin, 'The International Diffusion': see note 28 above.
31. There are tracts by G.A. Coleveldt, C. Velsen, S. Lakenman, J. van den Burchgraef and C.L. Ducrest, and material on J. Blanken and A. Blanken. On Blanken see *Jan Blanken. De Physique existentie deze lands. Catalogus Rijksmuseum*, 1987.
32. *Catalogus van boeken, der opendare Lees-Bibliotheek, Tot nut van 't Algemeen*, Middelburg, 1799; books in Dutch on mechanics were by Desaguliers, Steven, Derham, and the abbé Nollet. On the low state of experimental physics in the Republic, see J. Heilbron (1979), *Electricity in the 17th and 18th Centuries*, Berkeley: University of California Press, p. 138; and in France, p. 144.
33. For one man's struggle to change the situation see H.A.M. Snelders (1980), 'Lambertus Bicker (1732–1801). An Early Adherent of Lavoisier in The Netherlands', *Janus*, **67**, 103–22. For a contemporary critic of Dutch inventiveness, see Ijsbrand van Hamelsveld (1791), *De zedelijke toestand der Nederlandsche natie, op het eende der achttiende eeuw*, Amsterdam, esp. pp. 215–6 and p. 244. See also *De Koopman*, vol. I (1768), p. 40, pp. 332–3; vol. IV, p. 172, for explicit comparison with England. Early Dutch Newtonians like s'Gravesande were called upon to fulfil the tasks of civil engineers as early as the 1730s. See Royal Library, The Hague, MS 128 B 3 with his reports from the 1730s. S'Gravesande's successor in the chair of physics in Leiden, J. Allamand, appears to have done little of this sort of work.
34. For an overview of how the spying worked, see J.R. Harris (1996), 'Law, Espionage, and the Transfer of Technology from Eighteenth-Century Britain', in Robert Fox (ed.), *Technological Change. Methods and Themes in the History of Technology*, Amsterdam: Harwood Academic Publishers, pp. 122–35.
35. Archives nationales, Paris, MS F12 993, date 1747: much of this concerns the effort to grant John Kay an *arrêt*.
36. See Archives nationales, Paris, Marine G 106, for 1686–1779: f. 38 for a report of 1737 on an English company at Guadalcanal; f. 42 for a certificate from Desaguliers; f. 69 for an item dated 1736 where a pump is described as being able to elevate water 'que la manière des anglais'. In MSS Marine G 108, a mechanist is brought over from London to set up a factory for making pumps; and in Marine G 110, f. 138, an English engineer explains that Desaguliers was his master: letter from Thomas Stephens, 29 August, 1758.
37. B.H.V.P., Paris, call no. 951658; J.J.C. Farcot (1790), *Questions constitutionelles sur le commerce et l'industrie ... proposés à l'assemblée nationale par des negocians françois*, Paris.

38. B.H.V.P., Paris, call no. 15703, *Première exposition des produits de l'industrie française*, year 4. By the year 10 (1802) the number of industrial machines had increased markedly; see catalogue, call no. 910705, for guide book of exposition.
39. L.S. Le Normand and J.G.V. de Moléon (1824), *Description des expositions des produits de l'industrie française faites à Paris depuis leur origine jusqu'à celle de 1819 inclusivement* ..., 4 vols, Paris, vol. 1, p. 19: 'industrious men rarely search for fortune; they forget themselves and dream only of the progress of industry.' In 1819 the first prize at the exposition went to a student at the school at Saint-Quentin for constructing a steam engine.
40. Archives nationales, Paris, F12 1556, 'Rapport de l'agence des arts et manufactures à la Commission d'agriculture et des arts ... Atelier des Perfectionnement'.
41. Jean Dhombres (1980), 'L'enseignement des mathématiques par la "méthode révolutionnaire". Les leçons de Laplace à l'Ecole normale de l'an III', *Revue d'Histoire des Sciences*, **33**, 315–48.
42. Prefecture de Police, 30 May 1807, found in Archives nationales, Paris, F12 502; the survey from all over the country can be found in F12 2467.
43. See C. Stewart Gillmor (1971), *Coulomb and the Evolution of Physics and Engineering in Eighteenth Century France*, Princeton: Princeton University Press, pp. 48–62.
44. Archives nationales, Paris, F12 1556, 8 August 1793.
45. Archives nationales, Paris, F12 1556: this dossier covers material from 1778 to the year 10; a further report of the year 3 (18 Pluviose 1794) discusses how to use the cotton jennies and carding machines found in Brussels which had been there some ten years. In the year four a government agent in Aix was instructed to trace the designs of all the machines found there and to try to get workers to leave and go to France.
46. J.A.C. Chaptal (1893), *Mes Souvenirs sur Napoleon*, Paris: E. Plon, Nourrit et Cie, p. 99 (published posthumously).
47. M. Le Comte de Chaptal (1819), *De l'Industrie françoise*, 2 vols, Paris: Antoine-Augustin Renouard. For an early attempt to compare British and French industry, see John Nickolls (1754), *Remarques sur les Avantages et les Désavantages de la France et de la Gr. Bretagne*, Leyden.
48. J.A.C. Chaptal (1800), *Essai sur le Perfectionnement des Arts Chimiques en France*, Paris: Deterville; this work is devoted to education.
49. Archives nationales, Paris, F12 2200, 8 April 1817. In 1778 Watt claimed that he had installed 27 of his engines in England; for his list see Archives nationales, Marine G 110, f. 178.
50. J.T. Desaguliers (1744–45), *A Course of Experimental Philosophy*, 2nd edition, 2 vols, London: W. Innys, vol. ii, p. 412.
51. Biernacke, *The Fabrication of Labor*; see note 2.
52. Michael Adas (1989), *Machines as the Measure of Men. Science, Technology, and Ideologies of Western Dominance*, Ithaca: Cornell University Press, esp. chapter two.

APPENDIX

Sites visited for the project, 'The Cultural Foundations of Early Industrialization'

United Kingdom

Birmingham City Library, papers of the Boultons and the Watts.
Bristol Record Office and Bristol Central Library, papers relevant to harbour projects; the MSS of Richard Bright.
Sheffield Public Library, coal mine owners.
Derbyshire Public Record Office and Derby Public Library, the Strutts and Richard Arkwright.
Cambridge University Library, scientific notebooks and lectures; the Fitzwilliam Museum, the Strutts.
Manchester, The John Rylands Library.
Oxford, Manchester College, records from Warrington Academy.
London, Royal Society of London; British Library, MS of John Smeaton; The Institution of Civil Engineers, Smeaton, Yeoman; Dr. Williams' Library; House of Lords, MS archives for testimonies by engineers working for canal companies.
Spalding, Lincs., Records of the Spalding Gentlemen's Society.

France

Hérault, Archives départementales.
Lyon, Archives municipales and Archives départementales du Rhone, the silk industry and the dealings in particular of John Badger with agents of the state.
Nantes, Archives départementales du Loire Atlantique, records on the harbour, and records to do with M. Jary, who attempted in the 1780s to import Watt's engine. He had a concession to mine in the region.
Troyes, Bibliothèque municipale and archives départementales, the cotton industry and the issue of mechanization.
Paris, Archives nationales, F12 and F14; archives of the Marine, Papers of Piou, the Milne family, Overkampf, reports from academicians; Bibliothèque nationale, expositions; Préfecture de la Police, and Arsenal, industrial spies; Ecole nationale des ponts et chaussées; Ecole centrale des arts et métiers.

Belgium (all state archives)

Arlon, French revolutionaries and the attempt to transform local culture.
Saint Hubert, MS of its mid-eighteenth-century abbé.
Liège, local museum of industry.
Ghent, textiles.
Brussels, privy council records.
Mons, coal mining and steam.

The Netherlands

The Hague, national archives and library.
Amsterdam, University Library and the Gemeente Archief.
Rotterdam, Gemeente archief.
Tilburg, Gemeente archief.
Middelburg, Gemeente archief, Rijkswaterstaat and Provincial Archives.
Maastricht, Gemeente Archief and Rijksarchief Limburg.
Arnhem, Rijksarchief.
Leiden, University Library, for Fahrenheit, Marchand, s'Gravesande.
Friesland, Rijksarchiv Friesland, for Allamand.
Haarlem, Teyler's Museum, and private archives of the Hollandsche Maatschappij den Wetenschappen.
Gelderland, Rijksarchief.

U.S.A.

Delaware, the Hagley Museum, MSS of the Dupont.
Philadelphia, American Philosophical Society; Van Pelt Library of the University of Pennsylvania.
New York, Public Library.
Washington, D.C., Science Museum and Library of Congress

5. Science, Practice and Innovation in the Age of Natural Dyes, 1750–1860

Robert Fox

The belief that science was either potentially or in reality an important, even essential fount of technological improvement was already firmly established by the mid-eighteenth century. Some decades earlier in Paris, the engagement of members of the Académie Royale des Sciences as advisers in practical ventures had offered a Colbertian version of the belief,[1] and Larry Stewart's book, *The Rise of Public Science*, and *Public & Private Science* by Alan Morton and Jane Wess have shown how, in a very different form, the belief took root in early-eighteenth-century England, stimulating an unprecedented growth in the related activities of lecturing, semi-popular publishing, and instrument making.[2] In 1751, therefore, Diderot was expressing what most contemporaries would have perceived as a truism when he observed, in the article 'Art' in the *Encyclopédie*, that the success of any technology, or other practical activity, depended on the constant harmonious interaction between 'spéculation' and 'pratique' and hence between the world of the *savant* and that of the *artiste*.[3] A century later, a similar conviction underlay Lyon Playfair's warning that chemical manufacturing in Britain would fall behind its European competitors unless the interests of science, in its relations with industry, were promoted,[4] as it also underlay the lament that Sir Henry De la Beche voiced in his inaugural discourse at the opening of the School of Mines and of Science applied to the Arts (later to become the Royal School of Mines) in London in 1851. As De la Beche complained, all too many of his contemporaries in mining and geology tended to regard experience without scientific knowledge as 'more trustworthy than the like experience with it'.[5]

The persistence of the kind of views that Diderot, Playfair, and De la Beche expressed between the mid-eighteenth and mid-nineteenth centuries deserves attention in any study of technological revolutions, if only because the evidence for the truth of what was being alleged was exceedingly patchy.[6] A French debate about the efficiency of steam engines and other heat engines between 1815 and 1825 makes the point well.[7] In it, scientific exchanges of great sophistication on the conditions that would yield the greatest economy of fuel had as their scientifically finest product the brief, solitary masterpiece of Sadi Carnot, *Réflexions sur la Puissance Motrice du Feu* (1824), in which Carnot noted a twenty-fold disparity between the motive power delivered by an ideal engine (operating without friction or losses of heat by conduction and radiation) and that of the best working engine known at the time.[8] The gap between theory and practice was apparently of little concern to a former *polytechnicien* who had been fashioned not only by his two years at the Ecole Polytechnique but also by a father, Lazare, skilled in the mathematical theory of machines. Nevertheless, it pointed unerringly to the irrelevance of theory to practice at one of the cutting edges of power technology, as contemporary engineers, who almost without exception ignored the book, seem to have recognized.

It is not my intention to suggest that this degree of irrelevance was characteristic of all areas of industrial practice: studies of the substantial contribution of Lavoisier and other *savants* to the work of the French government's Régie des Poudres et Salpêtres after 1775 would themselves make such a view untenable.[9] The point is simply that what was routinely presented between 1750 and 1850 as a blanket confidence in the efficacy of science concealed at least some areas in which the confidence was manifestly unjustified and others in which the rôle of science in industry, though significant, was at best vaguely articulated. In focussing, in this chapter, on chemistry, and more specifically on the chemistry of natural dyes, I am examining one of the latter areas in order to illustrate, for this case, the complexities and elusive, evolving character of the relations between science and practice.

My starting point is a prominent paean to the contribution of science to manufacturing that placed special emphasis on chemistry. It appeared in J.A.C. Chaptal's two-volume *De l'Industrie Françoise*, published in 1819. As a chemist who had been actively involved in the heavy chemical industry while pursuing the early stages of an academic career in Montpellier, Chaptal looked back on three decades in which, as he saw it, science had moved to the heart of industrial activity. The context that he invoked was a broad one in which Lavoisier's contribution in raising chemistry to the status of a 'positive science' had been as important for the industrialist as it had been for the *savant*.[10] But with breadth there went an absence of differentiation. *De l'Industrie Françoise* made no serious attempt to distinguish between the diverse activities that could be embraced by the term 'science', or even by the more precise term 'science chimique'. Chaptal shifted, in fact, between two distinct positions: one according to which the chemical industry since 1770 had drawn on chemistry conceived rather narrowly as a body of theory, the other that saw chemistry as a source of a body of accumulated knowledge and of practices, such as systematic testing and experimental method, that were distinct from the particular theories that resulted from or informed them. It is seldom clear from Chaptal's work, or from most other polemics in a similar vein, how the interaction between science and practice occurred in particular instances, or how the nature of the interaction varied from one technological context to another or with time.

Such reservations about Chaptal's view of what he was concerned to present as a glorious period in the history of chemistry serve to remind us how cautious we need to be, and how far we still have to go, in our historical analysis of the theoretical underpinnings of technology in the Industrial Revolution. The taxonomy of the various relations between science and practice for which Barbara Whitney Keyser called in 1990 and towards which she delivered some illuminating opening shots (in the particular context of Claude-Louis Berthollet's work on bleaching and dyeing) has yet to be established.[11] Most of us, I think, would have little difficulty in accepting Charles Gillispie's persuasive argument of forty years ago that Nicolas Leblanc's process for the production of artificial soda was devised without the significant aid of contemporary theory, which was quite unable to account for the complex processes involved.[12] The impotence of theory is also demonstrated by the timing of some of the most important metallurgical innovations of the later eighteenth century, which predated or bypassed the new chemistry and rested, for such analysis as was made of them, on the doctrine of phlogiston or other conceptions that contributed little.[13] But these instances only bear on one particular form of interaction, in which the efficacy of science is equated with and judged by the efficacy of theory as a source either of understanding or of new practices. By this criterion, the chemistry of the mid- and later eighteenth century, whether new or old, failed the test.

What we know of the scientific background of the chemical communities of the main textile areas of this period and on into the first half of the nineteenth century tends to reinforce the point that Gillispie was making. Whether we look to Lancashire or to the Haut-Rhin, to Thessaly or to Catalonia, we find the majority of leading colourists engaging only in the most rudimentary way with contemporary theory.[14] The processes of dye-making and dyeing, like those involved in the production of most heavy chemicals, were simply too complicated for what theory could 'throw at them'.[15] Until the mid-nineteenth century, it was normal for a colourist to receive, at best, a brief, elementary grounding in 'academic' chemistry. Even in the technologically advanced area of Mulhouse, a visit of a few weeks to the laboratory of a Parisian chemist would be seen as a sufficient preparation for the training that really mattered, gained by poring over the vats of dye beside a master of the craft. Jean-Michel Haussmann, one of the greatest Alsatian colourists of the last quarter of the eighteenth century, is typical in this respect. The grounding he received as a young man during an extended stay in J.H. de Schulé's calico-printing works in Augsburg prepared him as no other initiation could have done for the distinguished tradition of research and development that he went on to pursue in his own factory near Colmar.[16] Learning on the job was equally important in Britain. There, John Mercer was a notable example of a colourist who built an outstanding career on the kind of information he could garner from James Parkinson's instructive but elementary *Chemical Pocket-Book or Memoranda Chemica* (London: H.D. Symons, four editions between 1800 and 1809). Despite his lack of a systematic foundation in science, between his first appointment as an 'experimental chemist' in the Oakenshaw print works in Lancashire in 1818 and his death in 1866, Mercer became a leader of his profession, a Fellow of the Royal Society, and a model of 'rags to riches' success worthy of the highest reaches of Samuel Smiles's pantheon of Victorian paragons.[17]

While Mercer's career illustrates the modest level of formal chemical training that was thought to be necessary until and even for some time after the introduction of synthetic dyes, it also exemplifies a trend towards greater specialization that became apparent in all the main areas of textile production from the 1820s. To return to Mulhouse, Haussmann had pursued his chemical activity in addition to the many other responsibilities he had as owner of the factory in which he worked. But, two generations on, the outstanding Mulhousian colour chemist of the second quarter of the nineteenth century, Henri Schlumberger, was a specialist who began his career as an apprentice colourist in 1818 at the age of fifteen, rising quickly to the status of an international authority and retaining that position until he ended his career, over thirty years later, as chief chemist with Dollfus-Mieg et Cie.[18]

Specialist colourists, such as Mercer and Schlumberger, constituted a small and generally rather closed aristocracy of competence. The sketchiness of their exposure to theory was compensated for by what they saw as far more important elements in their armoury. These included not just experience but more specifically, in most cases (here Mercer was an exception), experience acquired in a variety of locations. The Lancashire dynasty of the Lightfoots, who maintained close links with textile manufacturers in Rouen, Paris, and Alsace,[19] were by no means atypical; in fact, there were few leading colourists in either France or Britain who did not spend decisive periods across the Channel. This method of transmitting chemical knowledge and skills, comparable with the process by which skills in the operation of steam engines and then the new and more mechanized textile machinery of the early and mid-nineteenth century were passed on, was evidently far more effective than transmission by books.[20] It also reinforced the eminence that most colourists enjoyed and carefully cultivated in their immediate communities since it encouraged the establishment of local networks of expertise of the kind that existed in the Société Industrielle de Mulhouse. The details of the practices that were developed in these networks only entered the realm of public knowledge imperfectly, and although courses in or relevant to colour chemistry were beginning to be offered, especially in Germany, by the early years of the century and more ambitiously in most countries by the 1840s, the world of the leading colourists remained small, secretive, and hard to penetrate. Until well into the age of synthetic dyes, colourists tended to work in ones and twos, and they would often be the owners of a calico printing works or, as the process of specialization advanced, members, like Henri Schlumberger and other colourists in his family, of a textile dynasty. Their salaries reflected this closeness to the economic and social summits of the societies in which they worked and lived. At Dollfus-Mieg et Cie, for example, an annual salary of 12,000 francs (about £500) for a colourist was common, at a time when a French university professor in the provinces would be fortunate to receive half that amount.[21]

The appointment as 'chemical manager' that Lyon Playfair held briefly at the calico printing works of James Thomson in Clitheroe in 1841–42 could be seen, and clearly was seen by contemporaries, as a significant departure from the existing order, and even as a threat to it. Playfair had undergone a serious academic training in chemistry in Glasgow, London, and most importantly Giessen, where he had taken a Ph.D. for work performed under Justus von Liebig, and he brought to bear a command of theory and a competence in precise analysis of a kind that few, if any, other colourists possessed.[22] Mercer later recalled the scepticism with which the new generation of chemists, armed with Ph.D.s and other formal qualifications, were viewed by most of the traditional 'colour-men' and managers as they began to trickle into the textile industry, from the 1840s.[23] They were seen as ill-suited to the tasks to which they were put, which demanded a particular mixture of artistic and mechanical know-how, as well as chemical skills of a very specific, practical kind.

One justification for the scepticism lay in the speed with which techniques in the relevant areas of the textile industry were changing in the second quarter of the nineteenth century, placing new demands on the kind of experience that only a constant engagement with the day-to-day practices of production could refine. The spread of steam power, advances in cylinder printing, power looms, the self-acting mule, and improved combing techniques all demanded adjustments on the part of designers and colourists, coinciding in the 1830s and 1840s with a period of accelerated inventiveness in the production of new dyes that greatly extended the traditional repertoire.[24] Antimony orange, Prussian blue, chrome yellow, chrome orange, and manganese bronze were just some of the numerous mineral colours that were developed in this period, along with 'steam colours' offering subtly different shades;[25] and the decade of the 1850s was to see the development of one of the dyeing sensations of the mid-century, murexide, a dye that lay at the interface between the categories of the natural and the artificial.[26] This level of innovation reinforced the traditional colourists' sense of their success, making it hard to see what need there was for the sophisticated knowledge of the academically trained chemists. Yet at the same time it made colour chemistry an object of ever greater intellectual interest, outside as well as within the textile industry.

The interest was manifested, about the mid-century, by the growth in the number of textbooks treating colour chemistry, and further stimulated by the preparation of the first artificial dyestuff, mauve, by W.H. Perkin in London in 1856. Rudolf von Wagner's *Die chemische Technologie* (Leipzig: O. Wigand, 1850, and 12 subsequent editions, with minor changes of title, to 1892) and Paul Schützenberger's *Traité des Matières Colorantes* (2 volumes, Paris: Victor Masson et Fils, 1867) were typical literary products of this period that demonstrated the heightened sense, on the part of their authors, of the value of studies of dyeing that offered not only recipes and descriptions of processes, in the manner of Edward Parnell's *Dyeing and Calico-Printing* (London, 1849),[27] but also sustained theoretical analysis. A reading of these books would suggest, as was intended, that the worlds of the print works and the university laboratory were adjacent, even if they did not coincide, and that the future of dyeing and dyestuffs manufacture would depend on an opening of the realm of practice to the enlightenment of theory and to the refinement of advanced bench techniques of the kind that a university education or other form of advanced training alone could convey. Things, however, were not quite so, and, at least in the short run, they did not work out that way.

The evidence that the proud distinctiveness of the traditional colourists was as strong as ever at the mid-century is abundant. Here Playfair's assessment that Mercer's chemical language was 'powerful, though necessarily provincial' and the qualification that while Mercer had ideas that were 'original and full of sagacity', he was 'unable to work them out and explain them to others', is significant: it marks the gap which, for Playfair, still existed between him and a man whom he counted as a long-standing friend (not least as a fellow-member of a discussion group that Playfair organized at his home in south Lancashire in the early 1840s) and whom he genuinely respected.[28] Playfair's point was not that Mercer's method was based on uninformed trial and error, but that Mercer belonged to what, to use Larry Holmes's terminology, was a different 'chemical culture'.[29] It was a culture that fostered techniques and a kind of interaction between science and practice far removed from those emerging from the academic culture of *savants* who, for a century, had struggled to engage with such questions as how colours were bonded to the fibres to which they were applied (was the bond mechanical or chemical?) or why mordants effected a resistance to washing. As Luisa Dolza has shown, those questions, and much simpler ones, had defeated the learned committee that the Royal Academy of Science of Turin set up in 1789 to study the techniques of dyeing in Piedmont: the academicians found the world of the dyers almost impenetrable and never produced their promised report.[30] Even in the second half of the nineteenth century, the scientific understanding of the processes of dyeing was far from complete: the widely shared belief that both mechanical and chemical factors played a part still left these discussions far short of a state of 'closure'.[31]

The fact remains, however, that even if many of the most arcane facets of the science of dyeing remained obscure, there had been great progress in knowledge and understanding on the part of practising colourists by the 1850s, much of it achieved through an openness to academic chemistry. But the marks of an increasingly easy osmosis should not lead us to exaggerate the debt to the academic world or to see that world as the stimulus to improvement, still less as a source of 'ready-made' innovation. The fact that chemical concepts of unprecedented sophistication were in use by the mid-century cannot conceal the fact that most dyers and dye-makers remained rooted in cultures distinct from those of the academically trained chemical commentators who had viewed their work with such interest since the later eighteenth century.

In conclusion, I edge briefly into the age of synthetic dyes to make a point about continuity. From the late 1850s, in the wake of Perkin's discovery of mauve, most colourists simply adopted the new aniline dyes as an addition to their repertoire, and they continued to work in small, simple laboratories and with standards of precision little different from those that had sufficed when they were dealing with natural dyes. The departure from this state of affairs took place slowly and in two stages. The first stage began in the early 1860s, when chemists from academic backgrounds, many of them working in institutions of higher education, quite suddenly assumed a central rôle in the patent disputes that became, from the start, a feature of the age of synthetic dyes. In this development, the complex litigation associated with the notorious fuchsine affair in France in 1863 – an affair that turned on the contested rights to manufacture the new red aniline dye, fuchsine (also known as magenta) – was decisive.[32] Here at last was a matter that called unequivocally for laboratory skills and an authoritative command of theory of the kind that only leading chemists, such as Liebig's distinguished pupil A.W. Hofmann, with their intellectual roots outside industry, could dispense.[33]

Although legal disputes of the kind that erupted over fuchsine gave chemical science in the Liebig tradition a firm foothold in the dyestuffs industry, they left the main task of innovation in hands little different from those that had pursued it with mounting success for over a century. On this score, change only came in the second of the two stages to which I have referred. It began with the introduction of the alizarin family of dyes by Carl Graebe and Carl Liebermann in 1869, but was not fully consolidated until the proliferation of the azo-dyes (far more numerous than any previous category of dyestuffs) in the 1880s. It was in these later years, initially at BASF and Hoechst in the 1870s and then at Bayer, that the well-founded company laboratory finally assumed its position as an essential and formally designated element in the strategy of innovation.[34] In doing so, it at last made a concrete reality of the perception, which had fascinated observers since the eighteenth century, of science as a motor of technological advancement and economic success.

ACKNOWLEDGEMENTS

It is a pleasure to acknowledge the debt I owe to colleagues who have helped my thinking about the issues raised in this paper. My general approach and interpretations have been developed since the late 1980s in the collaboration with Anna Guagnini that has led to our forthcoming book, *Laboratories, Workshops, and Sites. Concepts and Practices of Research in Industrial Europe, 1800–1914*, to be published in 1998 by the Office of the History of Science and Technology, University of California at Berkeley. On the history of natural dyes, I am indebted above all to Dr Agustí Nieto-Galan, from whom I learned much during his stay in Oxford as a Spanish government post-doctoral fellow between 1993 and 1995. I am also grateful to the colleagues who attended an international workshop on the history of natural dyes, held at Oriel College, Oxford, from 4 to 6 January 1996. The papers from this workshop, which was financed by the European Science Foundation as part of its international project on the history of chemistry between 1789 and 1939, are now being prepared for publication.

NOTES

1. Robin Briggs (1991), 'The Académie Royale des Sciences and the Pursuit of Utility', *Past & Present*, **131**, 38–88.
2. Larry Stewart (1992), *The Rise of Public Science. Rhetoric, Technology, and Natural Philosophy in Britain, 1660–1750*, Cambridge: Cambridge University Press; Alan Q. Morton and Jane A. Wess (1993), *Public & Private Science. The King George III Collection*, Oxford: Oxford University Press in association with the Science Museum.
3. Denis Diderot and Jean Le Rond d'Alembert (eds) (1751–80), *Encyclopédie, ou Dictionnaire Raisonné des Sciences, des Arts et des Métiers*, 17 vols plus plates, supplements, and index, Paris, &c.: Briasson et al., &c., vol. 1, p. 714.
4. Lyon Playfair (1852), 'On the Chemical Principles involved in the Manufacture of the Exhibition', in *Lectures on the Results of the Great Exhibition of 1851, delivered before the Society of Arts, Manufactures, and Commerce*, London: David Bogue, pp. 189–208.
5. Sir Henry T. De la Beche (1852), 'Inaugural Discourse, delivered at the Opening of the School of Mines and of Science applied to the Arts, 6th November 1851', in *Records of the School of Mines and of Science applied to the Arts*, **1**, part 1, 1–22, p. 20.
6. The patchiness of the evidence is demonstrated in Robin Briggs's study of the utilitarian aspirations of the Académie Royale des Sciences from its foundation in 1666 until the Revolution. See Briggs, 'The Académie Royale des Sciences'.
7. Robert Fox (1976), 'The Challenge of a New Technology: Theorists and the High-Pressure Steam Engine before 1824', in *Actes de la Table Ronde 'Sadi Carnot et l'Essor de la Thermodynamique', Paris, Ecole Polytechnique (Juin 1974)*, Paris: Editions du Centre National de la Recherche Scientifique, pp. 149–67.
8. Nicolas Léonard Sadi Carnot (1824), *Réflexions sur la Puissance Motrice du Feu*, Paris: Bachelier, p. 117. For a comment on this disparity, see the modern English edition of the *Réflexions* – Carnot ([1824] 1986), *Reflexions on the Motive Power of Fire. A Critical Edition with the Surviving Manuscripts*, transl. and ed. by Robert Fox, Manchester: Manchester University Press, and New York: Lilian Barber Press, pp. 19 and 158.
9. Lucien Scheler (1973), 'Lavoisier et la Régie des Poudres', *Revue d'Histoire des Sciences*, **26**, 193–222; Charles Coulston Gillispie (1980), *Science and Polity at the End of the Ancien Regime*, Princeton, NJ: Princeton University Press, pp. 50–73; Seymour H. Mauskopf (1995), 'Lavoisier and the Improvement of Gunpowder Production', *Revue d'Histoire des Sciences*, **48**, 95–121; Patrice Bret (1996), 'Lavoisier et l'Apport de la Chimie Académique à l'Industrie des Poudres et Salpêtres', *Archives Internationales d'Histoire des Sciences*, **46**, 57–83.
10. J.A.C. Chaptal [M. le comte Chaptal] (1819), *De l'Industrie Françoise*, 2 vols, Paris: Antoine-Augustin Renouard, vol. 2, pp. 36–40.
11. Barbara Whitney Keyser (1990), 'Between Science and Craft: the Case of Berthollet and Dyeing', *Annals of Science*, **47**, 213–60.
12. Charles Coulston Gillispie (1957), 'The Discovery of the Leblanc Process', *Isis*, **48**, 152–70; Charles Coulston Gillispie (1957), 'The Natural History of Industry', *Isis*, **48**, 398–407, pp. 401–2.
13. Bertrand Gille (1947), *Les Origines de la Grande Industrie Métallurgique en France*, Paris: Editions Domat Montchrestien, pp. 85–100; Denis Woronoff (1984), *L'Industrie Sidérurgique en France pendant la Révolution et l'Empire*, Paris: Editions de l'Ecole des Hautes Etudes en Sciences Sociales, pp. 327–9.
14. This comment is made in the light of the contributions to the workshop on the history of natural dyes to which I refer in the Acknowledgements. The paper by Bensaude-Vincent and Nieto-Galan has been particularly helpful to me in thinking about the relations between theory and practice. See Bernadette Bensaude-Vincent and Agustí Nieto-Galan (1996), 'The Theories of Dyeing in Europe, 1750–1900: a View of a Long-Standing Controversy through the Works of Jean-François Persoz', paper given at the European Science Foundation workshop on the history of natural dyes, Oriel College, Oxford, 4–6 January 1996.
15. Archibald Clow and Nan L. Clow (1952), *The Chemical Revolution. A Contribution to Social Technology*, London: Bathworth Press; L.F. Haber (1958), *The Chemical Industry during the Nineteenth Century. A Study of the Economic Aspect of Applied Chemistry in Europe and North America*, Oxford: Clarendon Press, pp. 1–79; John Graham Smith (1979), *The Origins and Early Development of the Heavy Chemical Industry in France*, Oxford: Clarendon Press.

16. Robert Fox (1984), 'Science, Industry, and the Social Order in Mulhouse, 1798–1871', *The British Journal for the History of Science*, **17**, 127–68, pp. 139–40.
17. Edward Andrew Parnell (1886), *The Life and Labours of John Mercer, F.R.S., F.C.S. etc., the Self-Taught Chemical Philosopher*, London: Longmans, Green, & Co; Agustí Nieto-Galan (1997), 'Calico Printing and Chemical Knowledge in Lancashire in the Early Nineteenth Century: the Life and "Colours" of John Mercer', *Annals of Science*, **54**, 1–28.
18. Fox, 'Science, Industry, and the Social Order', p. 144.
19. Nieto-Galan, 'Calico Printing', p. 9.
20. André Brandt (1967), 'Apports Anglais à l'Industrialisation de l'Alsace au Début du XIXe Siècle', *Bulletin de la Société Industrielle de Mulhouse*, **1**, 27–41; André Brandt (1970), 'Travailleurs Anglais dans le Haut-Rhin dans la Première Moitié du XIXe Siècle', in *Actes du 92e Congrès National des Sociétés Savantes. Strasbourg et Colmar 1967. Section d'Histoire Moderne et Contemporaine*, 2 vols, Paris, vol. 1, pp. 297–312.
21. Ernst Homburg (1983), 'The Influence of Demand on the Emergence of the Dye Industry: the Roles of Chemists and Colourists', *Journal of the Society of Dyers and Colourists*, **99**, 325–34, p. 328; Fox, 'Science, Industry, and the Social Order', p. 144n.
22. T. Wemyss Reid (1899), *Memoirs and Correspondence of Lyon Playfair, First Lord Playfair of St. Andrews, P.C., G.C.B., LL.D., F.R.S. &c.*, London: Cassell and Company, Limited, pp. 42–52.
23. Parnell, *John Mercer*, p. 245–6.
24. Fox, 'Science, Industry, and the Social Order', pp. 155–6.
25. Nieto-Galan, 'Calico Printing', p. 12.
26. Homburg, 'The Influence of Demand', pp. 329–30; Anthony S. Travis (1993), *The Rainbow Makers. The Origins of the Synthetic Dyestuffs Industry in Western Europe*, Lehigh, PA: Lehigh University Press, pp. 43–5.
27. Edward Andrew Parnell (1849), *Dyeing and Calico-Printing*, London: Taylor, Walton, and Moberly [reprinted from Parnell's *Applied Chemistry in Manufactures, Arts, and Domestic Economy*, 1844].
28. Reid, *Lyon Playfair*, pp. 53–5; Robert H. Kargon (1977), *Science in Victorian Manchester. Enterprise and Expertise*, Manchester: Manchester University Press, pp. 89–90; Nieto-Galan, 'Calico Printing', p. 8.
29. Frederic Lawrence Holmes (1995), 'Beyond the Boundaries: Concluding Remarks on the Workshop', in Bernadette Bensaude-Vincent and Ferdinando Abbri (eds), *Lavoisier in European Context. Negotiating a New Language for Chemistry*, Canton, MA: Science History Publications, pp. 267–78, pp. 274–7; Nieto-Galan, 'Calico Printing', p. 2.
30. Luisa Dolza (1995), 'The Struggle for Technological Independence: Textiles and Dyeing in Eighteenth Century Piedmont', unpublished University of Oxford M.Litt. thesis; Dolza (1996), 'Dyeing in Piedmont in the late Eighteenth Century', *Archives Internationales d'Histoire des Sciences*, **46**, 75–83, pp. 78–81.
31. Bensaude-Vincent and Nieto-Galan, 'The Theories of Dyeing in Europe'.
32. Haber, *The Chemical Industry*, pp. 82, 112–14, and 201–2; Henk Van den Belt (1992), 'Why Monopoly Failed: the Rise and Fall of Société La Fuchsine', *The British Journal for the History of Science*, **25**, 45–63; Travis, *The Rainbow Makers*, pp. 67–87 and 104–37.
33. Anthony Travis has brought out clearly the importance of the litigation stimulated by the fuchsine affair, although he also stresses that Hofmann's detailed investigations of the aniline dyes had roots going back to the discovery of mauve by Perkin (who was his pupil at University College in London) in 1856. See Anthony S. Travis (1992), 'Science's Powerful Companion: A.W. Hofmann's Investigation of Aniline Red and its Derivatives', *The British Journal for the History of Science*, **25**, 27–44, and Travis, *The Rainbow Makers*, pp. 104–37.
34. Haber, *The Chemical Industry*, pp. 83–4; John Joseph Beer (1959), *The Emergence of the German Dye Industry* (Illinois Studies in the Social Sciences, vol. 44), Urbana, IL: Illinois University Press, pp. 70–93; Georg Meyer-Thurow (1982), 'The Industrialization of Invention: a Case Study from the German Chemical Industry', *Isis*, **73**, 363–81; Ernst Homburg (1992), 'The Emergence of Research Laboratories in the Dyestuffs Industry, 1870–1900', *The British Journal for the History of Science*, **25**, 91–111; Travis, *The Rainbow Makers*, pp. 163–230.

6. James Watt, Heroic Invention and the Idea of the Industrial Revolution

Christine MacLeod

In the nineteenth century, to be buried or commemorated in Westminster Abbey was an honour reserved for a select few, not least because space, both above and below ground, was becoming very short. By the middle of the century, complaints were regularly heard that the Abbey had become 'little more than a statuary's yard', where the 'illustrious dead' were so crowded that they could be 'seen climbing upon each other's shoulders'.[1] Chantrey's 'gigantic' statue of James Watt was perhaps the most controversial of these monuments. Dean Stanley described how, in 1834, with considerable difficulty (and barely concealed reservations), it had been manoeuvred into place.

> Of all the monuments in the Abbey, perhaps this is the one which provokes the loudest execrations from those who look for uniformity of design, or congeniality with the ancient architecture. Well may the pavement of the church have cracked and yawned, as the enormous monster moved into its place, and 'disclosed to the eyes of the astonished workmen rows upon rows of gilded coffins in the vaults beneath; into which, but for the precaution of planking the area, workmen and work must have descended, joining the dead in the chamber of death.' Well might the standard-bearer of Agincourt, and worthies of the Courts of Elizabeth and James, have started from their tombs in St Paul's Chapel, if they could have seen this colossal champion of a new plebian art enter their aristocratic restingplace, regardless of all proportion, or style, in all the surrounding objects. Yet, when we consider what this vast figure represents, what class of interests before unknown, what revolutions in the whole framework of modern society, equal to any that the Abbey walls have yet commemorated, there is surely a fitness even in its very incongruity.[2]

The commissioning of this colossal statue in 1824 and the equally massive public subscription which financed it are central to this chapter, but so is the sentiment expressed by Dean Stanley. Watt's memorial was both stylistically and culturally an intruder in this burial place of monarchs, courtiers and noble warriors, but, Stanley concedes, it was an appropriate one, as the harbinger of a new age, of 'revolutions in the whole framework of modern society'. James Watt had posthumously become the benign personification of the industrial revolution, the genius behind the engine which powered British manufacturing success, even the British empire and military predominance. He attained this popular image, I shall argue, largely through his adoption as the figurehead of an alliance of reforming 'men of science', manufacturers and economic liberals in the 1820s. The heroic depiction of Watt and the steam engine that gained ground in the second quarter of the nineteenth century moulded the concept of 'the industrial revolution', long before Arnold Toynbee famously (and grimly) promulgated the phrase in his lectures of 1882.

British historiography of the industrial revolution usually begins with Toynbee's *Lectures on the Industrial Revolution in England*.[3] Toynbee's highly influential book depicted the 'Industrial Revolution' as a cataclysmic discontinuity that, in the late eighteenth century, wrenched England from feudal, agrarian harmony, to a class society of mass 'pauperism' and 'degradation' co-existent with 'a great increase in wealth'. The free-market economics of Adam Smith in harness with technological change had swiftly and dramatically ushered in 'a new industrial age'.[4] This is a depiction of British industrialization which still retains – in whole or part – a strong hold on the popular imagination, reinforced by the mass media and by proliferating industrial museums.[5] Among professional historians, of course, it has been modified to varying degrees. Cliometricians, in particular, have flattened out a dramatic event into a gradual process, and toppled the steam engine from its pinnacle of importance: von Tunzelmann, for example, reaches the ultimate deflating conclusion that 'the level of national income reached on 1 January 1801 would not have been attained much before 1 February 1801 without James Watt'.[6] Others suggest that this goes too far.[7]

To a British readership in the 1880s, however, the broad outline of Toynbee's account was already very familiar. For over half a century Britons had been repeatedly told that national wealth and strength derived from a sudden upsurge in manufacturing output, beginning in the late eighteenth century and consequent primarily on steam-powered cotton machinery and a good transport system (at first canals, but, more recently, steam-driven railways). But this account, although emerging as dominant by the mid-nineteenth century, had been the subject of much political contention and fierce debate both in Parliament and in print. During the early nineteenth century, as Will Hardy argues, Parliament was forced increasingly to consider the condition of manufacturing industry. As free-traders and protectionists debated the corn laws and other economic issues, they appealed to the historical record to support their case. From this emerged a conceptualization of recent history which we can recognize as the traditional Industrial Revolution – the rapid expansion of manufacturing during the reign of George III. Its supposed causes were, of course, controversial; before the 1820s, however, improved machinery was only one strand in usually multi-causal explanations.[8] Thereafter, as I shall argue, veneration of Watt helped to identify the British industrial revolution increasingly with the steam engine and cotton factories.

The rise of manufactures also stimulated economic theory into new directions. Maxine Berg demonstrates how political economists cut their theoretical teeth on 'the machinery question' in the first half of the nineteenth century.[9] While conservatives dreaded the social consequences of mechanization and denied its supposed benefits, political economists expounded optimistically its rôle in both past and future economic growth. In 1824, when Watt and his steam engine were being eulogized, the 'machinery question' had become particularly intense: economic liberals were divided as to whether laws prohibiting the export of machinery should be repealed.[10] Manchester manufacturers, who normally campaigned for greater freedom of trade, balked at the threatened export of their machinery. They began to argue that Britain's lead in machinery was alone responsible for national prosperity and therefore must not be jeopardized by its export: the rhetoric of steam power was caught up in their case.

The reflection of industrialization in literature and painting has often been an anxious and, at best, an ambivalent one.[11] But the optimists and the beneficiaries of industrialization also made a cultural response, and if celebration of technological progress lacked the narrative tension required for a good story, it found its own medium in the public commemoration of inventors.[12] I shall argue that commemorative statuary and the fundraising efforts surrounding it both raised awareness of new technology and helped shape attitudes more positively towards it. Watt's admirers proffered an optimistic and patriotic interpretation of the terrifyingly rapid changes in livelihoods and landscapes which many of their contemporaries were witnessing – changes produced, they suggested, not by unintelligible, impersonal forces, but by a recognizable (if scarcely more intelligible) human being. Encomia of Watt also shaped our notion of inventors and inventions: to most British people, even in the late twentieth century, Watt remains the archetypal inventor, and great inventions are seen as the product of 'genius', itself unexplained and presumed inexplicable. At the time these encomia signalled a watershed in the way that inventors were regarded: veneration of Watt helped to raise the status of inventors as a whole and gradually to strengthen the protection of their intellectual property through the patent system. Although it was not until the third quarter of the nineteenth century that the commemoration of inventors became widespread,[13] the celebration of Watt and his steam engine blazed a trail that others followed.

WATT'S RISE TO FAME

Watt rose to fame on the back of his invention. The steam engine attracted numerous 'biographers' well before Watt did. The first (English) history of the steam engine appeared in the *Encyclopaedia Britannica* in 1797.[14] Its author was John Robison, a professor at the University of Edinburgh and a close friend of Watt's. In many ways Robison set the pattern which nineteenth-century British books on the steam engine followed.[15] He asserted the pre-eminence of the steam engine among all inventions throughout history; he provided an historical account of its technical development, but paid little attention to its socio-economic context; he gave it an exclusively British pedigree; and he allowed Watt the crowning glory of achievement, praising his uniquely 'fertile genius' and 'truly philosophical mind'.[16]

Hugh Torrens has demonstrated how Watt and his supporters, including Robison, wrote his rivals out of the history of the steam engine.[17] Watt was not satisfied with Robison's glowing account of his achievements, since it also gave (very limited) credit to an engineer whom Watt believed to have infringed his patent. He privately chided Robison into making some 'corrections' to his entry when it was republished.[18] Several subsequent authors were also pressured to amend their texts, not least by Watt's friends writing for *The Edinburgh Review*.[19] Watt's tenure of the summit of British steam-engine development was thus established during his lifetime, in a somewhat less than honourable fashion. His reputation was boosted to new heights by the obituary written in 1819 by another friend, the editor of the *Edinburgh Review*, Francis Jeffrey. This was frequently reprinted and set the hyperbolic tone for much that was to follow.[20] Jeffrey lauded Watt as a saviour of his country and a benefactor of mankind. Not only were all branches of industry indebted to Watt's invention, but, wrote Jeffrey,

> It is our improved steam-engine that has fought the battles of Europe, and exalted and sustained, through the late tremendous contest, the political greatness of our land ... It has increased indefinitely the mass of human comforts and enjoyments, and rendered cheap and accessible, all over the world, the materials of wealth and prosperity. It has armed the feeble hand of man ... with a power to which no limits can be assigned, completed the dominion of mind over the most refractory qualities of matter, and laid a sure foundation for all those future miracles of mechanic power which are to aid and reward the labours of after-generations. It is to the genius of one man, too, that all this is mainly owing; and certainly no man ever before bestowed such a gift on his kind.[21]

Watt was honoured more in his own time than any previous British inventor or engineer, except perhaps Edward Jenner, who was lionized (but also vilified) for his invention of smallpox vaccination. Some of this he owed to his son, James Watt junior, who, following his father's death in

1819, began freely spending both money and time to perpetuate his memory. At a cost of approximately £4,000 he donated marble statues or busts by Chantrey to Watt's friends; to Handsworth parish church, where Watt was buried; to Greenock, where he had been born in 1736 (and which was later given a library to house it properly); and to Glasgow University, where he had made his most famous invention in the 1760s.[22] In 1823 the younger Watt anonymously contributed the laudatory entry 'Watt' to the *Encyclopaedia Britannica*, and ten years later he kept a close eye on the wording of Brougham's celebrated epitaph for the Westminster Abbey monument.[23]

Watt's fame was not, however, the product solely of filial piety; it owed much to his friends and admirers, some of whom also saw him as an ideal figurehead for their cause. These causes were several; at their head were scientists, promoters of working-men's education and economic liberals, both in Parliament and the provinces. Many of them took their cue from the elevation of Watt to the status of a national hero in 1824.

WATT IN WESTMINSTER ABBEY

In June 1824 the Prime Minister, Lord Liverpool, wrote to George IV, suggesting that he should volunteer a contribution of £500 to the impending public subscription to commemorate James Watt, 'as the Inventor of the Steam Engine, the greatest and most useful Invention of Modern Times'. It was an opportunity, said Liverpool, for the King to demonstrate his support for 'the important interests of science and literature'.[24] Some of Watt's friends had sought a Parliamentary vote for a monument in either Westminster Abbey or St Paul's Cathedral.[25] The Government, however, feared to establish a new precedent, such that Parliament would be constantly besieged by claims on behalf of 'eminent men'. Instead, ministers would attend a public meeting in their private capacity.[26] The resultant gathering of 'the great and the good' in the Freemasons' Hall, Westminster, on 18 June 1824 was not only highly distinguished, but also apparently diverse: it encompassed both Whigs and Tories, Royal Society reformers and die-hards, and spokesmen for, as well as critics of, the factory system.[27] Liverpool had been approached by Edward Littleton, M.P., on behalf of 'several of the principal manufacturers of the kingdom' and with the support of some of 'the most influential persons in the House of Commons'.[28] The Prime Minister in turn informed the King that the initiative stemmed from 'all the persons connected with Science, and likewise by many Members of both Houses of Parliament'.[29] At a time when Royal Society membership was not reserved for a scientific élite and many gentlemen pursued astronomy, chemistry or other scientific interests, these circles mixed socially and were by no means mutually exclusive. As we shall see, men of science and liberal politics, and Watt's personal friends, largely account for the membership of the committee of 78 which emerged from the meeting and which eventually raised over £6000 to fund Chantrey's colossal and controversial statue.[30]

Several overlapping networks of friendship and interest can be detected. The largest comprised members of the Athenaeum, the London club founded in 1823 expressly to cater for literary and scientific men. Nearly half (36) of the Watt monument committee were members of the Athenaeum, including four of the Club's five trustees.[31] At least 52 members of the Athenaeum subscribed to the monument, together contributing over £1300, or approximately one-fifth of the total cost.[32] Nine of the twelve speakers on 18 June were members. The Athenaeum brought Lord Liverpool and his Cabinet colleagues together not only with some of their Whig opponents, but also with literary and scientific figures, including several of Watt's closest friends and admirers – Sir Humphry Davy, Sir Walter Scott, Francis Jeffrey, Matthew Robinson Boulton (son of Watt's late business partner), George Rennie and Thomas Telford, the engineers, and a closely knit group of Whig merchants (Samuel Boddington, George Philips, M.P., Richard Sharp, Charles Hampden Turner) who had been involved in the Limehouse Rope Manufactory with Watt's friend and fellow inventor, the late Captain Joseph Huddart. There can be no doubting the part played by Watt's personal friends in promoting the monument: they provided at least 35 members of the monument committee and subscribed generously.[33] Boulton gave the largest private donation (£200), and Miss Boulton gave a further £100; Turner, Benjamin Gott (the Leeds wool spinner), and Peter Ewart (the Manchester cotton spinner and one-time Boulton & Watt engine erector) each gave £100; Sir Joseph Huddart and John Woolmore (both of the Limehouse ropery), Telford and Sir Isaac Coffin, M.P., were among those giving £50. Most donations were between five and ten guineas.

Prominent among the speakers was Sir Humphry Davy, a long-time friend of the Watt family. As President of the Royal Society of London, Davy claimed Watt for the cause of science. Not just 'a great practical mechanic', said Davy, Watt 'was equally distinguished as a natural philosopher and a chemist'; his improvements were no accident, but 'founded on delicate and refined experiments'. Moreover, Watt's inventions demonstrated the great practical utility of science:

> our modern Archimedes ... has permanently elevated the strength and wealth of this great empire: and during the last long war, his inventions and their applications were amongst the great means which enabled Britain to display power and resources so infinitely above what might have been expected from the numerical strength of her population. Archimedes valued principally abstract science; James Watt, on the contrary, brought every principle to some practical use.[34]

Davy's words doubtless found approval with two of the Royal Society's 'Young Turks' on the committee, Charles Babbage and John Herschel, and helped to reconcile them for the moment to his presidency.[35] In 1830, Babbage was notoriously to write a scorching critique of the Royal Society's gentlemanly inactivity and the state's neglect of science which, he believed, would ultimately have disastrous consequences for national wealth and strength. Since scientific research had evident economic and strategic benefits, it merited Government support.[36] Most sympathetic to their concerns in the Cabinet was probably the Home Secretary, Robert Peel. Peel was in regular contact with Davy and other fellows of the Royal Society, securing financial support for them where he could.[37] In 1826 he persuaded George IV to create the Royal Medals for men of science and literature, and was largely responsible for the honours and civil list pensions awarded for scientific achievements by William IV in the 1830s.[38]

The utility of science was usually canvassed most explicitly at a far less exalted level, in the Mechanics' Institute Movement, which also sought to diffuse the 'science' of political economy.[39] Henry Brougham, the Whig leader and friend of Watt, took the opportunity to promote it: Watt's achievements could be emulated by all working men through the Mechanics' Institutes, where a scientific education would equip 'many a man of genius ... to follow in the footsteps of him whose name, once so obscure, now shines forth with so brilliant and so useful a lustre'.[40] In his inaugural speech as Rector of Glasgow University the following year, Brougham again called on the name of Watt, 'to demonstrate that the highest intellectual cultivation is perfectly compatible with the daily cares and toils of working men'.[41] Indeed Watt's name was regularly invoked for this purpose by both the Mechanics' Institute Movement and its offshoot, the Society for the Diffusion of Useful Knowledge.[42] Several members of the monument committee shared Brougham's educational concerns, including eight who became local representatives of the S.D.U.K.[43]

Politically and financially, Watt's friends had chosen their time well. The economy was reviving. The long years of postwar adjustment, widespread distress and uncertainty seemed to be over, and the threat of unemployment and civil unrest was receding. Furthermore, liberal Tories in Liverpool's government were querying the wisdom of the corn law that excluded foreign grain: in the future grain imports would be unavoidable – to be financed by the export earnings of a buoyant industrial sector. Although wary of the social implications of industrialization, they were forced to appreciate the economic significance of manufactures.[44] And, thanks to rising revenues, it was possible to assist manufacturers by reducing tariffs, beginning with the budget of February 1824.[45] Increasingly the Whigs on the opposition front benches found themselves approving and supporting Government fiscal policies; inter-party hostility in general was dissipating – to the point where Parliament was accused of dullness.[46] The House of Commons was in a good mood in the summer of 1824 and the manufacturing districts were prospering: the time was ripe for a cross-party celebration among occupants of the centre ground.[47] Watt's friends from the worlds of science, trade and manufacturing suggested the occasion: here was a secular saint they could all comfortably worship.

Approximately one-third of the monument committee consisted of politicians, divided almost equally between moderate Whigs and liberal Tories, the two centrist wings of each party, with a sprinkling of independents. Lord Liverpool himself, accompanied by three senior members of his Cabinet, chaired the meeting. Taking his cue from Jeffrey's eulogistic obituary, he spoke fulsomely of the great advantages of steam power in both manufacturing and navigation, as well as of Watt's personal qualities, and opened the subscription list with the King's £500 and his own £100 donation.[48] Liverpool's estimation of Watt's importance was not purely for the occasion: as early as 1820 he had rebutted a call for more agrarian protection and a fiscal brake on mechanization with the retort that 'England was indebted for its present greatness' to men such as Watt, Boulton, and Arkwright.[49] The Prime Minister's sentiments were echoed by William Huskisson, President of the Board of Trade and the Government's chief proponent of economic liberalism. To Huskisson, the steam engine was not only 'the most powerful instrument in the hands of men, to alter the face of the physical world ... [but also] a powerful moral lever in forwarding the great cause of civilization' – British manufactures, he implied, were at least as effective as Christian missionaries in civilizing the cannibals of the Pacific. At home, steam power had advanced 'personal comfort and public wealth', had diffused 'a spirit of improvement, [and] a thirst for instruction', and, by boosting Britain's resources at a crucial time, had been instrumental in defeating Napoleon.[50] Huskisson was enunciating claims common among economic liberals: commercial expansion would promote international peace and extend 'civilization'. Now he implicitly recruited Watt to the cause: free trade enabled commercial expansion; steam power provided its motive force.

Both parties could be seen making their obeisance to the commercial classes. For the Whigs, Sir James Mackintosh, M.P., Scots lawyer and contributor to the *Edinburgh Review*, expressed his pleasure 'at seeing here ... many of that enlightened, ingenious, independent, and upright class of men, the manufacturers of England'; the occasion was a tribute not only to Watt but also to them and to 'the useful arts'.[51] For the Tories, however, Robert Peel had the advantage, drawing much acclaim by acknowledging his own commercial roots: 'I feel the class of society from which I derive my origin exalted and honoured, by possessing such a man [as Watt] among its ranks.' It was fitting, observed Peel, 'that the same dome which now covers the monuments of the warrior, and of the poet, and of the statesman ... will protect also the memorials which a grateful nation may raise to the men who have perfected the arts of civilized life'.[52]

Trade and manufacturing were well represented on the monument committee. The cotton industry provided ten names, including Peel and his father, a Strutt, an Arkwright, and the Glasgow cotton magnate, Kirkman Finlay. There were also two ironmasters, a prominent woollen manufacturer (Gott), and a member of the Wedgwood pottery dynasty, as well as six partners in the rope manufactory previously mentioned. Many of them had known Watt personally, some as customers for his steam engines. It was also perhaps this direct business link with the firm of Boulton and Watt that prompted subscriptions from at least four London brewers, thirteen London merchant houses and the West India Dock Company,[53] three steam-ferry companies, nine Welsh ironmasters (listed in sequence, five guineas each), and a Cornish mines developer, Lord de Dunstanville – a generous £50.[54] Most Birmingham subscribers had been personal friends of Watt; the Birmingham Canal Navigation proprietors voted £100.[55]

The largest provincial group on the committee was from Manchester. Here it is especially difficult to separate scientific (and educational) interests from economic ones. John Dalton, the distinguished chemist who was President of the Manchester Literary and Philosophical Society from 1817 to 1844, headed the list of ten, all of them members of that Society and all of them cotton spinners, except Dalton himself and William Henry, who was an industrial chemist and a close friend of James Watt junior.[56] Friendship with Watt or his son was common to most of them. Seven were Unitarians, members of the heterodox but wealthy and well-educated sect which had first founded the Manchester 'Lit. and Phil.' in 1781, and which notoriously shared 'a more or less intense political radicalism ... and a more or less explicit commitment to the dictates of political economy'.[57] The other three were Tories: the two Peels and H.H. Birley, Chairman of the Manchester Chamber of Commerce. Two of the ten, Peter Ewart and John Kennedy, had been in London during the spring of 1824 to put before the Commons' Select Committee the Chamber of Commerce's case for retaining the ban on exports of machinery.[58] They had lobbied ministers hard, and had been rewarded in May 1824 with a deferral of the decision pending further investigation.[59] Ewart and Kennedy shared another immediate interest with Birley, as members of the committee planning to link Manchester and Liverpool by rail.[60]

Most noticeable by their absence from both the monument committee and the 'London' subscription list were the mechanical engineers. Partners and senior employees of the firm of Boulton and Watt were present, naturally, as were three prominent civil engineers, Telford, Rennie and James Walker, and three further 'civils' subscribed. Apart from Henry Maudslay, who subscribed ten guineas, the engine makers stayed away in droves. Perhaps the rancorous patent disputes and Watt's exclusion of his competitors from both profit and historical recognition had taken their toll. Alternatively, more immediate politicking may have been critical: the London engineers, who favoured repealing the ban on machinery exports, had lost the argument to Manchester's representatives, who were now celebrating Watt with the very ministers they had convinced to retreat from freer trade.[61]

WATT OUTWITH WESTMINSTER

London's lead was quickly followed, with varying degrees of success, in Manchester, Glasgow and Edinburgh. Almost certainly Manchester made the largest provincial contribution to the Westminster Abbey monument – at least £870.[62] Yet Manchester was neither Watt's birthplace, nor ever his home. It was, however, the centre of the English cotton industry, dependent for its motive power on the steam engine, and it prided itself on its scientific culture. Later, it became the acknowledged capital of free trade (the heartland of the Anti-Corn-Law League in the late 1830s and 1840s). In 1824, however, the immediate issues exercising Mancunian minds included the prospect of unrestricted exports of cotton machinery, the projected Liverpool and Manchester Railway on which George Stephenson intended to run steam locomotives, and the proposal to open a Mechanics' Institute. Again we find genuine admiration and gratitude to Watt. He was being co-opted to various causes: Watt the man was becoming Watt the icon.

On returning to Manchester, members of the delegation quickly organized a public meeting to launch a subscription toward the Westminster Abbey memorial. Opening the proceedings, George Philips, M.P., continued in the grandiloquent and liberal tones he had heard at the London meeting: through 'the genius of Mr Watt' Britain 'had acquired a mastery over the winds and waves; the space which separated nations was contracted, and they were brought into closer contact with each other ... [promoting] peace and good will among men'.[63] He was met by a counter-proposal that Manchester should have its own statue. Its unsuccessful proponents stressed the city's particular obligation: Manchester 'owed all its importance to Mr Watt's inventions'.[64] Manchester's dependence on steam marked it out from the rest of the Lancashire cotton industry, which had risen first on water power.

The 36-member committee which supervised the Manchester subscription overwhelmingly comprised representatives of the cotton trade. At least 13 were Unitarians, at least 15 (with considerable overlap) members of the 'Lit. and Phil.'.[65] Twelve of the 36 names reappear in the 22-strong committee that was currently establishing the Manchester Mechanics' Institution,[66] to whose members Watt was again presented as a rôle model – a fellow working-man who had achieved 'opulence and honour' through the study of science.[67] Contributions followed a similar pattern: among 71 known subscribers, firms in the cotton trades predominated – 41 from spinners, manufacturers, calico printers, merchants, and 9 from machine makers.[68] Not only was Cottonopolis recognizing its peculiar debt to the steam engine: to liberal Mancunians Watt's steam engine symbolized both the power of scientific knowledge and the wider benefits of unfettered, worldwide trade.[69]

In Scotland the preference for local monuments prevailed. Glasgow felt that it had a special claim to celebrate Watt. At a meeting in the Town Hall in November 1824 the Lord Provost suggested that citizens might prefer to support 'a monument, which would at once perpetuate Mr Watt's memory, and adorn the city which gave birth to those mighty efforts of his genius'.[70] A committee of 90, dominated by a wide spread of Glasgow's commercial classes, but also including city officials, members of the University professoriate and other professional men, raised nearly £3500 from their own kind to commission a 'colossal' bronze statue from Chantrey; the cotton trade was again conspicuously generous and several groups of mechanics gave substantial sums.[71] The small town of Greenock supported its belief that Scotland should have its own monument, there in Watt's birthplace, with an impressive public subscription of £1703. Topped up by James Watt junior, this commissioned a marble statue from Chantrey in 1835.[72] Edinburgh, lacking a local connection, was less successful. At a poorly attended meeting ('principally ... men of science, and those connected with the arts'), Sir Walter Scott and Francis Jeffrey carried a resolution for the Westminster Abbey project, but could not prevent the calling of a second meeting by the advocate, Henry Cockburn. This favoured a Scottish national monument, 'because', as Cockburn argued, 'this was his own country, and it was their duty to encourage similar exertions, by shewing to Scotchmen the ultimate rewards which genius and industry would meet with'.[73] In Edinburgh's largely service economy, a committee dominated by city officials and professional men raised a creditable £1265 in three weeks, but it was 1851 before their goal of a new 'Watt' building to accommodate the School of Arts was achieved – complemented by a statue of Watt in 1854.[74]

INVENTORS AND INVENTIONS

The most immediate beneficiaries of Watt's rocketing reputation were probably the railway promoters, not least the Liverpool and Manchester Rail Road Company, then preparing its Parliamentary bill. Only two months after the Freemasons' Hall meeting, a supportive piece in the *Morning Chronicle* triumphantly laid claim to its rhetorical influence:

> The attempt on the part of Canal Projectors to prevent their adoption is utterly hopeless. The discovery of the Loco Motive Engine will be almost as important to the trade and commerce of this country as the discovery of the Steam Engine itself ... A few weeks since we witnessed the Prime Minister and his coadjutators, surrounded by men distinguished for their scientific and patriotic pursuits, assembled for the purpose of voting a Monument to the Memory of Mr Watt, for the distinguished services he conferred on his country and on mankind. Surely, then, Parliament will never refuse to permit the country to reap the full benefit of his discovery, by throwing out the Bills intended to be applied for.[75]

The current rage for speculation in railways, surmised the *Quarterly Review* a few months later, owed not a little 'to the great encouragement held out by the first Minister of the crown and his colleagues (at a meeting for considering the proposal of a monument to the late James Watt) to those who should effect a further improvement of this mighty engine, and to the promoters and discoverers of other useful inventions'.[76]

Canny publishers identified and fed an upsurge of interest in the steam engine. Where previously there had been a trickle of encyclopaedia entries and periodical articles, the late 1820s and 1830s saw the publication of a regular stream of books.[77] John Farey, for example, offered nearly 700 pages of steam-engine history and technical detail, prefaced by an encomium on steam's wealth-creating and civilizing powers.[78] Full-length biographies of Watt were slower to appear, beginning in 1839 with two translations of Arago's *Eloge* – written for the Académie des Sciences, of which Watt had been a foreign associate.[79] Meanwhile shorter biographical sketches in the steam-engine literature and in 'improving' texts like Craik's *Pursuit of Knowledge Under Difficulties* kept him constantly before the public.[80]

Watt would surely have been delighted that the high regard professed for him apparently helped to raise the status of inventors as a body and to prompt concern for reform of the antiquated patent system, for which he had campaigned in vain.[81] There was a well-established tendency to dismiss inventors as 'projectors' or 'schemers', men whose ingenuity would lead them, and anyone foolish enough to invest in their ideas, to financial ruin. But now the climate was changing. The *Quarterly Review*, for example, preached a new tolerance: the public should not be too quick to judge an invention, since one 'which at first blush may appear frivolous, and even ludicrous, turns out in the end to be extremely useful ... every invention should have fair play'.[82] *The Times*, going so far as to describe inventors as 'the elect of the human race', gave its full support to a call to reduce the cost of patents.[83] Demands began to be heard more often for a better deal for inventors; they routinely invoked the name of Watt. Responding to one such plea in 1825, the Attorney General affirmed Parliament's duty to facilitate the work of inventors, for it was 'likely to become useful to their country, and beneficial to mankind'.[84] In 1829 Parliament held its first-ever investigation into the operation of the patent system. Although wholesale reform was delayed for a further two decades, Lord Brougham was able to instigate a minor legislative change to benefit patentees in 1835.[85] There was also a marked alteration in the attitude of judges and juries towards patentees during the 1830s: the balance of success in litigation shifted towards the prosecutors of infringements, as patentees began to be regarded less as grasping monopolists, and more as national benefactors.[86]

The eulogizing of Watt and the steam engine was also a turning-point in the 'metaphysics' of invention. Until the late eighteenth century, invention was generally conceived in providentialist terms: successful inventors (that is, those who were not dangerous 'projectors') were mere instruments of the divine plan.[87] The invention of printing, for instance, had been part of God's preparations for the Reformation.[88] The growth of the patent system, which rewarded individual achievements, offered an implicit challenge to this viewpoint. By 1800 a more secular account was emerging, in which human agency was recognized, but little attention was yet paid to particular inventors and most inventions remained effectively anonymous.[89] With Watt the dominant discourse of invention became not just individualistic, but heroic. Invention was assimilated into the wider Romantic belief in creative imagination: it came to be seen as the product of individual 'genius', an innate capacity fostered by other personal qualities – especially perseverence and hard work.[90] During the third quarter of the nineteenth century this heroic model was extended to a larger pantheon of inventors and deployed in defence of the then embattled patent system.[91]

Popular concepts of invention came to be shaped increasingly by biography. Histories of the steam engine recounted (and disputed) earlier individual contributions, before culminating in the uncontested achievements of Watt.[92] His invention of the separate condenser was implicitly explained through his biography, starting with his early interest in things mechanical. It was the exogenous product of one man's 'genius' or 'happy idea', prompted by his perception of the technical shortcomings of the model Newcomen engine he was repairing.[93] This flash of inspiration was converted into the working steam engine through a combination of Watt's personal traits of character, his outstanding mental abilities and a measure of manual dexterity – with a little help from the chemistry of Black and the capital of Roebuck and Boulton. The wider context of the growing need for a more fuel-efficient means of draining mines is absent from these accounts. Yet this technology which sprang from one man's mind was accredited with single-handedly changing the world out of all recognition. As we have seen, the steam engine was allegedly shaping the fortunes not just of Manchester but of the globe as a whole: it had determined the outcome of war and raised Britain to international pre-eminence; in the future, through trade, it would promote peace among nations; even now it was the herald of civilization and Christianity throughout the world. The mode of thought we now term 'technological determinism' – technology conceived as an exogenous and independent force which shapes society – was evidently gaining ground.

THE INDUSTRIAL REVOLUTION

A new history of Britain had also been sanctioned. Steam's rôle in Britain's rising prosperity and power began to be regularly canvassed, not least in the burgeoning literature devoted to explaining steam power. But it was also made topical by the debate over the legislative ban on machinery exports. A pivotal question was the extent to which Britain's manufacturing lead rested on its superiority in machinery, as Manchester manufacturers maintained.[94] J.R. McCulloch, the political economist, calmly insisted that Britain enjoyed many advantages, so that preserving a British monopoly in machinery was unnecessary, and in 1825 the Commons' Select Committee, heavily influenced by the London engineers and committed free traders, concurred.[95] However, the steam engine now enjoyed totemic status: having captured the popular imagination as a symbol of British inventive genius and industrial success, in less politicized explanations of British predominance than these it moved to centre stage.

In the 1830s, indeed, steam power became a feature of more general works on British history and contemporary society. Both celebrants and critics of industrialization emphasised its rôle. Watt's inventions were variously described as 'the foundation ... for the prodigious advance in wealth and population which marked the reign of George III',[96] as the cause of 'one of the most striking revolutions ever produced [for the worse] in the moral and social condition of a moiety of a great nation',[97] and as 'a power which has effected a complete revolution in manufacturing operations'.[98] The latter authors commented on the contrast between Watt's struggles to maintain his patents and 'the bursts of enthusiastic admiration which, after [his] death ...

seemed to exhaust the powers of language in eulogizing the man and his imperishable inventions'.[99] Perhaps the strongest testimony to this intrusion of the steam-powered industrial revolution into British history comes from those who feared its imminent hegemony. In 1836 William Fitton, a leading Lancashire radical, reportedly argued 'that without shorter hours the factory system would become a bitter curse and proceeded to deny strenuously that "England owes its greatness to the modern discoveries of steam and machinery"'.[100]

That Fitton's protests were in vain is demonstrated by Brougham's speech against the corn law in 1839, where he confidently deployed this image of British industrialization as evidence for his case:

> The [export] bounty was taken off in 1774, and the same period was the most remarkable, the most brilliant, and I will say the proudest in the history of the English nation. ... The whole face of nature was changed; the whole aspect of the earth in this country became ... one large, wealthy, industrious, expert and skilful workshop. These were the miracles of manufacturing industry. This was the period of manufactories.

At its heart, said Brougham, was the development of mining and machinery, especially 'the new power of steam, ... enlarging the sphere of human potency and giving man a new existence and a new dominion on the earth'.[101] Significantly, Brougham's protectionist opponents did not dispute either this periodization or his emphasis on manufactures; instead Sir Archibald Alison, for example, strove to redefine the period as one not of relative free trade but of effective protection.[102]

The short, sharp, steam-driven revolution was now common currency. If anything, claims for the historical rôle of steam only became more exaggerated with time; it was celebrated rather than scrutinized. It is a mark of Toynbee's relative sophistication that he, at least, set the steam engine in a broader context than most commentators. Nonetheless, in condensing his account for a popular audience, Toynbee reverted to the heroic genre: James Watt and Adam Smith were 'the two men who did most to bring ... about' the industrial revolution, 'for the *Wealth of Nations* and the steam-engine (with the great inventions, like the spinning-jenny and the power-loom, ...) destroyed the old world and built a new one'.[103] Free trade and technology triumphant – for Toynbee, a catastrophic partnership. This concept of the 'Industrial Revolution' had been current for half a century; Toynbee gave it a name and an academic seal of (dis)approval.

We have seen how Watt had been made a hero. In the process, nothing less than a new conceptualization of invention and a new myth of British history had been forged.[104] The celebration of Watt by manufacturers who sought recognition for their own 'standard-bearer', by politicians who favoured freer trade and by scientists who sought funding from politicians helped to promote a new orthodoxy of British history. No other invention or inventor of the period captured the historical record in this way. Balloon flight seized the popular imagination; smallpox vaccination (and Jenner) were celebrated for saving countless lives. Both helped to raise the status of inventors and promote technological optimism; neither, unlike Watt's steam engine, was seen as having changed the course of history.

NOTES

1. *The Times*, 23 August 1844, p. 5f; 24 August 1844, p. 4e.
2. Arthur Penrhyn Stanley (1869), *Historical Memorials of Westminster Abbey*, 3rd edition, London: John Murray, pp. 349–50. See also the criticism of Watt's statue in *The Times*, 23 August 1844, p. 5f; 26 August 1844, p. 3d, and in *The Builder*, **44** (8 September 1883), p. 308. Chantrey's statue of Watt is now in the Scottish National Portrait Gallery, Edinburgh, awaiting a suitable site in Heriot-Watt University.
3. David S. Landes (1993), 'The Fable of the Dead Horse: or, the Industrial Revolution Revisited', in Joel Mokyr (ed.), *The British Industrial Revolution: An Economic Perspective*, Boulder, Oxford: Westview Press, p. 133; D.C. Coleman (1992), *Myth, History and the Industrial Revolution*, London and Rio Grande: The Hambledon Press, pp. 3–22.
4. Arnold Toynbee (1884), *Lectures on the Industrial Revolution in England*, London: Rivingtons, pp. 84–93, 189–93.
5. Coleman, *Myth, History and the Industrial Revolution*, pp. 1–3, 22–36, 42; Bob West (1988), 'The Making of the English Working Past: a Critical View of the Ironbridge Gorge Museum', in Robert Lumley (ed.), *The Museum Time-Machine*, London and New York: Routledge, pp. 36–62.
6. G.N. von Tunzelmann (1978), *Steam Power and British Industrialization to 1860*, Oxford: Clarendon Press, pp. 286–7.
7. Coleman, *Myth, History and the Industrial Revolution*, pp. 36–42.
8. William Hardy (1994), 'Conceptions of Manufacturing Advance in British Politics, *c.* 1800–1847, with Special Reference to Parliament, Governments and their Advisers', unpublished D.Phil. thesis, University of Oxford, pp. 4–5, 46, 132–321.
9. Maxine Berg (1980), *The Machinery Question and the Making of Political Economy, 1815–1848*, Cambridge: Cambridge University Press, pp. 1–19.
10. Berg, *The Machinery Question*, pp. 203–25.
11. Francis D. Klingender ([1947], 1972), *Art and the Industrial Revolution*, ed. Arthur Elton, London: Paladin; Raymond Williams (1961), *Culture and Society, 1780–1950*, Harmondsworth: Penguin Books; Leo Marx (1964), *The Machine in the Garden: Technology and the Pastoral Ideal in America*, New York: Oxford University Press; Herbert Sussman (1968), *Victorians and the Machine: The Literary Response to Technology*, Cambridge, MA: Harvard University Press.
12. For a catalogue of Watt portraiture, see H.W. Dickinson and Rhys Jenkins (1927), *James Watt and the Steam Engine*, 2nd edition with a new introduction by Jennifer Tann (1981), Ashbourne: Moorland Publishing Co., pp. 81–9. Portraits of Watt were frequently reproduced: William Coolidge Lane and Nina E. Browne (eds) (1906), *A.L.A. Portrait Index: Index to Portraits contained in Printed Books and Periodicals*, Washington: Government Printing Office, p. 1531. They usually accompanied accounts of his life or of the steam engine, which were also celebratory in tone.
13. Christine MacLeod (1996), 'Concepts of Invention and the Patent Controversy in Victorian Britain', in Robert Fox (ed.), *Technological Change: Methods and Themes in the History of Technology*, Amsterdam: Harwood Academic Press, pp. 137–40.
14. A French book was first in the field: G.C.F.M.R. de Prony (1796), *Nouvelle Architecture Hydraulique*, Paris.
15. For a bibliography of early works on the steam engine, see Dickinson and Jenkins, *James Watt and the Steam Engine*, pp. xix–xx, 360–71.

16. [John Robison] (1797), 'Steam Engine', in G. Gleig (ed.), *Encyclopaedia Britannica*, vol. 17, Edinburgh: Bell and Macfarquhar, pp. 743–72.
17. Hugh Torrens (1994), 'Jonathan Hornblower (1753–1815) and the Steam Engine: an Historical Analysis', in Denis Smith (ed.), *Perceptions of Great Engineers: Fact and Fantasy*, London: Science Museum for the Newcomen Society, National Museums and Galleries on Merseyside and the University of Liverpool, pp. 23–34.
18. Eric Robinson and Douglas McKie (eds) (1970), *Partners in Science: Letters of James Watt and Joseph Black*, London: Constable & Co. Ltd., pp. 232–5; [John Robison] (1801), 'Steam Engine', in G. Gleig (ed.), *Supplement to the Third Edition of the Encyclopaedia Britannica*, vol. 2, Edinburgh: Bonar, pp. 522–4.
19. Torrens, 'Jonathan Hornblower', pp. 26–8. These were sins of omission; Watt did not lay claim to other people's inventions. Patent litigation, while ultimately successful, had caused him serious anxiety and so much expense that he did not forgive alleged 'pirates'.
20. Compare Jeffrey's hyperbole with the more measured tones of the earlier, anonymous biography of Watt, in *British Public Characters of 1802–3* (1803), London: R. Phillips, p. 512. I am grateful to Dr David Bryden for this reference.
21. Francis Jeffrey (1819), 'Character of Mr Watt', *The Scotsman*, 4 September 1819; reprinted in *Edinburgh Magazine*, September 1819, 203.
22. R. Gunnis (n.d.), *Dictionary of British Sculptors, 1660–1851*, new edition, London: The Abbey Library, p. 33; Alison Yarrington, Ilene D. Lieberman, Alex Potts, and Malcolm Baker (1994), *An Edition of the Ledger of Sir Francis Chantrey, R.A., at the Royal Academy, 1809–1841*, London: the Walpole Society, pp. 52–4, 143–4, 159, 213, 217.
23. British Library, London (hereafter B.L.), Additional MS 34613, ff. 178, 204, 209, 315; University College, London, Brougham MSS 27513, 27743, 27778, 42555. For the text of the epitaph, see Yarrington et al., *The Ledger of Sir Francis Chantrey*, pp. 213–4.
24. B.L. Add. MS 38576, f. 141.
25. [Charles Hampden Turner] (ed.) (1824), *Proceedings of the Public Meeting held at Freemasons' Hall, on the 18th June, 1824, for Erecting a Monument to the Late James Watt*, London: John Murray, pp. iii–ix.
26. Turner, ibidem, p. vi. This remained a cause of complaint for some, e.g. [David Brewster] (1840), 'Life and Discoveries of James Watt', *Edinburgh Review*, **70**, 499; Dionysius Lardner (1840), *The Steam Engine Explained and Illustrated; with an Account of its Invention and Progressive Improvement, and its Application to Navigation and Railways; Including also a Memoir of Watt*, 7th edition, London: Taylor & Walton, pp. 318–9.
27. Turner, *Proceedings of the Public Meeting*, pp. 93–7; *The Times*, 19 June 1824, p. 6c.
28. B.L. Add. MS 38298, f. 316.
29. B.L. Add. MS 38576, f. 141.
30. See Appendix; Yarrington et al., *The Ledger of Sir Francis Chantrey*, pp. 214–5.
31. [Athenaeum] (1826), *An Alphabetical List of the Members, with the Rules and Regulations of the Athenaeum*, London.
32. Birmingham Reference Library (hereafter B.R.L.), Boulton and Watt MSS, Timmins vol. 2, f. 44. This printed list contained 201 individual and corporate names, the majority of them from London; their subscriptions totalled £3794 (including the King's £500). At least 40 of them knew Watt personally.
33. A valuable list of Watt's friends at the time of his death in 1819, made by James Watt junior, is in B.R.L., Boulton and Watt MSS, MI/5/14 (1)a; other friends have been identified from Watt's correspondence in B.R.L., James Watt MSS, Letterbooks 1–5. I am grateful to Dr Richard Hills for his references to these.
34. Turner, *Proceedings of the Public Meeting*, pp. 8–9, 13. Watt's friends were equally happy to have 'science' support their case for Watt, which stressed his merits as a 'philosopher', no mere artisanal tinkerer: for example, Turner, ibidem, p. 16; Robison, 'Steam Engine' (1797), p. 762; John Farey (1827), *A Treatise on the Steam Engine, Historical, Practical, and Descriptive*, reprinted (1971), Newton Abbot: David & Charles Reprints, p. 309.
35. David Philip Miller (1983), 'Between Hostile Camps: Sir Humphry Davy's Presidency of the Royal Society of London, 1820–1827', *British Journal for the History of Science*, **16**, 13–19, 26–35.
36. Charles Babbage (1830), *Reflections on the Decline of Science in England and on some of its Causes*, London: B. Fellowes.

37. John Wilson Croker (1884), *Correspondence and Diaries*, ed. Louis J. Jennings, London: John Murray, vol. 1, p. 263; Charles Stuart Parker (1891–9), *Sir Robert Peel*, London: John Murray, vol. 1, pp. 363–5.
38. Roy MacLeod (1970), 'Science and the Civil List, 1824–1914', *Technology and Society*, **6**, 47–55.
39. Berg, *The Machinery Question*, pp. 145–66.
40. Turner, *Proceedings of the Public Meeting*, pp. 53–4.
41. Henry, Lord Brougham (1838), 'Inaugural Discourse on being Installed Lord Rector of the University of Glasgow, April 6 1825', in *Speeches of Henry Lord Brougham*, vol. 3, Edinburgh: Adam and Charles Black, p. 95.
42. See below, p. 106, and, for example, the S.D.U.K.'s publishing intentions, appended to [Henry Brougham] (1827), *A Discourse of the Objects, Advantages, and Pleasures of Science*, London: Baldwin, Cradock & Joy, where Watt, Rennie, Smeaton, Arkwright, Franklin and Wren are listed in the series of 'Self-exalted Men'. For the S.D.U.K., see Berg, *The Machinery Question*, pp. 292–3, and J.N. Hays (1964), 'Science and Lord Brougham's Society', *Annals of Science*, **20**, 227–41.
43. These eight are listed in [George L. Craik] (1830–1), *The Pursuit of Knowledge under Difficulties*, London: Charles Knight, title page.
44. Boyd Hilton (1977), *Corn, Cash, Commerce: The Economic Policies of Tory Governments, 1815–1830*, Oxford: Oxford University Press, pp. 303–7.
45. Norman Gash (1984), *Lord Liverpool: The Life and Political Career of Robert Banks Jenkinson, Second Earl of Liverpool 1770–1828*, London: Weidenfeld & Nicolson, pp. 217–9.
46. Austin Mitchell (1967), *The Whigs in Opposition, 1815–1830*, Oxford: Clarendon Press, p. 183; Chester W. New (1961), *The Life of Henry Brougham to 1830*, Oxford: Clarendon Press, pp. 266–8.
47. Gash, *Lord Liverpool*, p. 221.
48. Turner, *Proceedings of the Public Meeting*, p. 92.
49. *Hansard's Parliamentary Debates*, 1820, 2nd series, vol. 1, col. 421. See also ibidem, cols. 577, 593. For Liverpool's economic views, see Gash, *Lord Liverpool*, pp. 116–9, 157–8, 184, 200–1, 217–20; Hardy, 'Conceptions of Manufacturing Advance', pp. 187–98.
50. Turner, *Proceedings of the Public Meeting*, pp. 23–33.
51. Turner, ibidem, pp. 51–2.
52. Turner, ibidem, pp. 70–1.
53. West Indian sugar plantations provided the second largest group of customers for Watt engines after the cotton industry, many of them exported through London: Jennifer Tann (1997), 'Steam and Sugar: the Diffusion of the Stationary Steam Engine to the Caribbean Sugar Industry, 1770–1840', *History of Technology* (forthcoming).
54. B.R.L., Boulton and Watt MSS, Timmins vol. 2, f. 44.
55. *The Times*, 1 October 1824, p. 2d.
56. Turner, *Proceedings of the Public Meeting*, pp. 94–7; A.E. Musson and Eric Robinson (1969), *Science and Technology in the Industrial Revolution*, Manchester: Manchester University Press, pp. 238–9.
57. V.A.C. Gatrell (1982), 'Incorporation and the Pursuit of Liberal Hegemony in Manchester, 1790–1839', in Derek Fraser (ed.), *Municipal Reform and the Industrial City*, Leicester: Leicester University Press and New York: St Martin's Press, p. 27; see also John Seed (1982), 'Unitarianism, Political Economy, and the Antinomies of Liberal Culture in Manchester, 1830–50', *Social History*, **7**, 1–25; Robert Kargon (1977), *Science in Victorian Manchester: Enterprise and Expertise*, Manchester: Manchester University Press, pp. 5–14.
58. A.E. Musson (1972), 'The "Manchester School" and Exportation of Machinery', *Business History*, **14**, p. 27.
59. Musson, ibidem, pp. 22–9.
60. *Morning Chronicle*, 20 November 1824, p. 1a.
61. Musson, 'The "Manchester School"', pp. 28–9. In 1826 James Watt junior offered his help to the Manchester Chamber of Commerce in its continuing resistance to the repeal of this legislation: Musson, ibidem, p. 34 n.
62. *Manchester Guardian*, 31 July 1824, p. 2a. This includes the £100 from Ewart and £20 from Sir Robert Peel, Bart., given through the 'London' subscription: B.R.L., Boulton and Watt MSS, Timmins vol. 2, f. 44. It should probably be raised a further £50 by contrib-

utions of £25 each from George Philips, M.P., and G.A. Lee which were included in both the Manchester and the 'London' lists, but have been counted only once here.

63. *Manchester Guardian*, 10 July 1824, p. 3c. Philips' speech was mocked by 'Timotheus' in a letter to a Manchester newspaper, which was eagerly reprinted by William Cobbett, who was already satirizing the occasion in order to attack both cotton lords and Government, demanding to know 'What's Watt?': *Cobbett's Political Register* (1824), vol. 51, cols. 325–7, 371–4; vol. 52, cols. 297–304, 803–7.
64. *Manchester Guardian*, 10 July 1824, p. 3e. Manchester finally raised its own statue of Watt in 1857; Birmingham did so in 1868.
65. *Manchester Guardian*, 10 July 1824, p. 1a. For membership lists, see *Memoirs of the Manchester Literary and Philosophical Society*, 2nd series, vols. 2 (1813) and 6 (1842).
66. *Manchester Guardian*, 10 April 1824, p. 1a.
67. Sir Benjamin Heywood (1843), *Addresses Delivered at the Manchester Mechanics' Institution*: London, pp. 6–7, 11–13.
68. *Manchester Guardian*, 31 July 1824, p. 2a. In 1824, with trade booming, Manchester's machine makers did not dispute the Chamber of Commerce's position on machinery exports. It was only when domestic demand for machinery declined in the depression of 1825–26 that the machine makers broke away and started petitioning Parliament for the opportunity (legally) to supply overseas markets: Musson, 'The "Manchester School"', 32–3.
69. Manchester did not celebrate the famous inventors of cotton machinery – Hargreaves, Arkwright, or Crompton – whose patents it had fought or whose unpatented inventions it had used without payment. In 1890 Ford Maddox Brown depicted John Kay in one his murals for Manchester Town Hall – fleeing angry weavers who feared for their livelihoods: Charles Dellheim (1982), *The Face of the Past: the Preservation of the Medieval Inheritance in Victorian England*, Cambridge: Cambridge University Press, p. 173.
70. James Cleland (1825), *Historical Account of the Steam Engine, and its Application in Propelling Vessels, with an Account of the Proceedings at London, Glasgow, and Greenock for Erecting Monuments to Mr Watt*, Glasgow: Khull, Blackie & Co., p. 30.
71. Yarrington et al., *The Ledger of Sir Francis Chantrey*, pp. 215–6; *Glasgow Mechanics' Magazine*, **2** (48), 27 November 1824, pp. 303–4.
72. Cleland, *Historical Account of the Steam Engine*, pp. 65–6; *The Times*, 6 October 1826, p. 3c; Yarrington et al., *The Ledger of Sir Francis Chantrey*, p. 217.
73. *The Scotsman*, 10 July 1824, p. 529; 24 July 1824, p. 557.
74. *The Scotsman*, 24 July 1824, p. 559; 14 August 1824, p. 606; 16 December 1840, p.3f; William Lee Bride (1970), *James Watt – His Inventions and His Connections with Heriot-Watt University*, Edinburgh: Heriot-Watt University, p. 26; Dickinson and Jenkins, *James Watt and the Steam Engine*, p. 87. Dr David Bryden, who is researching the commemoration of Watt in Scotland, very kindly gave me copies of the Glasgow and Edinburgh subscription lists. Subscribers have been identified through the *Glasgow Post-Office Directory for 1828–9* (1828), Glasgow: John Graham & Co., and through *Pigot & Co.'s New Commercial Directory of Scotland for 1825–6* (1826), London.
75. *Morning Chronicle*, 18 August 1824, p. 4a.
76. [John Barrow] (March 1825), 'Canals and Rail-Roads', *Quarterly Review*, **31**, 358–9.
77. Dickinson and Jenkins, *James Watt and the Steam Engine*, pp. xix–xx, 360–71.
78. Farey, *A Treatise on the Steam Engine*, p. v.
79. Dickinson and Jenkins, *James Watt and the Steam Engine*, pp. xix, 367; D.F.J. Arago (1839), *Life of James Watt*, Edinburgh: Adam & Charles Black; 3rd edition, trans. J.P. Muirhead, London: John Murray.
80. [Craik], *The Pursuit of Knowledge under Difficulties*, vol. 2, pp. 295–323.
81. H.I. Dutton (1984), *The Patent System and Inventive Activity during the Industrial Revolution, 1750–1852*, Manchester: Manchester University Press, pp. 36–41.
82. Barrow, 'Canals and Rail-Roads', p. 355.
83. *The Times*, 22 April 1826, p. 3a.
84. *Hansard*, 1825, 2nd series, vol. 15, cols. 72–3.
85. Dutton, *The Patent System*, pp. 13–14, 18.
86. Dutton, ibidem, pp. 77–9. Lubar notes a similar phenomenon at the same time in the United States, then at an earlier stage of industrialization: Steven Lubar (1991), 'The Transformation of Antebellum Patent Law', *Technology and Culture*, **32**, 938–40. Is it

plausible that the growing respect for inventors, aided by Watt's posthumous influence, crossed the Atlantic, or was it just a coincidence?

87. For the continuation of providentialist and more secular deterministic accounts of invention through the nineteenth century, see MacLeod, 'Concepts of Invention', pp. 137–53. In the 1950s Archbishop Fisher could still suggest that nuclear weapons might be part of the divine plan.
88. Christine MacLeod (1988), *Inventing the Industrial Revolution: the English Patent System, 1660–1800*, Cambridge: Cambridge University Press, pp. 202–4.
89. MacLeod, ibidem, pp. 219–22.
90. See, for example, Farey, *A Treatise on the Steam Engine*, p. 395; Robert Stuart (1829), *Historical and Descriptive Anecdotes of Steam Engines and of their Inventors and Improvers*, London, pp. 258–9.
91. MacLeod, 'Concepts of Invention', pp. 143–5.
92. Disputes were dominated by national jealousies, principally the debate concerning the contribution of the Frenchman, Denis Papin. Histories of the cotton industry, which also began to appear in the 1820s, were similarly organized by inventions attributed to individuals, with disputes focusing on Arkwright's entitlement to his patents. Among his supporters Arkwright achieved heroic status similar to Watt's, but he had too many detractors alleging plagiarism to allow a similar secular canonization: see David J. Jeremy (1991), 'British and American Entrepreneurial Values in the Early Nineteenth Century: a Parting of the Ways?', in R.A. Burchell (ed.), *The End of Anglo-America*, Manchester, Manchester University Press, pp. 34–40.
93. See, for example, Lardner, *The Steam Engine Explained*, p. 120; Robert L. Galloway (1881), *The Steam Engine and its Inventors: a Historical Sketch*, London: Macmillan & Co., p. 142.
94. Musson, 'The "Manchester School"', pp. 25–6.
95. British Parliamentary Papers, 1824, V, *First Report from the Select Committee on Artisans and Machinery*, p. 619; British Parliamentary Papers, 1825, V, *Report from the Select Committee on the Laws relating to the Export of Tools and Machinery*, p. 129.
96. [John Wade] (1833), *History of the Middle and Working Classes*, London: Effringham Wilson, p. 82.
97. P. Gaskell (1833), *The Manufacturing Population of England, its Moral, Social, and Physical Conditions, and the Changes which have Arisen from the Use of Steam Machinery; with an Examination of Infant Labour*, London: Baldwin and Cradock, p. 52.
98. George L. Craik and Charles MacFarlane (1841–44), *The Pictorial History of England during the Reign of George the Third*, 3 vols, London: Charles Knight, vol. 1, p. 582.
99. Craik and MacFarlane, ibidem, vol. 3, p. 674.
100. D.S. Gadian (1978), 'Class Consciousness in Oldham and Other North-Western Industrial Towns', *Historical Journal*, **21**, p. 166.
101. *Hansard*, 1839, 3rd series, vol. 45, cols. 542–3.
102. [Sir Archibald Alison] (1844), 'Free Trade and Protection', *Blackwood's Edinburgh Magazine*, **55**, p. 399.
103. Toynbee, *Lectures on the Industrial Revolution*, p. 189.
104. I use the term 'myth' here not to deny that the industrial revolution happened, but, rather, to echo Coleman's argument that the industrial revolution has an emotive, explanatory function as part of British national identity: Coleman, *Myth, History and the Industrial Revolution*, pp. 1–3, 36–8.

APPENDIX

Names of the Committee for the Erection of a Monument to the late James Watt

Earl of Liverpool, K.G.
Rt Hon. George Canning, M.P.
Rt Hon. Robert Peel, M.P.
Hon. Heneage Legge, M.P.
Sir Humphry Davy, Bart., P.R.S.
Sir Robert Peel, Bart.
Sir John Wrottesley, Bart. M.P.
Sir James Mackintosh, M.P.
Alexander Baring, M.P.
J.W. Croker, M.P.
Davies Gilbert, M.P.
E.J. Littleton, M.P.
Francis Lawley, M.P.
George Philips junior, M.P.
William Wilberforce, M.P.
Samuel Boddington
H.H. Birley
M.R. Boulton
W.T. Brande, Sec. R.S.
Rev. John Corrie
James Davies
Peter Ewart
Francis Freeling
Rev. T. Lane Freer
Charles Hatchett
J.F.W. Herschel, Sec. R.S.
John Kennedy
Thomas Murdoch
George Rennie
Richard Sharp
Joshua Smith Simons Smith
Thomas Telford
John Woolmore
James Walker

Earl of Aberdeen, P.A.S.S.
Rt Hon. Frederick Robinson, M.P.
Rt Hon. William Huskisson, M.P.
Sir Isaac Coffin, Bart. M.P.
Sir James Graham, Bart. M.P.
Sir Walter Scott, Bart.
Sir Thomas Lawrence, P.R.A.
Matthias Attwood, M.P.
Henry Brougham, M.P.
D.S. Dugdale, M.P.
T.F. Kennedy, M.P.
T. Frankland Lewis, M.P.
George Philips, M.P.
J.H. Tremayne, M.P.
Richard Arkwright
Charles Babbage
George H. Barker
John Bolton
William Clayfield
John Dalton
William Cotton
Kirkman Finlay
George H. Freeling
Benjamin Gott
William Henry, M.D.
Francis Jeffrey
George A. Lee
William Murdock
Joseph Reynolds
William Strutt
C.H. Turner
John Vivian
Josiah Wedgwood
James Watt

(Turner, *Proceedings of the Public Meeting*, pp. 93–5)

PART 3

Inventors and Products

7. The Curious History and Imminent Demise of the Challenge and Response Model

Trevor Griffiths, Philip Hunt and Patrick O'Brien

> such a mass of inaccuracies, contradictions, and blunders, have been piled together, respecting the spinning inventions, as I think can scarcely be paralleled, within the same compass, on any other subject.[1]

Recent research into the British Industrial Revolution has been primarily concerned with the issue of growth, in particular with estimating its pace and extent. Yet, however the Industrial Revolution is viewed, whether as a broadly-based sequence of change or as a concentrated burst of creativity centred on one or two leading sectors, the process of technical change and the experience of the cotton textile industry remain of fundamental importance. Even if it was not the only manufacturing sector, cotton was certainly the most dynamic one, a point which all growth indices have tended to confirm.[2] This preoccupation with the nature of the Industrial Revolution means that historians have, of late, been less concerned to explain the changes which their findings seek to describe. As a result, any explanations which have been offered for the British experience of industrialization have been informed by models long rooted in the historiography of that experience. The most popular, and the most frequently repeated, of these explanations is that encapsulated by the term 'challenge and response'. Invention, this hypothesis suggests, was driven by the need to overcome productivity imbalances, innovation in one stage in the manufacturing process generating pressures for change at points earlier or later in the production sequence.[3]

Although often expressed in generalized terms, the 'challenge and response' thesis is most frequently validated by reference to the sequence of invention in cotton textiles during the eighteenth century. Briefly stated, the model posits that the invention of the flying shuttle in 1733 created, or in other versions exacerbated, a productivity imbalance between weavers and spinners. The resulting growth in demand for yarn made more urgent the need to mechanize preparatory processes, including spinning. However, the argument proceeds, the spinning machines which emerged in the 1760s and 1770s themselves created a new imbalance, flooding the market with surplus yarn, the absorption of which could most effectively be achieved through the development of powered forms of weaving.[4] Variants of this model have appeared over time. However, it is in this 'classical' form that it has come to exercise an enduring influence over the historical consciousness, so much so that two of the most recent overviews of British industrialization in general and the development of the cotton industry in particular have both given credence to such a sequence of events.[5] The attraction of the 'challenge and response' model is obvious, imposing, as it does, an order and logic on an otherwise apparently random and chaotic process. Yet its very attraction has ensured that the model has rarely, if ever, been subjected to critical evaluation. This essay thus seeks to plug an important historiographical gap, exposing the 'classical' version of the 'challenge and response' hypothesis to close empirical scrutiny.

The approach adopted here is twofold. Firstly, we present a brief review of the historiography, to identify the point at which the model achieved its final form. The hypothesis that has informed many an undergraduate essay emerges as a belated product of a flawed historiographical tradition. We then proceed to offer a more direct, statistical evaluation of the model and, in particular, its central tenet that the 'challenge' posed by growing productivity imbalances necessitated a redirection of inventive effort into improving complementary processes. In other words, the pressures created or exacerbated by Kay's flying shuttle should, if the model is correct, have stimulated research into the development of spinning and other preparatory innovations. A clustering of inventive activity might be expected to follow, as prospective inventors recognized the process most likely to generate a profit.[6] The extent to which any such clustering occurred, or to which any shift in inventive effort can be observed, is gauged from changes in the broad pattern of recorded innovation over time, as reflected in data sets intended to encompass the bulk of inventive effort in textiles in the century between the invention of the flying shuttle and the self-acting mule. The material comprises registered patents of invention, along with innovations sponsored by agencies including the Irish Linen Board, the Commissioners and Trustees for Improving Fisheries and Manufactures in Scotland, and the Society for the Encouragement of Arts, Manufactures and Commerce, and various private inventions which figure in the literature for the period.[7] The conclusions suggested by this arithmetical exercise are inevitably impressionistic, as the references extend across a wide range of industries and processes. However, they can be tested further by the application of a more traditional mode of analysis, isolating and describing particular sequences of invention. A pattern emerges which is fundamentally at variance from that suggested by the 'challenge and response' model.

COMPOUNDING AN ERROR: HISTORIANS AND THE CHALLENGE AND RESPONSE MODEL

The notion of invention as a response to a clearly defined set of challenges is not wholly absent from the contemporary literature. As early as 1760, the Society of Arts had noted that

> it is notorious that a Scarcity of Hands raises the Price of Labour so high as, by rendering our Manufactures too dear for Foreign Markets, threatens Ruin to many Branches of our Trade, and also as it appears at this time, that in some Branches the want of Workmen is so great that some Millions-worth of Manufactures are imported from foreign Nations, the Rivals to our own[.][8]

This observation provided the preamble to a proposed premium of £200 for improvements to the hand loom, which 'will enable one Man to work as much as two have done hitherto'.[9] It reflected a broad concern, current across textile-producing areas in the middle decades of the eighteenth century, that labour was in short supply and sufficiently expensive in comparative terms to threaten competitiveness.[10] Yet such fears were not confined to particular regions or industrial sectors. Lacking in the Society of Arts' proposal is any sense of problems arising from technological imbalances. Indeed, it is significant that the Society was endeavouring to promote improvements in more than one process. Premiums were considered for both an improved hand-loom and a spinning machine capable of working six threads simultaneously.[11] Such initiatives do not fit readily into any recognizable 'challenge and response' framework.

Subsequent authors identified more clearly the problem of productivity imbalances. Writing in the 1780s, the Manchester fustian cutter, James Ogden, recalled a time when a shortage of workable yarn rendered weavers in the fustian and calico trades vulnerable to the demands of their spinners. Relief came with the invention of the jenny, which was capable of being worked by children and which ensured that 'many who had been insolent before, were glad to be employed in carding and slubbing cotton for these engines'.[12] Yarn shortages and inelasticities in the supply of spinning labour are themes which figure prominently in the literature on the textile trade produced in the early nineteenth century. For example, Ure, in 1836, remarked that 'It was found more easy to multiply weavers than spinsters, and hence looms were often at a stand for want of yarn'.[13] Significantly, however, at no stage was an explicit link drawn between such problems and the advent of the flying shuttle. Indeed, in this as in most accounts, Kay's invention is notable for its absence. Instead, the growth in demand for yarn is linked to broader market trends, in particular the expansion in calico exports.[14] Even where the shuttle is deemed to merit a mention, no con-nection is made between its development and subsequent sequences of invention. For example, Guest's *Compendious History*, first published in 1823, acknowledges the shuttle's importance, arguing that it 'enabled the weaver to make nearly double the quantities of cloth he could have made on the old system'.[15] Yet the prime source of the yarn shortages which precipitated the spinning inventions of the 1760s is still seen to have been the growth in exports. No broader significance was attached to the diffusion of the flying shuttle in the cotton trade from 1760.[16] Guest's work thus provides no corroboration for the 'challenge and response' model as set out above. However, his statement regarding the productivity implications of the shuttle would influence profoundly the elaboration of that hypothesis by future historians.

By the middle decades of the nineteenth century, an abbreviated version of the 'challenge and response' story was in circulation. This concerned the development of powered weaving by the Reverend Edmund Cartwright, work stimulated, according to Cartwright's own account, by an encounter with Manchester manufacturers at Matlock Spa. The latter were anxious that the general adoption of Arkwright's machines would mean that

> so much more yarn would be manufactured than our own weavers could work up, that the consequence would be a considerable export to the Continent, where it might be woven into cloth so cheaply, as to injure the trade in England. Mr. Cartwright replied to this observation, that the only remedy for such an evil would be to apply the power of machinery to the art of weaving as well as to that of spinning, by contriving looms to work up the yarn as fast as it was produced by the spindle.[17]

As late as the 1870s, this Matlock episode, and the circumstances surrounding it, was the most commonly propounded version of the 'challenge and response' thesis.[18] For over one hundred years after the successful mechanization of cotton spinning, few made the connection between it and the earlier development of the flying shuttle.

The 'challenge and response' model only entered the historiographical mainstream in its final, mature form in the 1880s. In 1886, Thomas Ellison produced a book which, while directed at the business community, included a brief historical overview of technical progress within the cotton trade.[19] For information, Ellison drew substantially on the standard early-nineteenth-century histories of Guest, Ure and Baines. However, he included an observation to be found in none of these works. According to Ellison, the productivity imbalance between hand spinning and hand weaving worsened significantly 'after the invention of the fly-shuttle by Mr. John Kay, of Bury, in 1738, a contrivance which enabled the weaver to turn out twice as much cloth as before in a given space of time'.[20] Although repeating points made half a century earlier, including the mis-dating of Kay's innovation,[21] Ellison departed from conventional accounts in making explicit the link between the advent of the shuttle and the 'challenge' of production imbalances. He was not quite the first author to do so. Gilbert French, in his life of Samuel Crompton, first published in 1859, had made a similar point.[22] In both cases the principal source employed appears to have been Guest's *Compendious History* and the claim made therein that the shuttle doubled weaving productivity.

To gauge the accuracy of this assertion, some account needs to be taken of the particular context in which Kay's improvement was first applied. It found favour most immediately in the Colchester bays trade, an industry then experiencing decline in its principal export markets.[23] The advantage of the fly shuttle was that it would enable one weaver to undertake work which usually required two. Rather than the shuttle being thrown across the loom by hand, the movement of the weft was controlled by a single operator.[24] A significant cost saving was promised, boosting the industry's capacity to recover its export potential. Yet the gain in labour productivity was transformed by Guest and the authors who drew on his work into a doubling in weaving output, an unlikely prospect in the context of the bays trade. Furthermore, figures prepared by Kay's son, Robert, and submitted to the Society of Arts in 1764 suggest that, by the most generous estimates, use of the fly shuttle boosted output on broad looms by only 30 per cent, while the gain on narrower cloths, including cottons, was even more modest at 10 to 17 per cent.[25] Ellison's assertions, and the arguments which have been derived from them, appear fundamentally flawed, based as they are on a mis-reading of the evidence.

Subsequent texts have compounded the error, identifying, without attempting to investigate, a link between the diffusion of the flying shuttle and the mechanization of cotton spinning. The precise significance given to this point varies between authors. Wadsworth and Mann are characteristically fastidious, noting the pressures created by market growth, while arguing that any scarcity in yarn supplies 'was not felt acutely until after the introduction of the fly shuttle'.[26] Others are less cautious. Edwards, for example, is clear that 'The balance between the weaving and spinning sections had been disturbed by the introduction in the mixtures trade of

Kay's "flying shuttle", which by the 1750s had increased the productivity of the weavers'.[27] The principal source for this claim is Guest, which indicates the extent to which earlier perspectives have continued to inform historical analysis of technical change. Two years later, David Landes incorporated this example into a broader theory of invention, which remains the fullest exposition of the 'challenge and response' model.[28] The notion of inventive effort driven by a succession of technological disequilibria has become a commonplace of histories of technology, and features in many economic histories of the period.[29] Geoffrey Timmins is merely the latest to pursue this line, arguing that, as a result of the shuttle's adoption, 'A technological inbalance [sic] was ... created in the industry, with the result that, all too often, weavers were unable to obtain the amounts of yarn they needed'.[30]

The historiographical tradition of which this is the latest expression dates back just over one hundred years. For something like a century after the inventions of Hargreaves, Arkwright and Crompton, the 'challenge and response' thesis was largely absent from the literature. Its subsequent emergence was the result of a mis-reading of an existing body of evidence. Although frequently reiterated since, it lacks validation in contemporary sources. However, before we deliver a verdict, it remains important to subject the model to direct statistical evaluation, by examining broad changes in the pattern of invention over time.

WEIGHED IN THE BALANCE AND FOUND WANTING: A STATISTICAL EVALUATION OF THE MODEL

Utilizing the sources outlined above, we can reconstruct fluctuations in the direction of inventive effort by categorizing textile innovations under three broad headings: 'preparatory', involving all processes up to and including the spinning of yarn; 'manufacturing', covering the transformation of yarn or thread into cloth; and 'finishing', which determined the final appearance and texture of the end product. The data cover the period which encompasses the development and diffusion of acknowledged macro-inventions,[31] from John Kay's flying shuttle (1733) to Richard Roberts's self-acting mule (1825). The period as a whole is divided into sub-periods, so as to isolate the effects of particular innovations. In each case, twenty-year periods are taken, to allow for the full effects of an invention's diffusion to work themselves through. The data-set, it must be acknowledged, is unweighted and includes inventions and developments which found no subsequent commercial application. Nevertheless, the figures can be taken to be approximate measures of the overall direction of research energies in the textile industry within the specified period. At all points, two sets of figures are provided. Estimates of total recorded invention are supplemented by statistics relating solely to patented innovations. The latter subset may be thought to have certain advantages: firstly, it is free of institutional biases, many sponsoring bodies having been keen to stimulate research into designated areas; and secondly, because market signals are not thereby distorted, it is preferred by many historians for testing economic theories of invention.[32] Thus, if the patent series fails to validate the 'challenge and response' model, that model must be on shaky foundations indeed.

The earliest sub-period covers the decade or so prior to the patenting of the flying shuttle. Although it would appear from Table 7.1 (below) that inventive activity was primarily geared towards the improvement of preparatory processes, this should not be taken to indicate the existence of a technological imbalance to which Kay's shuttle was a response. Three years before patenting the shuttle, Kay developed a machine to spin and dress worsted thread (patent number 515), which suggests a desire to address mechanical problems in general, rather than a concern to relieve particular production bottlenecks.[33]

Table 7.1. Pattern of Textile Innovation before the Flying Shuttle, 1720–33

	Preparatory	Manufacturing	Finishing	Not Known	Total
Number of innovations	10.5	7	5.5	0	23
%	45.6	30.4	23.9	0	100
Number of Patents	7.5	4	3.5	0	15
%	50.0	26.7	23.3	0	100

Sources: Bennet Woodcroft (1854), *Titles of Patents of Invention, Chronologically Arranged from March 2, 1617 (14 James I) to Oct. 1, 1852 (16 Victoriae)*, London: Patents Office of Great Britain; *The Gentleman's Magazine* (1731–1800), London; Royal Society of London (1720–1800), *Philosophical Transactions*, London; John Apletre [and Henry Barham] (1720), *The Produce of India, Italy, and France, Raised in England, by the Silk Manufactures*, London; Richard V. Saville (1979), 'Some Aspects of the Role of Government in the Industrial Development of England, 1686–1720', unpublished Ph.D. thesis, University of Sheffield, pp. 422–33; Gravenor Henson (1831, reprinted 1970), *History of the Framework Knitters*, Newton Abbot: David and Charles, pp. 257–8; Alexander J. Warden (1864), *The Linen Trade, Ancient and Modern*, London: Longman, Green, Longman, Roberts & Green, pp. 448–9.

From that point, the 'challenge and response' thesis would suggest that inventive effort should have been channelled into the improvement of preparatory processes, to meet the problem of productivity imbalances. In fact, in the 20 years after the shuttle's appearance, a perceptible shift occurred away from 'preparatory' innovations, a trend which is even more obvious if attention is confined to patented inventions alone (see Table 7.2, below). The share of 'manufacturing' improvements rose slightly, but the most marked increase occurred in 'finishing' inventions, with the patent series again emphasizing the shift. Such figures reflect the importance of product innovations, in an age when market growth was most effectively achieved by diversifying the range of outputs.[34]

Table 7.2. Pattern of Textile Innovation after the Flying Shuttle, 1734–53

	Preparatory	Manufacturing	Finishing	Not Known	Total
Number of Innovations	12.5	19.5	21	2	55
%	22.7	35.4	38.2	3.7	100
Number of Patents	2	5.5	5.5	14	27
%	20.4	20.4	51.8	7.4	100

Sources: Woodcroft, *Titles of Patents of Invention*; *The Gentleman's Magazine* (1731–1800); Royal Society of London (1720–1800), *Philosophical Transactions*; Warden, *The Linen Trade*, pp. 450–2.

Yet the altered balance between earlier production stages should also occasion little surprise. The immediate context in which the shuttle was first applied ensured that any pressures created in complementary processes would be slight.[35] Preparatory innovations can be traced to these years. Paul and Wyatt's roller spinning machine (patent number 562) and Adrianus van den Bommanaer's machine to twist yarn for the manufacture of lace and cambric (number 563) both appeared in 1738. Yet, despite their chronological propinquity, there is no evidence to link either invention with Kay's. While the shuttle was, in its early years, confined to the plain bays trade, roller spinning was first applied to the cotton manufacture, while van den Bommanaer's machine produced yarn intended for lace making rather than for the loom.[36]

No other 'preparatory' inventions designed to raise productivity were recorded until 1748, when two carding engines were patented, one by Daniel Bourn (number 628) and the other by Lewis Paul (number 636). Yet the intervening years had witnessed a variety of 'manufacturing' innovations, including Kearney's improved sleay on the loom (not patented, 1742), Patrick McGillevrie's warping machine (not patented, 1743) and John Johnstone's improved method of throwing the shuttle (not patented, *c.* 1746).[37] Inventive activity in the two decades after Kay patented the shuttle thus appeared to be as much concerned with improving it and developing alternatives to it as it was with promoting advances in related processes.

A note of qualification must be entered here. The shuttle's diffusion into the cotton manufacture was delayed until about 1760, so that any clear 'challenge and response' pattern might only be expected to become apparent after that date.[38] Even then, the narrative of invention is at odds with the basic model. Contrary to its central premise, diffusion of the shuttle was *preceded* by a wave of spinning inventions, including those of Lawrence Earnshaw (not patented) and James Taylor (patent number 693) in 1753 and 1754 respectively. Both originated in the cotton- and linen-producing area of south-east Lancashire and north Cheshire.[39] Neither can be accommodated within any plausible 'challenge and response' sequence, as both emerged before any pressures to boost productivity could have been felt. This point, it must be conceded, is not in itself decisive, as neither invention was taken up commercially.

Of greater significance is the fact that the broader pattern of invention in the two decades following the shuttle's diffusion provides equally little support for the 'challenge and response' hypothesis. As Table 7.3 below shows, the slight growth in the proportion of 'preparatory' inventions, exaggerated in the overall figures by applications for the Society of Arts' premium for an improved spinning wheel, is comfortably outweighed by the increase in 'manufacturing' innovations. The latter is accounted for, in large part, by improvements and elaborations on the stocking frame, representing a sustained burst of activity in the framework knitting trade following Strutt's invention of the Derby rib frame in 1758.[40]

Such qualifications aside, it must be acknowledged that within a decade of the shuttle's diffusion in the cotton manufacture, two fundamental breakthroughs had been achieved in spinning processes: Hargreaves' invention of the spinning jenny and Arkwright's development of the water frame.[41] Although the chronological coincidence is suggestive, more detailed investigation casts doubt on any causal connection between the various macro-inventions. As above, any productivity gains achieved by the shuttle on narrow cloths were slight. Potentially of greater significance in promoting its adoption by cotton manufacturers were the gains which were anticipated in the quality of output. According to Robert Kay, the greater quantity of yarn conveyed in the shuttle reduced the number of 'piecings' required in producing a given length of cloth. Furthermore, the shuttle's more direct line across the warp made for a neater, more uniform weave.[42]

Table 7.3. Pattern of Textile Innovation after the Diffusion of the Flying Shuttle, 1760–79

	Preparatory	Manufacturing	Finishing	Not Known	Total
Number of Innovations	86	110.5	99.5	0	296
%	29.1	37.3	33.6	0	100
Number of Patents	20	40.5	31.5	0	92
%	21.7	44	34.2	0	100

Sources: Woodcroft, *Titles of Patents of Invention*; *The Gentleman's Magazine* (1731–1800); Royal Society of London (1720–1800), *Philosophical Transactions*; Royal Society of Arts, Ad Hoc Committee Minutes, 1758–63, Minutes, 1760–1850, Minutes of Standing Premium Committees, 1760–1850; Alastair J. Durie (1979), *The Scottish Linen Industry in the Eighteenth Century*, p. 84; Richard Guest (1823), *A Compendious History of the Cotton-Manufacture*, pp. 12–13; *The Trial of a Cause Instituted by Richard Pepper Ardern, Esq., His Majesty's Attorney General, by Writ of Scire Facias, to Repeal a Patent Granted on the Sixteenth of December 1775, to Mr Richard Arkwright* (1785), pp. 52–7, 67–8; Pierre J. Macquer (1763), *Art de la Teinture en Soie*; Edward Baines, Junior (1835), *History of the Cotton Manufacture in Great Britain*, p. 229; Andrew B. Humphries (1983), *Twelve Bobbins Spinning*; Jean Hellot (1767), *The Art of Dying Wool and Woollen Stuffs*; Seymour R. Broadbridge (ed.) (1980), 'British Industry in 1767: Extracts from a Travel Journal of Joseph Banks', p. 135; *Journals of the House of Commons* (1700–1800), vols 13–55; Sydney H. Higgins (1924), *A History of Bleaching*, p. 15; William E.A. Axon (ed.) (1886), *The Annals of Manchester*, p. 99; John Wilson (1786), *An Essay on Light and Colours*; Thomas K. Derry and Trevor I. Williams (1960), *A Short History of Technology*, p. 560; Stanley D. Chapman (1974), 'Enterprise and Innovation in the British Hosiery Industry, 1750–1850', *Textile History*, **5**, 14–37; William Felkin (1867, reprinted 1967), *History of the Machine-Wrought Hosiery and Lace Manufactures*; Gravenor Henson (1831, reprinted 1970), *History of the Framework Knitters*.[43]

Quality considerations were crucial for Lancastrian producers anxious to retain and to build upon the foothold gained in the lucrative Guinea trade, and may have been decisive in encouraging the take-up of the shuttle.[44] It is conceivable that market trends rather than technological imbalances better explain the pattern of invention in cotton textiles in the third quarter of the eighteenth century. To that extent, the 'challenge and response' model merely serves to obscure reality (Table 7.3).

The emergence of the spinning jenny and the water frame cannot convincingly be linked to the diffusion of the flying shuttle. Equally, neither machine appears to have altered the subsequent pattern of invention to any significant degree. If anything, the effect was to shift activity towards 'preparatory' inventions, rather than towards the improvement of complementary processes (Tables 7.4 and 7.5 below).

Table 7.4. Pattern of Textile Innovation after the Spinning Jenny, 1765–84

	Preparatory	Manufacturing	Finishing	Not Known	Total
Number of Innovations	83	94.5	109.5	0	287
%	28.9	32.9	38.2	0	100
Number of Patents	27	50	43	0	120
%	22.5	41.7	35.8	0	100

Sources: as Table 7.3, above, plus: Gilbert J. French (1859, reprinted 1970), *Life and Times of Samuel Crompton*, Bath: Adams & Dart; Warden, *The Linen Trade*.

Table 7.5. Pattern of Textile Innovation after the Water Frame, 1769–88

	Preparatory	Manufacturing	Finishing	Not Known	Total
Number of Innovations	88	92.5	121.5	1	303
%	29.1	30.5	40.1	0.3	100
Number of Patents	31	57	45	0	133
%	23.3	42.9	33.8	0	100

Sources: as Table 7.4, above, plus M.S. [Mary Strickland] (1843, reprinted 1971), *A Memoir of Edmund Cartwright*, Bath: Adams & Dart.

This point is confirmed if we turn to examine the emergence and subsequent improvement of the mule from 1779. Crompton was driven to develop a new spinning machine because of the problems encountered in producing yarn of sufficient quality on the jenny.[45] His work may be seen as part of an entirely familiar process of improving upon established methods of manufacture, and coincided with a modest but perceptible shift in favour of 'preparatory' inventions from the mid-1770s (Table 7.6 below).

Table 7.6. Pattern of Textile Innovation after the Mule, 1779–98

	Preparatory	Manufacturing	Finishing	Not Known	Total
Number of Innovations	117	114	143.5	3	377
%	31.1	30.2	37.9	0.8	100
Number of Patents	52.5	67	76.5	1	197
%	26.6	34.1	38.8	0.5	100

Sources: as Table 7.5, above, plus Enid Gauldie (1966–7), 'Water-Powered Beetling Machines', *Transactions of the Newcomen Society*, **39**, 125–8.

Overall statistical trends should not obscure the fact that attempts to improve weaving processes were made soon after the emergence of the early spinning machines. Prior to Edmund Cartwright's first power-loom patent in 1785 (number 1470), mechanical improvements were registered by William Cheape (number 1237, in 1779) and William Betts (number 1265, in 1780). However, a large number of spinning inventions can also be traced to this period, starting with the improved jennies of Thomas Highs (not patented, 1771) and Coniah Wood (number 1018, in 1772).[46] Of more immediate significance, a concerted surge of spinning inventions occurred later in the decade. This may be seen as a response to Arkwright's attempt to extend his control over the textile trade as a whole through his inclusive carding and spinning patent of 1775 (number 1111).[47] No fewer than 15 spinning machines appeared in the four years which followed, eight of which were patented: 1776 – Henry Marsland, number 1126; Thomas Wood, number 1130; Thomas Dewhurst, number 1131; 1777 – Edward Hallam and Thomas Heys, number 1151; William Vickers, number 1168; 1778 – Charles Mordaunt, number 1198; 1779 – Robert Peel, number 1212; Richard Marsh, number 1236. The desire to evade Arkwright's monopoly rights appears to have provided the principal impetus behind this particular wave of invention.[48] Once again, a macro-invention had the effect of stimulating research into the same rather than complementary processes.

The pattern was repeated after 1785, when Edmund Cartwright's patent for the power loom was followed within ten years by 16 weaving improvements, eight of which were patented.[49] In aggregate terms, a slight shift occurred in favour of 'manufacturing' inventions. In part, this was driven by the perceived threat posed to domestic producers by the export to the Continent of cheap 'surplus' yarn, produced in ever larger quantities on the new spinning machines (Table 7.7 below).

Table 7.7. Pattern of Textile Innovation after Cartwright's Power Loom, 1786–1805

	Preparatory	Manufacturing	Finishing	Not Known	Total
Number of Innovations	123	148	163	3	437
%	28.1	33.9	37.3	0.7	100
Number of Patents	70.5	96	104.5	1	272
%	25.9	35.3	38.4	0.4	100

Sources: as Table 7.6, above.

William Radcliffe's dressing frame was one such invention and was an important improvement in promoting the economic viability of powered weaving. Yet its adoption signalled no marked change in favour of 'preparatory' innovation.[50] Rather, the share of 'manufacturing' inventions continued to grow, more especially at the expense of 'finishing' improvements (Table 7.8 below).

As we move from the era of prototype discoveries to the age of 'improvement',[51] so the notion of 'challenge and response' has far less purchase on the historical imagination. This is, in many respects, entirely appropriate. Invention in this latter period was an increasingly specialized and 'professional' activity, engaging both general mechanics, such as Richard Roberts, who strove to overcome problems across a number of

Table 7.8. Pattern of Textile Innovation after the Dressing Frame, 1803–22

	Prepar-atory	Manufac-turing	Finish-ing	Not Known	Total
Number of Innovations	97.5	154.5	151	1	404
%	24.1	38.3	37.4	0.2	100
Number of Patents	86	125.5	117.5	0	329
%	26.1	38.2	35.7	0	100

Sources: Woodcroft, *Titles of Patents of Invention*; Royal Society of Arts, Minutes, 1760–1850, Minutes of Standing Premium Committees, 1760–1850.

Table 7.9. Pattern of Textile Innovation after the Self-Acting Mule, 1826–45

	Prepar-atory	Manufac-turing	Finish-ing	Not Known	Total
Number of Innovations	346	504.5	343.5	1	1195
%	28.9	42.3	28.7	0.1	100
Number of Patents	342.5	456	320.5	0	1119
%	30.6	40.8	28.6	0	100

Sources: as Table 7.8, above.

processes, and more specialist manufacturers, who worked to achieve incremental improvements in their own sector of the industry.[52] At various stages in his career, Roberts was active in promoting both 'preparatory' and 'manufacturing' inventions. Yet, as is well known, his decision to shift from power-loom production to work on the mechanization of mule-spinning operations was prompted by developments within the spinning trade itself, arising from disputes involving local labour, rather than from pressures induced by more productive weaving techniques.[53] As might have been predicted from the foregoing, the 20 years following Roberts's first self-actor patent were marked by no sudden discontinuity in the pattern of invention. Rather, they witnessed a continuation of the gradual decline in the proportion of inventive activity concerned with the appearance of the final product and further growth in that accounted for by 'preparatory' and 'manufacturing' improvements (Table 7.9 above).

The application of statistical and narrative methods of analysis reveals that the 'challenge and response' model, as specified in much of the literature over the last one hundred years or so, lacks convincing empirical support. Inventors' entirely rational quest for profit led them to work in areas already defined by existing macro-inventions. Breakthroughs, such as the spinning jenny and the water frame, stimulated research into the same process, with the aim of improving on existing techniques or evading monopoly rights. No unambiguous shift into related up- or downstream processes was apparent at any stage in the period separating the flying shuttle from the self-acting mule. One of the very few direct references to such a sequence remains Cartwright's visit to Matlock Spa, an episode which must be judged a flimsy foundation on which to construct a general theory of invention.

CONCLUSION: ASSESSING THE DAMAGE

The 'challenge and response' model has proved, through its logic and simplicity, an enduring attraction for generations of historians and their students. It remains one of the quoted theories designed to explain technological progress. Yet it is, in large measure, a late-nineteenth-century confection on which twentieth-century authors have drawn to impose an order and unity on a complex sequence of events. This order, it must now be acknowledged, is largely spurious, finding no validation in either the contemporary literature or the statistics of inventive activity. In utilizing the 'challenge and response' model, historians and students should be aware that they are handling damaged goods. Its reiteration merely serves to obscure the reality of technical change in the British Industrial Revolution and to divert research from potentially more profitable avenues. Attention should now be focussed on tracing the origins of specific macro-inventions which transformed textile production in the later eighteenth century. Such research must be founded on a close appreciation of market trends, locating the cotton industry within the broader economic, political and social development of Hanoverian Britain. This, at least, is the agenda suggested by an examination of the contemporary literature. It is one in which the 'challenge and response' model has little or no part to play. It is time to bury an old friend.

NOTES

1. Richard Guest (1828), *The British Cotton Manufactures, and a Reply to an Article on the Spinning Machinery Contained in a Recent Number of the Edinburgh Review*, Manchester, p. 157.
2. Nicholas F.R. Crafts (1985), *British Economic Growth during the Industrial Revolution*, Oxford: Clarendon Press; Nicholas F.R. Crafts and C.K. Harley (1992), 'Output Growth and the British Industrial Revolution: a Restatement of the Crafts-Harley View', *Economic History Review*, 2nd series, **45**, 703–30; Thomas S. Ashton (1948), *The Industrial Revolution, 1760–1830*, London: Oxford University Press; Maxine Berg and P. Hudson (1992), 'Rehabilitating the Industrial Revolution', *Economic History Review*, 2nd series, **45**, 24–50; Peter Temin (1997), 'Two Views of the British Industrial Revolution', *Journal of Economic History*, **57**, 63–82.
3. David S. Landes (1969), *The Unbound Prometheus. Technological Change and Industrial Development in Western Europe from 1750 to the Present*, Cambridge: Cambridge University Press, pp. 84–7.
4. Landes, *The Unbound Prometheus*, pp. 84–6; Paul Mantoux (1928), *The Industrial Revolution in the Eighteenth Century. An Outline of the Beginnings of the Modern Factory System in England*. New and revised edition (1961) with a preface by T.S. Ashton, London: Jonathan Cape, pp. 206–9, 241; Phyllis Deane (1979), *The First Industrial Revolution*, Cambridge: Cambridge University Press, second edition, pp. 88–90; Eric Pawson (1979), *The Early Industrial Revolution. Britain in the Eighteenth Century*, London: Batsford, pp. 86–7.
5. Martin J. Daunton (1995), *Progress and Poverty. An Economic and Social History of Britain, 1700–1850*, Oxford: Oxford University Press, p. 187; Geoffrey Timmins (1996), 'Technological Change', in Mary B. Rose (ed.), *The Lancashire Cotton Industry. A History Since 1700*, Preston: Lancashire County Books, pp. 29–62, p. 39.
6. Mantoux, *The Industrial Revolution*, pp. 208–9; Patrick K. O'Brien (1991), 'The Mainsprings of Technological Progress in Western Europe, 1750–1850', in Peter Mathias and John A. Davis (eds), *Innovation and Technology in Europe from the Eighteenth Century to the Present Day*, Oxford: Blackwell, pp. 6–17; Nathan Rosenberg (1982), *Inside the Black Box: Technology and Economics*, Cambridge: Cambridge University Press, pp. 4–27.
7. A card index of inventions for the period, 1660–1850, is available for consultation at the Institute of Historical Research, Senate House, Malet Street, London WC1E 7HU. See Trevor Griffiths, Philip A. Hunt and Patrick K. O'Brien (1992), 'Inventive Activity in the British Textile Industry, 1700–1800', *Journal of Economic History*, **52**, 881–906.
8. Royal Society of Arts, Ad Hoc Committee Minutes 1758–60, 6 May 1760.
9. Ibidem.
10. *Journals of the House of Commons*, London, vol. 15, p. 544; vol. 16, p. 137; vol. 18, p. 686; John Smith (1747, reprinted 1969), *Chronicon Rusticum-Commerciale or Memoirs of Wool being A Collection of History and Argument Concerning the Woolen Manufacture and Woolen Trade in general*, 2 volumes, New York: Augustus M. Kelley, vol. II, pp. 18, 30–1, 180–1, 291, 315.
11. Royal Society of Arts, Ad Hoc Committee Minutes 1758–60, 1 May 1760, 6 May 1760.
12. William E.A. Axon (ed.) (1887), *Manchester 100 Years Ago. A Description of Manchester by a Native of the Town, James Ogden, Published in 1783*, reprinted in William E.A. Axon (ed.) (1983), *Manchester 200 Years Ago*, Radcliffe: Neil Richardson, p. 28; John Aikin (1795, reprinted 1968), *A Description of the Country from Thirty to Forty Miles round Manchester*, New York: Augustus M. Kelley, p. 167.
13. Andrew Ure (1836), *The Cotton Manufacture of Great Britain*, 2 volumes, London: Charles Knight, p. 192; see also James Montgomery (1833), *The Theory and Practice of Cotton Spinning; or, the Carding and Spinning Master's Assistant*, Glasgow: John Niven junior, p. 260.
14. Aikin, *A Description of the Country*, pp. 161–6; Ure, *The Cotton Manufacture*, p. 192; George W. Daniels (1920), *The Early English Cotton Industry. With some unpublished letters of Samuel Crompton*, Manchester: Manchester University Press, p. 60.
15. Richard Guest (1823), *A Compendious History of the Cotton-Manufacture; with a disproval of the claim of Sir Richard Arkwright to the Invention of its Ingenious Machinery*, Manchester: Joseph Pratt, p. 8; see also T. Sutcliffe (1843), *An Exposition of Facts relating*

to the Rise and Progress of the Woollen, Linen and Cotton Manufactures of Great Britain, Manchester, p. 5.

16. Guest, *A Compendious History of the Cotton-Manufacture*, p. 12.
17. M.S. [Mary Strickland] (1843, reprinted 1971), *A Memoir of Edmund Cartwright*, Bath: Adams & Dart, p. 56.
18. Alfred Barlow (1878), *The History and Principles of Weaving by hand and by power*, London: Sampson Low, Marston, Searle, and Rivington, p. 233; also Edward Baines junior (1835), *History of the Cotton Manufacture in Great Britain*, London: H. Fisher, R. Fisher, and P. Jackson, p. 229; James Bischoff (1842), *A Comprehensive History of the Woollen and Worsted Manufactures, and the Natural and Commercial History of Sheep, from the Earliest Records to the Present Period*, 2 volumes, London: Smith, Elder and Co., p. 317; Thomas Ellison (1858), *A Hand-Book of the Cotton Trade; or, A Glance at the Past History, Present Condition, and Future Prospects of the Cotton Commerce of the World*, London: Longman, Brown, Green, Longman, and Roberts, pp. 127–9.
19. Douglas A. Farnie (1978), 'Three Historians of the Cotton Industry: Thomas Ellison, Gerhart von Schulze-Gaevernitz, and Sydney Chapman', *Textile History*, **9**, 75–89, pp. 76–7.
20. Thomas Ellison (1886, reprinted 1968), *The Cotton Trade of Great Britain*, London: Frank Cass; New York: Augustus M. Kelley, p. 15.
21. Guest, *A Compendious History of the Cotton-Manufacture*, p. 8.
22. Gilbert J. French (1859, reprinted 1970), *Life and Times of Samuel Crompton*, Bath: Adams & Dart, pp. 20–1.
23. A.F.J. Brown (1968), 'Colchester in the 18th Century', in L.M. Munby (ed.), *East Anglian Studies*, Cambridge: Heffer, pp. 146–73, pp. 148–9; Eric Kerridge (1985), *Textile Manufactures in Early Modern England*, Manchester: Manchester University Press, p. 172; W.H. Chaloner (1965–66), 'The Textile Inventor John Kay', *Bulletin of the John Rylands Library*, **48**, 9–12, p. 11; Alfred P. Wadsworth (1926–28), 'Sidelights on Rochdale History', *Transactions of the Rochdale Literary and Scientific Society*, **16**, 55–73, p. 72.
24. Alfred P. Wadsworth and Julia de Lacy Mann (1931), *The Cotton Trade and Industrial Lancashire, 1600–1780*, Manchester: Manchester University Press, p. 450.
25. Sir H.T. Wood (1911–12), 'The Inventions of John Kay, 1704–70', *Journal of the Royal Society of Arts*, **60** (3081), 73–86, pp. 83–4; A. Paulinyi (1986), 'John Kay's Flying Shuttle: some Considerations on its Technical Capacity and Economic Impact', *Textile History*, **17**, 149–66, p. 154.
26. Wadsworth and Mann, *The Cotton Trade*, p. 472.
27. Michael M. Edwards (1967), *The Growth of the British Cotton Trade, 1780–1815*, Manchester: Manchester University Press, and New York: Augustus M. Kelley, p. 3.
28. Landes, *The Unbound Prometheus*, pp. 84–7.
29. Jacques Payen and Jean Pilisi (1980), 'Weaving and Mechanical Finishing', in Maurice Dumas (ed.), *A History of Technology and Invention. Progress Through the Ages. Vol. III. The Expansion of Mechanization, 1725–1860*, London: John Murray, pp. 599–618, pp. 601–2; Bertrand Gille (1986), *The History of Techniques. Volume 1. Techniques of Civilization*, London: Gordon and Breach, pp. 624–30; Donald Cardwell (1994), *The Fontana History of Technology*, London: Fontana Press, pp. 141–5; Nathan Rosenberg (1996), 'Uncertainty and Technological Change', in R. Landau, T. Taylor and G. Wright (eds), *The Mosaic of Economic Growth*, Stanford, California: Stanford University Press, pp. 334–53, p. 345.
30. Timmins, 'Technological Change', p. 39.
31. Joel Mokyr (1990), *The Lever of Riches: Technological Creativity and Economic Progress*, Oxford: Oxford University Press; Joel Mokyr (1994), 'Technological Change, 1700–1830', in Roderick Floud and Donald McCloskey (eds), *The Economic History of Britain since 1700. Second Edition. Volume 1: 1700–1860*, Cambridge: Cambridge University Press, pp. 12–43, pp. 15–16.
32. Kenneth L. Sokoloff (1988), 'Inventive Activity in Early Industrial America: Evidence from Patent Records, 1790–1846', *Journal of Economic History*, **48**, 813–50; Richard J. Sullivan (1989), 'England's "Age of Invention": the Acceleration of Patents and Patentable Invention during the Industrial Revolution', *Explorations in Economic History*, **26**, 424–52; Richard J. Sullivan (1990), 'The Revolution of Ideas: Widespread Patenting and Invention during the English Industrial Revolution', *Journal of Economic History*, **50**, 349–62; Richard J. Sullivan (1995), 'Patent Counts and Textile Invention: A Comment on Griffiths,

Hunt and O'Brien', *Journal of Economic History*, **55**, 666–70; although see Christine MacLeod (1988), *Inventing the Industrial Revolution. The English Patent System, 1660–1800*, Cambridge: Cambridge University Press, p. 157; Griffiths, Hunt and O'Brien, 'Inventive Activity'; Patrick K. O'Brien, Trevor Griffiths and Philip A. Hunt (1995), 'There Is Nothing Outside The Text, and There Is No Safety In Numbers: A Reply to Sullivan', *Journal of Economic History*, **55**, 671–2.

33. Bennet Woodcroft (1863), *Brief Biographies of Inventors of Machines for the Manufacture of Textile Fabrics*, London: Longman, Green, Longman, Roberts and Green, pp. 1–5.
34. Griffiths, Hunt and O'Brien, 'Inventive Activity'; Patrick K. O'Brien, Trevor Griffiths and Philip A. Hunt (1996), 'Theories of Technological Progress and the British Textile Industry from Kay to Cartwright', *Revista de Historia Economica*, **14**, 533–55, pp. 543–4.
35. Brown, 'Colchester in the 18th Century'; G.H. Tupling (1927), *The Economic History of Rossendale*, Manchester: Chetham Society, new series, vol. 86, pp. 181–2.
36. Kerridge, *Textile Manufactures*, p. 172; Wadsworth and Mann, *The Cotton Trade*, pp. 419–55.
37. Alexander J. Warden (1864), *The Linen Trade, Ancient and Modern*, London: Longman, Green, Longman, Roberts & Green, pp. 451–2.
38. Wood, 'The Inventions of John Kay', p. 75; Baines, *History of the Cotton Manufacture*, p. 117.
39. *The Gentleman's Magazine* (1787), pp. 1165–6, 1200; W. Chadwick (reprinted 1972), *Reminiscences of Mottram*, Longdendale, pp. 24–9; F. Espinasse (1874), *Lancashire Worthies*, London: Simpkin, Marshall, & Co., and Manchester, Abel Heywood & Son, pp. 315–6; Aikin, *A Description of the Country*, pp. 466–8; Wadsworth and Mann, *The Cotton Trade*, p. 473.
40. R.S. Fitton and Alfred P. Wadsworth (1958), *The Strutts and the Arkwrights, 1758–1830*, Manchester: Manchester University Press, pp. 24–46; Gravenor Henson (1831, reprinted 1970), *History of the Framework Knitters*, Newton Abbot: David and Charles, pp. 259–76; Stanley D. Chapman (1974), 'Enterprise and Innovation in the British Hosiery Industry, 1750–1850', *Textile History*, **5**, 14–37; Alan Rogers (1981), 'Rural Industries and Social Structure: The Framework Knitting Industry of South Nottinghamshire, 1670–1840', *Textile History*, **12**, 7–36, pp. 10–11.
41. W.A. Hunter (1951–52 and 1952–53), 'James Hargreaves and the Invention of the Spinning Jenny', *Transactions of the Newcomen Society*, **26**, 141–51; Christopher Aspin and Stanley D. Chapman (1964), *James Hargreaves and the Spinning Jenny*, Helmshore: Helmshore Local History Society; Christopher Aspin (1968), 'New Evidence on James Hargreaves and the Spinning Jenny', *Textile History*, **1**, 119–21; R.S. Fitton (1989), *The Arkwrights. Spinners of Fortune*, Manchester: Manchester University Press, pp. 22–49.
42. Wood, 'The Inventions of John Kay', pp. 83–4; J. Lord (1899), 'John Kay, the Inventor of the Fly Shuttle', pt. iv, *The East Lancashire Review*, **3**, 331–4, p. 333; Paulinyi, 'John Kay's Flying Shuttle', p. 154.
43. Alastair J. Durie (1979), *The Scottish Linen Industry in the Eighteenth Century*, Edinburgh: John Donald Publishers Ltd; Guest, *A Compendious History of the Cotton-Manufacture*; *The Trial of a Cause Instituted by Richard Pepper Ardern, Esq., His Majesty's Attorney General, by Writ of Scire Facias, to Repeal a Patent Granted on the Sixteenth of December 1775, to Mr Richard Arkwright* (1785), London; Pierre J. Macquer (1763), *Art de la Teinture en Soie*, Paris: Academie des Sciences; Baines, *History of the Cotton Manufacture*; Andrew B. Humphries (1983), *Twelve Bobbins Spinning (the Story Surrounding a Horizontal Flax Wheel Invented by Bernard Barton c. 1766)*, Ivegill: A.B. Humphries; Jean Hellot (1767), *The Art of Dying Wool and Woollen Stuffs*, Dublin; Seymour R. Broadbridge (ed.) (1980), 'British Industry in 1767: Extracts from a Travel Journal of Joseph Banks', in A. Rupert Hall and Norman Smith (eds) (1980), *History of Technology*, vol. 5, London: Mansell, pp. 119–42; *Journals of the House of Commons* (1700–1800), vols 13–55, London; Sydney H. Higgins (1924), *A History of Bleaching*, London: Longmans & Co.; William E.A. Axon (ed.) (1886), *The Annals of Manchester*, Manchester: John Heywood; John Wilson (1786), *An Essay on Light and Colours, and What Colouring Matters Are That Dye Cotton and Linen*, Manchester; Thomas K. Derry and Trevor I. Williams (1960), *A Short History of Technology*, Oxford: Clarendon Press; Chapman, 'Enterprise and Innovation'; William Felkin (1867, reprinted 1967), *History of the Machine-Wrought Hosiery and Lace Manufactures*, Newton Abbot: David and Charles; Henson, *History of the Framework Knitters*.

44. Wadsworth and Mann, *The Cotton Trade*, pp. 148–61.
45. John Kennedy (1830), *A Brief Memoir of Samuel Crompton; with a Description of his Machine called the Mule and of the subsequent Improvement of the Machine by Others*, Manchester, p. 4; T. Midgeley (1927), *Samuel Crompton, 1753–1827 – a Life of Tragedy and Service*, Bolton, pp. 7–8; Michael E. Rose (1965–66), 'Samuel Crompton (1753–1827), Inventor of the Spinning Mule: a Reconsideration', *Transactions of the Lancashire and Cheshire Antiquarian Society*, **75–76**, 11–32, p. 15.
46. *Leary's Directory* (1788) (Manchester Central Reference Library, Archives Department).
47. *A State of Facts respecting the case of Mr. Richard Arkwright of Cromford, in the county of Derby, Cotton Manufacturer* (1782), Special Collections, Columbia University Library; J. Hewish (1985), *Prejudicial and Inconvenient?, a Study of the Arkwright Patent Trials, 1781 and 1785*, London: British Library.
48. Stanley D. Chapman and S. Chassagne (1981), *European Textile Printers in the Eighteenth Century* (Pasold Studies in Textile History, 1), London: Heinemann, p. 48.
49. Bennet Woodcroft (1861), *Patents for Invention, Abridgements of Specifications Relating to Weaving*, London: Patent Office; G. White (1851), *A Practical Treatise on Weaving, by Hand and Power Looms*, London: Whittaker & Co., and Glasgow: John Niven, pp. 94–105; Warden, *The Linen Trade*, pp. 456–60; R. Marsden (1895), *Cotton Weaving: its Development, Principles, and Practice*, London: G. Bell & Sons, pp. 61–71; W. Wilkinson (1927), 'Power Loom Developments', in *Official Record of the Annual Conference of the Textile Institute, held at Bolton, 7th, 8th and 9th June 1927*, London: Journal of the Textile Institute, pp. 122–48.
50. William Radcliffe (1828), *Origin of the new System of Manufacture, commonly called 'Power-Loom Weaving', and the purposes for which this system was invented and brought into use*, Stockport: James Lomax, pp. 11–29; Woodcroft, *Brief Biographies*, pp. 32–7; White, *A Practical Treatise*, pp. 124–8.
51. O'Brien, Griffiths and Hunt, 'Theories of Technological Progress', pp. 543–5.
52. Harold I. Dutton (1984), *The Patent System and Inventive Activity During the Industrial Revolution, 1750–1852*, Manchester: Manchester University Press, p. 114; H.W. Dickinson (1945–46 and 1946–47), 'Richard Roberts, his Life and Inventions', *Transactions of the Newcomen Society*, **25**, 123–37; W.H. Chaloner (1968–69), 'New Light on Richard Roberts, Textile Engineer (1789–1864)', *Transactions of the Newcomen Society*, **41**, 27–44; Marsden, *Cotton Weaving*, pp. 71–3.
53. Andrew Ure (1835, reprinted 1967), *The Philosophy of Manufactures: or, An Exposition of the Scientific, Moral, and Commercial Economy of the Factory System of Great Britain*, London: Frank Cass & Co. Ltd., pp. 364–8; K. Bruland (1982), 'Industrial Conflict as a Source of Technical Innovation', *Economy and Society*, **11**, 91–121, pp. 100–3.

8. Product Innovation in Core Consumer Industries in Eighteenth-Century Britain

Maxine Berg

Since the nineteenth century the industrial revolution has been explained by the technological breakthrough of the Watt steam engine.[1] As Christine MacLeod has recently argued, however, the popularity of this explanation always owed much to the ease with which Watt and his engine could be rendered the subjects of myth.[2] Much less frequently discussed, by contrast, were the concurrent product innovations – inventions of new commodities within the framework of international markets. The mechanical and process innovations were important – they shifted production functions and raised productivity in altogether new ways. But the reason they took place at the time they did, and the explanation for their widespread impact, must lie in the expansion of markets and the proliferation of a whole range of new products which made the use of these techniques viable.

We need to look at industrial expansion in eighteenth-century England not just in terms of a few major industries and technologies, but in terms of the goods produced. This calls, I think, for some lateral thinking which takes us away from the enclosed worlds of the textile historian, the engineering historian, or the historian of the iron industry. We need to think across commodities which were used in a complementary fashion. I would thus argue that it is important to set out the place played in the eighteenth-century economy by a set of new consumer industries which included fashion articles of clothing, new fabrics and patterns as well as domestic commodities and ornamental and semi-luxury commodities. Producers of these goods frequently drew on similar ideas and skills; they were frequently organized in similar ways; and their perceptions of their markets were similar.

In this chapter I look at invention and innovation in the context of the product, the commodity. I have chosen to focus on a core group of commodities – light metalwares, ceramics, glass, clocks and watches and furniture. I seek to argue that a framework of innovation was developed on the complementarities among these commodities, and on a dynamic market among the middling classes which pushed producers to create new products, finishes and tools. New commodities in this core were developed using new materials – these included earthenwares and flint and cut glass, the chemical development of new metal alloys or finishes such as gilt and silver-plate, stamped brassware, japanned tinware and papier mâché, ormolu and cut steel. Networks among producers and merchants acted as conduits of information and connection in product development, design and technology – and helped to make these industries centres for innovation.

Investigation into the commodities leads into an area rather unfamiliar to the economic historian: that of luxury, semi-luxury and new consumer goods. These manufactures were located partly in the world of metropolitan luxuries – the goldsmiths and silversmiths, fine furniture makers, and colonial finishing trades. But they were also part of a world of provincial manufacturing regions – the watch and clock part makers, fine tool makers and steel filemakers filling burgeoning proto-industrial villages between Prescot and Warrington; the toymakers and japanners of Birmingham with its hinterland of lockmakers in Wolverhampton and flint glass makers and glasscutters in Stourbridge; the potters of Staffordshire and the cutlers, silversmiths and instrument makers of Sheffield. Did these industries have a greater part to play in Britain's industrial transformation in the eighteenth century than economic historians, traditionally dismissive of the luxury trades, have hitherto allowed?

Two of the great centres of manufacture in early industrial Britain were Birmingham and Sheffield. Luxury and semi-luxury goods were a mainstay of their manufactures from early in the eighteenth century. Medallions, buttons and buckles, ornamental boxes, jewellery, buttons, table-ware, cutlery, silver and silver-plated ware, candelabra and small furnishings and ornaments, made of brass, plated ware, light metal alloys, ormolu and japanned ware, and cut and coloured glassware: these provided the household decencies and semi-luxuries of the expanding middle classes of Britain and her settler colonies. Contemporaries were impressed with Birmingham as a manufacturing showplace of Europe, though some were uncomfortable with its products; the debate on luxury coloured their perceptions. Adam Smith remarked on the technical progress in both towns, and the great reductions in prices in Birmingham and Sheffield ware. But he noticed that wages were higher in Birmingham, and that its advantages were similar to those metropolises and maritime towns trading on luxury manufactures.

> Manufactures for which the demand arises altogether from fashion and fancy, are continually changing and seldom last long enough to be considered as old established manufactures ... Birmingham deals chiefly in manufacture of the former kind; Sheffield in those of the latter; the wages of labour in those two different places, are said to be suitable to this difference in the nature of their manufactures.[3]

The diversity of the products of Birmingham and (to a lesser degree) Sheffield was regarded at the time as a form of innovation; ingenuity, taste and workmanship were praised by James Walker in 1799.[4]

PRODUCT INNOVATION

Incorporating product innovation into the analysis of the industrial revolution, and relating this to current analyses of the rôle played by technological change, is a difficult task.

First, our current economic histories of the period have played down the significance of the eighteenth-century sense of novelty, innovation, new workshops and factories, engines and tools. Secondly, technological innovation has been depicted as a supply-side phenomenon – as Mokyr has argued, macro-inventions were random shocks, and the sustaining micro-inventions, improvements and learning by doing were the results of an élite core of mechanically skilled workers.[5] As Mokyr has also concluded, however, Britain's success lay in turning new ideas into economically viable projects.[6] The products and a sense of their markets were integral to the success of the new techniques. This is where the recent social history of consumerism might be expected to help. This research has identified a flowering of material culture in early modern and eighteenth-century Europe, and has claimed an expansion of consumption among bourgeois and ordinary people in eighteenth-century Britain.[7] The historians of consumption, material culture and the decorative arts have examined the luxury and semi-luxury trades, but they have shown little interest in the way these trades have operated as industries. They have looked at possessions, advertising, and, as Michael Sonenscher has argued,[8] at ornate objects and personal creations, not at the economy of luxury production.

Economic historians have not compensated for this gap, for they have not collected output data for many of the industries producing such new consumer ware.[9] Metalwares are a good example. There appears to have been no attempt among collectors of data for the eighteenth and even the nineteenth century to separate out metal products from iron, and to distinguish decorative and consumer goods from basic hardware, armaments and engineering goods.[10]

A study of product innovation during the industrial revolution provides a way of bringing together the findings of historians of technology and historians of consumerism. It also provides a radical approach to reconsidering the current impasse between supply-side and demand-side explanations of innovation. Where Mokyr and O'Brien have argued against the rôle of a bigger market, or of the impact of material culture,[11] Kenneth Sokoloff has consistently argued the case for demand-side motivations to innovation in the ante-bellum U.S. economy. There he found the sources of invention in the characteristics of demand – the structure of the market, domestic competition, relatively small firms and a relatively prosperous pre-industrial population with a working knowledge of current technology and a lively response to market opportunities.[12]

The study of product innovation, however, poses many problems to economic historians, some of which account for its relative neglect. The importance of this type of innovation, especially as revealed in patent records, has been singled out by MacLeod and O'Brien. During the eighteenth century, as MacLeod has pointed out, just as many patents claimed to improve quality as to save capital. Furthermore, she estimates that about one-third of late eighteenth-century patents were for items of personal and domestic consumption.[13] O'Brien estimates that almost 34 per cent of manufacturing inventions, both patented and private, in textiles during the eighteenth century were product innovations: 'until 1790, output diversification and quality improvements remained the principal catalysts of growth.'[14]

The analysis of this innovation, however, is less clearly identified. Historians of technology traditionally try to set this innovation in a separate sphere. MacLeod (in an important article on invention and the capital goods sector) distinguishes machine users from machine makers, in parallel with the divide between process and product innovation. To the user of consumer goods, any new machine he invented was a process invention; to the capital goods manufacturer, a new machine was a product invention. The institutional separation of the producers of machinery from its users at the end of the eighteenth century and in the early nineteenth century speeded up the process of diffusion.[15] This division is also followed by O'Brien in his analysis of textile patenting. The earlier period, the century after the Restoration, was dominated by product innovation, that is, new mixes of yarns woven into cloth that was bleached, dyed and printed in new ways. This was, he argues, simply a response to the widening of domestic and foreign markets, Smithian growth. It formed the precondition for a new phase of technological innovation, one which became endogenous to particular sections of the textile industry.[16] It is thus that general accounts tend to by-pass product innovation, or to treat it, as does von Tunzelmann, as an early phase of innovation antecedent to the real and by imputation more important process inventions.[17]

Product innovation, discussed by historians of material culture, takes on a different meaning: the marketing of products. As John Styles has argued, the goods that made up the material culture of early modern England were a mixture of entirely new products that the English had invented or drastically remodelled, such as pocket microscopes and watches, of novelties that originated elsewhere in Europe, like delftware plates and Venetian glass, and of others like mahogany, lacquered cabinets and printed calico that arrived from Asia and the Americas. Product innovation was about tailoring products for new markets.[18]

The failure by economic historians seriously to incorporate product development and quality changes has much to do with the inability of economists to measure the impact of such changes. Attempts to identify sources of economic growth during the industrial revolution have relied on measures of total factor productivity growth. These measures cannot take account of innovation in the nature of outputs or of changes in the quality of inputs.[19] The national accounts framework used by economists measures the replication of goods and services, but cannot easily incorporate either the appearance of entirely new goods not present at the start of a time series, or improvements over time in the quality of goods or services. There is a gap between empirical evidence for the part played by new commodities, and the traditional theory of consumer behaviour, which ignores the impact of the physical properties of goods upon demand. Some economists have pointed to the ramifications of this gap. A theory which omits the technical characteristics of goods leaves it incapable of handling product variations and new goods.[20] Economists have accounted for this theoretical deficiency by claiming that the problem is a new one, contingent upon new historical conditions which have emerged in recent decades. Before this, it is argued, innovation and technological change was largely confined to the means of production rather than extending to the means of consumption. Inventors and innovators had been concerned with making existing commodities more efficiently, rather than with making more efficient commodities. It has been argued further that earlier economists who laid the foundation of the marginal utility theory of consumer behaviour were, therefore, justified in confining their examples of invention to producer goods, and in ignoring changing characteristics of consumer products.[21] Yet we have seen that there is no historical basis for this world of simple commodities in either the nineteenth century or the eighteenth.

The implications of the failure to consider quality adjustments and new products for estimates of economic growth during the eighteenth century must be parallel to the effects recently measured by R.J. Gordon for the U.S.

economy after the Second World War. He found large quality biases in the official price indexes for both producer and consumer durables, and raised the question of the extent to which this is a recent problem, or an age-old one.[22] He drew attention to those quality changes stemming from technical change which reduced the costs of purchasing a given quantity of characteristics, and to techniques which reduced the costs of operation, as well as to totally new products such as televisions and the automobile.[23] He estimated a doubling of former estimates of the growth of producer durables, and a smaller though significant adjustment for consumer durables.[24] Even his own estimates, he argued, understated other quality changes such as improvements in the reliability of photocopying machines, reduced vibration and noise of jet planes, longer lifetimes of diesel engines, better sound quality in radios and better picture quality in televisions.[25]

PRODUCT INNOVATION: CASE STUDIES FROM CORE CONSUMER INDUSTRIES

This recent recognition of the significance of quality differences and of new products in the twentieth century challenges us to enquire into the extent and characteristics of product innovation during the industrial revolution. How extensive was product innovation in the core consumer industries I have selected? Were there complementarities in product development and design among these industries? For evidence on this, I will draw on patent records and trade catalogues.

The first source is the patent records. The limitations of this source as an overall indicator of technological change in general have been well rehearsed elsewhere. We can re-examine these records, however, in the light of consumer culture rather than from the traditional perspectives of the historian of technology. Indeed some of these latter analyses have divided up this patent data into anachronistic categories which hold little meaning for the motivations of patentees or the description of the records. One such example is Richard Sullivan's selection of 'six technologically progressive industries', and his selection of 'useful inventions', and 'machine inventions.'[26] Categories such as these distort the patent evidence, and the criteria for selection of individual patents into any of these categories seem mysterious.

Patents represented only a minimalist picture of eighteenth-century invention and product innovation. Birmingham, one of the places with the most prolific invention of goods, gadgets, new materials and machinery, claimed only 102 patents over the whole period 1680–1800.[27] Patents cost between £120 and £350 depending on their coverage between England and the British Isles.[28] In luxury and new consumer industries with high mobility of skilled workers, and rapid international communication of new designs and techniques based on the international connections of court cities, it was difficult indeed to protect invention from European competitors. Many of the inventors, as revealed through the patent specifications, emanated from a world of natural philosophy, mechanics, automata and technological fantasy among the upper and middle ranks of British society. O'Brien has argued similarly for textile innovations. This was a world not far removed from that of John Joseph Merlin and James Cox's Museum of Automata.[29] These middle ranks sometimes included goldsmiths, silversmiths and watchmakers, who might turn their taste for invention to a variety of schemes. John Tuite, a silversmith, patented several inventions in civil engineering from the 1730s to 1750s, related to land drainage. Lewis Pantin, a goldsmith, worked in partnership with James Valoue, a watchmaker, and patented processes for raising ballast from river bottoms and for driving piles in the water.[30] Or these ranks might include the occasional well-educated spinster such as Elizabeth Perryman from Soho, who invented a street lamp. In her patent specifications she explained the drawbacks of existing street lighting in clear scientific language, and moved on to set out the nature of her own experiments, her use of the theory of optics and the law of catoptrics, and her own invention accompanied with wonderfully detailed drawings.[31]

For a number of these inventors, the patent was itself a consumer good, displaying their place in enlightened society, as creators of novelties, as making advances based on scientific principles, as part of the world of the arts. For many of the patent specifications were contributions to the development of the graphic arts in much the same way as were military topographies. Patent drawings and numbers were also used by manufacturers as part of the advertising in their trade cards, directories and trade catalogues, lending greater credibility to the claims they made for a whole range of other products.

I have collected a set of patents from the period 1627–1825 covering the group of metalwares, glass, ceramics, furniture and clocks and watches, as well as more general finishing techniques on a range of consumer goods.[32] This yielded 1610 patents, most of these concentrated in the eighteenth century; they made up approximately 30 per cent of the total patents taken out over this period.[33] This select group of consumer goods industries reveals a range of process and product inventions. Innovation in metalwares were central, accounting for 65 per cent of the patents in my consumer goods selection. Patents dealing with glass made up another 10 per cent of the group, while there were only small numbers of patents dealing with ceramics. Many of the innovations were of three basic types. Firstly, there were new chemical processes for making new metals, or for using coal-fired techniques in flint glass ware and earthenware. Secondly, new mechanical implements and automated processes were developed alongside batch production of different types of goods. Thirdly, and above all, new products were invented by transferring materials and designs across industries. Most of the patents were indeed for new products or for variations within those products. Equally important among product developments were patents directed to the use of different or substitute materials. These patents crossed the borderland into process innovation, as new materials were also developed. This fashion for substitution ran parallel to the application of specific techniques to different materials or goods.

The consumer orientation of much of this patenting stands out. The focus on new goods, on the ornamenting or finishing of a range of existing goods, on extending the qualities and varieties within ranges of goods, and on novelties and showy luxuries, raises a number of questions about the use of patent records as evidence of technological innovation. One popular focus of patenting activity, for example, was carriages and their ornamentation. Twelve such patents were taken out over the period 1769–1799. John Pickering took out a patent in 1769 for chasing in gold, silver and other metals, coffin furniture, ornaments for coaches, chariots, sedans and other carriages.[34] Are patents such as these to be ignored, or, alternatively, cited as evidence that patenting activity tells us very little about the course of technological change? Or do they challenge us to consider the part played by product innovation, by the proliferation of consumer goods during the eighteenth century, and by the connections between the patenting of these and broader technological innovation?

Let us consider a number of groups taken from the range of consumer-goods orientated patents selected as outlined above (see Table 8.1). The large number of patents within these groups for new products, for improvements and variations on these, and for ornamenting, painting, printing, inlaying and imitating, stands out. The largest numbers of patents in these groups were for cloth printing and printing and stamping paper, followed by patents for painting, paints and varnishes. After these, patents involving metal products are prominent. Plating, covering and ornamenting metals are closely followed by patents for buckles and fastenings, and by cutlery.

Patents for product changes and finishing processes were not random, but concentrated around specific problems or subjects. Product changes appeared as new objects or new shapes and sizes. Ornamentation and portability were key motivations. Patents for various types of applied ornamentation became important from 1740; these included various mouldings, ornaments for glass, chasing on metals, and ornamenting steel for toys.[35] Five of the patent groups in the table below specifically include ornamenting in their titles.

Other patent groups, including embossing, engraving, painting and printing, were primarily focussed on ornamentation.

Table 8.1. Patents for New Products, Ornamenting and Finishing, U.K., 1627–1825

Buckles and fastenings	36
Cutlery: knives, forks, razors, sword cutlery	27
Glass, ivory and bone	4
Drawing and photography: exhibiting prints and painted scenery	16
Earthenware and porcelain: glazing, painting, gilding, printing and ornamenting	8
Embossing, gilding, damasking	15
Engraving, etching and chasing	12
Making and ornamenting frames for pictures and looking glasses	7
Workboxes, music stands, dressing boxes and firescreens	4
Castors, knobs and handles	10
Cabinet and other furniture	14
Furniture:	
Tables	17
Chairs, sofas and similar articles	8
Bedsteads and couches	16
Glass: painting, gilding, ornamenting	10
Metals and metallic substances:	
Plating, tinning, lining, covering	35
Ornamenting, inlaying and polishing	9
Moulding and ornaments for buildings, coaches and furniture	18
Papier mâché and japanned ware	5
Painting, paints and varnishes	56
Pearl, ivory and bone	3
Printing on cloth	67
Printing and stamping paper	19

Source: Patents selected on core consumer industries from Bennet Woodcroft (1857), *Subject Matter Index of Patents of Invention from March 2, 1617 to October 1, 1852*, London: Patent Office; Woodcroft (1854), *Alphabetical Index of Patentees of Invention from October 2, 1617 to October 1, 1852*, London: Patent Office; Public Record Office C/210, Petty Bag Office, *Specification and Surrender Rolls*; and Public Record Office D/79, *Calendar of Specification Rolls*.

A good example was Thomas Rodgers, a London glass cutter, who in 1786 patented a method of ornamenting with carved or moulded glass with application to chimney pieces, looking glasses, lustre, pictures, frames, doors, ceilings, wardrobes, coaches, tea urns, caddies and a whole range of other listed furnishings.[36] The invention of ornamenting, executed in a variety of new materials, applied to new products or made and applied with new techniques, was furthermore closely tied to a very specific sense of novelty, one which drew for design principles on a 'recombination of old ornament'.[37] A whole range of items including lamps, stoves, kitchen equipment and mangles were introduced from the 1760s in portable dimensions. Samuel Roberts patented a retractable toast rack in 1807, and a year later David Thomas patented a perforated vessel and portable percolator frame for preparing coffee.[38]

Inventors patented ways of making existing goods in different and new materials. The development of new materials using chemical processes and mixing base and luxury metals, and the invention of entirely new materials such as 'japan', were usually associated with a proliferation of consumer items. Patents were frequently taken out for materials and ornamentation in 'imitation' of more expensive or foreign materials and processes. Patents from my selected subgroups specifically mentioning imitation are included in Table 8.2.

Table 8.2. Patents Specifying Imitations, U.K., 1627–1825

Metal imitations for buttons and shoe buckles	5
Plating and tinning	16
Imitation and ornamentation for ceramics	8
Imitation inlaying or stamping on marble, metals and varnish	6
Stamping and printing on cloth – imitations	5
Stove ornaments of plated, enamel or lapis lazuli	2
Glass imitation	2
Paper ornaments resembling wood carvings	2
Imitation of marble	1

Source: As for Table 8.1.

Substitution for foreign luxuries was included in patent descriptions such as that of William Bayley in 1692 for 'a new sort of glazed printed hangings, made of cotton, worsted, or woollen yarn, of all sorts of curious figures and landscapes, which for beauty of colours, exactness of figures, strength and gloss, is hard to be distinguished from the finest silk tapestry hangings brought from foreign parts.' Jane Tasker in 1709 patented flask cases in imitation of those from Venice, and John Peltaw and William Naylor in 1781 patented 'dyeing spots of silk handkerchiefs in imitation of India'. Other processes imitated classical or Renaissance arts, such as Wedgwood's patent in 1769 for ornamenting earthenware to imitate encaustic painting, or Robert Redrich and Thomas Jones' patent in 1724 for 'staining, spotting, veining and clouding on earthenware to imitate marble, porphyry and other stones, and tortoiseshell'. Other patents developed base metals or decorating techniques to imitate precious materials, such as George Ravenscroft's patent in 1674 for the 'manufacture of crystalline glass resembling rock-crystal', or Richard Dovey's patent of 1773 to 'make a metal of the colour and in all respects resembling gold'.[39]

Products well-known in one material might be made in an entirely new material, perhaps initially as an imitation, but soon simply as a novelty. Henry Clay, the Birmingham japanner, patented buttons in such a new material, described as japanned and dyed materials, in 1778, 1786 and 1790. William Williams, a London buttonmaker, similarly patented 'a composition of gold, silver and metal for making coat and waistcoat buttons entirely new and not before practised'; it entailed a combination of mixing copper and brass, silver-plating and gilding.[40]

Japanned ware was inspired by oriental lacquered furnishings imported into Europe in the seventeenth and eighteenth centuries. A European varnish was initially substituted, and with it a European luxury industry developed. But soon an altogether new technique including new materials and products was developed. Papier mâché and tin-plated ware were much cheaper materials, and were amenable to the production of all manner of small items, from teaboards and trays to vases and canisters, buttons and buckles as well as large items such as coach panels. All could be japanned and decorated. The new processes were developed in Bilston in Staffordshire, and in Pontypool in South Wales. Improvements were patented in Birmingham and elsewhere; though only five patents were recorded in the category of Papier Mâché and Japanned Ware, at least 15 were taken out for processes involving or applying to japanned ware between 1757 and 1825.[41]

The most widely applied process innovations among this group of manufactures were methods of rolling and stamping various materials. Techniques for rolling silver, copper and fine alloys were patented from the 1740s, and there were 115 patents over the period for stamping, pressing and embossing metals. These ranged from making hat and cloak pins, to metal handles, firefronts and coins.[42]

If patents across the core group are analysed by industry, metalwares were absolutely central, with product-related patents making up the highest proportion. These included items such as locks, buttons, buckles, forks, braces, scissors, fireplaces and stoves and stove furniture. Many of the process patents were focussed on casting, stamping and polishing to produce specific consumer goods.

TRADE CATALOGUES

The patent records reveal a great range of new commodities and variations on these, along with the development of finishing, ornamenting and imitation processes with applications across broad groups of commodities. The evidence of these records is complemented by that of trade catalogues produced to advertise and sell the goods. The catalogues were where the new commodities were graphically displayed in all their different dimensions and ornamentation. These catalogues, produced by some manufacturers and merchants, bear some similarities to the much more extensive marketing catalogues of the twentieth century, but there are also major differences.

R.J. Gordon drew the data for his new price index for the U.S. economy in the postwar period from the Sears Catalogue. Richard Sears began a mail-order watch business in 1887, assisted by Alvah Curtis Roebuck, a watch repairer who organized an assembly line to assemble inexpensive watches. The Sears, Roebuck and Co. catalogue for a wide range of mail-order goods began in 1893, and by 1895 numbered 507 pages. The print run of the catalogues grew from 318,000 in 1897 to 3.6 million in 1908, and sales grew from $1.4 million in 1896 to $50 million in 1907. The success of the catalogue was based on a rural market, the bypassing of high-mark-up middlemen and the high prices at traditional country stores, and the improved mail service. The catalogues provided data on the largest range of commodities priced at average or below average rates, and the closest index of the appearance and disappearance of individual products.[43]

There were certainly no catalogues during the eighteenth century to match the range of consumer goods covered by the Sears catalogue. But from the last third of the eighteenth century, manufacturers and merchants used trade catalogues as the means of advertising and selling their goods, especially in overseas markets. The eighteenth-century catalogues were a development on the pattern and sample books kept by any manufacturer as a record of his patterns and as a guide for orders. The sample books were rarely printed or published; the trade catalogues were. The trade catalogues are highly visual forms of evidence, and as such have only rarely been used by economic historians. The catalogues frequently combined engravings, either full-size or to scale, and prices. The largest surviving collection of these catalogues is for various lines of metalwares. Among the earliest of the printed pattern or 'sample books' was the tool catalogue for the clock and watch trade issued by John Wycke of Liverpool. Matthew Boulton's Soho Pattern Books of 1762 are the earliest to survive for the plated metalware trade. These were followed by a number of brassware, general metalwork, silver-plated ware, and ironwork catalogues in the later 1760s and 1770s. There are 68 of these catalogues, covering the eighteenth and early years of the nineteenth century, in the Victoria and Albert Museum, and an equally large collection at the Henry Francis du Pont Winterthur Library. The greatest concentration of catalogues is for brassware, and nearly all of these are from Birmingham. The catalogues indicate the huge success of the brassfounders in export markets; much of the trade was transacted through factors or agents, and prices and annotations were sometimes in foreign languages.[44] In many cases, the catalogues were anonymous, for they were used by agents who did not want their customers to discover the sources of their wares. American merchants made numerous journeys to England from the 1760s. One such, Samuel Rowland Fisher, visited factories and workshops, and, on one of his later trips in 1783–4, acquired in Birmingham an extensive metalwork catalogue, one of the largest (146 pages) still existing. He bypassed London purchasing agents, and preferred to choose his goods himself at lower prices 'direct from the factory'.[45]

The illustrations in the catalogues were printed from copper-plate engravings, showing the lines of manufactured products. The artistic illustrations, which were also very realistic, indicated that the engravers had a knowledge of the designing or manufacturing of the products. A set of engraved copper plates was a substantial investment. Old patterns were frequently repeated in new editions of catalogues, to accommodate conservative tastes.[46] Some of the pattern books were sold to agents or clients. Prices were often written in by hand, or set out in separate hand-written or printed lists. Prices of a range of goods such as ornate brass furniture handles appear to have varied little over the period, with standard-size ornate handles selling for 4 to 5 shillings a dozen, and large-size handles for 6 to 7 shillings a dozen. Prices for goods like these appeared fairly standard across the period, but varieties of ornamentation increased. The big difference between these catalogues and a twentieth-century example such as Sears Roebuck lies in the cost of their production, and therefore the extent to which they could act as an index of change in products.

What the catalogues do illustrate is the process, also clear in the patent records, of developing materials and ornamentation with applications across a range of related products. Some of the early catalogues were, to be sure, highly specialist, focussed (as was that of John Barker of Birmingham in 1772) on furniture hardware. But others, such as that of another Birmingham firm, Jee, Eginton & Co., in 1790, listed a whole variety of metal objects, including frames for looking glasses, urns, moulding, friezes and decorative ornaments. John Ryland, another Birmingham firm, sold wire fenders, sieves, wire dish covers, chains, escutcheon pins and 'all sorts of wire and forged beading shanks for coach furniture'.[47] John Taylor, whose trade emblem described himself as a working goldsmith and silversmith, jeweller, tortoiseshell & ivory box, gilt & general toy manufacturer, had a catalogue containing all manner of gold and silver jewellery, buckles, buttons, snuff boxes, small tableware and gilt and glass decorative items.[48]

The trade catalogues not only provide design and product information and price data; they demonstrate complementarities and connections within the group of metalware, glass and ceramics manufacturers, and watchmakers. A number of the catalogues contained designs and products for plated ware and glass, or for silver and plated ware and ceramics.[49] The catalogue owned by Rowland Fisher advertised brass furniture ware alongside rococo watch frames and stands;[50] Benjamin Hadley's catalogue advertised tea and coffee urns, trays, candlesticks, glass stands, bottles and pitchers.[51] James Whitehead's catalogue set out his earthenware designs together with his plated ware.[52]

A PRODUCT AND DESIGN CENTRE FOR INNOVATION?

The clear connections among the products in my core consumer group, as revealed in the trade catalogues, must lead us to consider more widespread connections, and the part played by these connections in the making of centres for innovation. We can ask to what extent the producers of semi-luxury and new consumer goods provided a precedent for the later engineers, toolmakers and machine makers of the capital goods sector. Historians of technology have investigated the development of the capital goods sector in the nineteenth century as an institution whose own economic development also generated domestic diffusion of technology across the economy as a whole. As Rosenberg and, more recently, MacLeod have outlined, improvements made within this sector were sold as products to users, but also improved their processes.[53]

In the century before this development of a capital goods industry there were parallel connections across a set of semi-luxury industries. Networks based on marketing and personal technical expertise provided conduits of information and connection in product development, design and technology. There developed a type of product and technical interrelatedness which made these industries centres for innovation. The product complementarities were also ones of technique and skill. These were not new developments, but grew out of the methods of the luxury trades. We have perhaps been too quick as economic historians to divide off manufacture for luxury markets from that for mass production. David Landes, for example, has compared the design for an ornament for narrow luxury markets with the design for performance, efficiency and cost. 'It was this precocious, rapidly growing, and socially diversified demand that made the eighteenth-century Britain's period of mastery in the history of watch manufacture.' It was the 'sale of watches and clocks as commodities rather than as personal creations' that opened the door to the 'pure trader'.[54] The incentive for most of this manufacture, though now made available to a wider range of the population, was still the same as that for luxury production. Contemporaries such as Bernard Mandeville were absolutely clear about this. 'By identifying the "restless desire after Changes and Novelty" as the psychologically activating mechanism of commercial societies, Mandeville placed the issue of wealth creation within the enlarged and precarious context of markets for rapidly changing symbols of esteem, the demand for which ironically dominated the demand for articles of use.'[55]

The precedent for connections across industries in product development was long established in the luxury trades. Michael Sonenscher in his work on the Paris luxury trades demonstrates the wide impact of the Martin japanning firm. He argues that the hallmark of Parisian design from the later seventeenth century was the versatility with which designs could be adapted to multiple materials and products.[56] These luxury trades deployed the division of labour, subcontracting and the retailing of other producers' work. There was no sharp divide, as was once believed, between the craft sector and the modern manufacturing sector. Indeed Matthew Boulton and Josiah Wedgwood were in many senses simply following in the traditions of these luxury manufactories. Boulton's manufacture of metal 'toys', buckles and buttons, expanded into silver and silver-plated ware and ormolu; alongside his steam engine enterprise with Watt he moved into the manufacture of steel toys, glass lamps and coining.[57] The departments of his works relied on key specialists, some of whom were working under contract or consultancy for Boulton, or in partnership with him.[58] The range of products produced by Matthew Boulton was matched at a higher level at the end of the eighteenth century and in the early years of the nineteenth century by that classic producer of luxury goods, Benjamin Vulliamy, who occupied in London a parallel position to that of the Martins in eighteenth-century Paris. Known as a firm of luxury clock and watchmakers, the Vulliamys also advertised ornamental plate, silver, ornamental work in metal and the sale of diamonds and pearls; 'they were prepared to supply their customers with anything from a chimney-piece to a door handle, from a piano to a button'.[59] Few of these items, even the clocks, were produced by Vulliamy, but instead he subcontracted out and divided up the parts and processes.

The semi-luxury producers used the example set by the luxury trades, but the difference was that they had to go much further. Their networks in developing a design and product milieu were not just ones of personal connection and reputation, but ones spanning industrial regions and drawing in large numbers of unknown small-scale producers. The London watchmakers and finishers of Clerkenwell dealt with a classical proto-industrial region – 'From Prescot to Liverpool, eight miles as the crow flies', in David Landes's words, 'the countryside was dotted with the cottages of springmakers, wheel cutters, chainmakers, casemakers, dialmakers – every specialty that went into the making of a watch.' Matthew Boulton and Josiah Wedgwood shadowed each other closely, but both were ensconced in rapidly-growing specialist regions. The pottery manufacture which had doubled in size in the mid-eighteenth century was specialized, subcontracted and served by ancillary trades in flint grinding, packing and merchant activity. Birmingham held a 'swarm' of specialized toy, jewellery, and instrument makers, glass cutters and japanners, many producing components for each other, and densely packed into the streets and courts between the churches of St. Martin and St. Philip.[60]

The close interaction between Boulton and Wedgwood is demonstrated clearly in their products, and in the techniques they used. The best example is provided by Wedgwood's engine-turning lathe which he developed using a specialist lathe developed for him by Matthew Boulton, and adapted with the help of the Liverpool watch- and toolmaker, John Wycke. The products showed the close interdependence of Birmingham and Staffordshire. As the trade catalogues and the patents demonstrated, designs, finishes and new products brought the metalware, earthenware and glassmaking regions into close contact. The particularly vital part played by metalwares and by their Birmingham producers in product innovation in consumer wares led on to their part in toolmaking and engineering innovation in capital goods production.

This semi-luxury consumer goods sector also produced the skills which contributed to the development of later technologies. Much product innovation also took place in the interstices of the small producer economies of regions which also had close connections through their skilled workforces. This is the world of Mokyr's micro-inventions: improvements made by an innovative, flexible and skilled workforce. These semi-luxury industries relied on a world of apprentices and journeymen who knew the value of their skills to a wide range of producers – it was also a world of counterfeiters, producers of fakes and other low-quality goods.

The craft training and apprenticeship which still took place in various forms in the luxury and semi-luxury trades carried benefits for the introduction of new products and industries.[61] There was a pool of interchangeable skilled labour: in the metal trades, workers in gold and silver now intermeshed with workers in fine cut steel, glass cutters and wire- and toolmakers. Britain's skilled workforce has been credited with much of the technological innovation of the eighteenth century, but has been seen as a workforce skilled in coal-using technologies, in metals, and mainly in engineering – a small élite of capital goods producers, acting as conduits of innovation to the rest of the economy.

If, however, we consider Sokoloff's democracy of invention as a result of the demand for new consumer commodities, this élite of skills broadens out. It also leads us to consider the connections between those capital goods producers and producers in the new consumer and semi-luxury trades. Scientific instruments were ornaments of consumption as much as they were tools of engineering. The 'machines' beloved of eighteenth-century patentees were the contraptions and devices of new consumer ware. The metal compositions and the rolling and stamping techniques were vital to the new materials, new finishes and ornamentation, and new goods of luxury and consumer ware. These in turn were adapted to heavy industrial uses. Producers of small tools as well as of complex lathes and engines were often the same individuals producing ornamental stamped brassware, medallions and mechanical toys.

A world of new commodities as revealed in the patent records and the trade catalogues displayed industry as developing the arts and the sciences. Products combined the useful and the decorative. Their complementarities across industries and markets nurtured key technological skills, tools and machines whose effects would ripple outward to the wider economy.

The evidence left by eighteenth-century records may not be either extensive or systematic enough to estimate the quantitative significance of quality differences and product development on productivity growth. Historians have long attributed this productivity growth to technological change inspired by inventive activity. A new look at the proliferation of new consumer goods during the later seventeenth and eighteenth centuries must at least press us to consider the part played by product innovation. It must also lead us to consider the extent to which the innovations of eighteenth-century manufacture constituted not a sharp break from the world of luxury producers, but a development based in this framework.

NOTES

1. Maxine Berg (1980), *The Machinery Question and the Making of Political Economy 1815–1848*, Cambridge: Cambridge University Press, pp. 90–92, 130–32.
2. Christine MacLeod, 'James Watt, Heroic Invention and the Idea of the Industrial Revolution', Chapter 6 of this volume.
3. Adam Smith (1776), *An Inquiry into the Nature and Causes of the Wealth of Nations*, reprinted in W.B. Todd (ed.) (1976), *Glasgow Edition of the Works and Correspondence of Adam Smith*, vol. 1, Oxford: Oxford University Press, p. 131.
4. James Walker (1792, 1794), *The Copperplate Magazine or Monthly Cabinet of Picturesque Prints consisting of Sublime and Interesting Views in Great Britain and Ireland Beautifully*

Engraved by the Most Eminent Artists from the Paintings and Drawings of the First Masters, London: Harrison & Co.

5. Joel Mokyr (1990), *The Lever of Riches*, New York and Oxford: Oxford University Press, pp. 291–99; Joel Mokyr (1994), 'Technological Change, 1700–1830', in Roderick Floud and Donald McCloskey (eds), *The Economic History of Britain since 1700*, Vol. I, Cambridge: Cambridge University Press, pp. 12–43, p. 43.
6. Mokyr, 'Technological Change, 1700–1830', p. 43.
7. E.L. Jones (1973), 'The Fashion Manipulators: Consumer Tastes and British Industries, 1660–1800', in L.P. Cain and P.J. Uselding (eds), *Business Enterprise and Economic Change*, Ohio: Ohio University Press, pp. 198–226; N. McKendrick, J. Brewer and J.H. Plumb (1982), *The Birth of a Consumer Society*, London: Europa; Lorna Weatherill (1988), *Consumer Behaviour and Material Culture in Britain 1660–1760*, London: Methuen; Beverley Lemire (1991), *Fashion's Favourite: the Cotton Trade and the Consumer in Britain 1660–1800*, Oxford: Oxford University Press; John Brewer and Roy Porter (eds) (1993), *Consumption and the World of Goods*, London: Routledge; Jan De Vries (1994), 'The Industrial Revolution and the Industrious Revolution', *Journal of Economic History*, **54**, 249–70.
8. Michael Sonenscher (1989), *Work and Wages in Eighteenth-Century France*, Cambridge: Cambridge University Press, pp. 210–43.
9. Maxine Berg and Pat Hudson (1992), 'Rehabilitating the Industrial Revolution', *Economic History Review*, **45**, 24–50; J. Hoppit (1990), 'Counting the Industrial Revolution', *Economic History Review*, **43**, 173–93.
10. Crafts gives output figures only for pig iron over the whole period, and for copper ore only from 1771. Elizabeth Schumpeter estimated exports of wrought iron, of iron nails, of wrought copper and wrought brass. Ralph Davis provided export figures, broken down to a greater degree for the early nineteenth century showing exports of hardware and cutlery. This category of metalwares is here distinguished from hand guns and swords, and from brass, copper and tin wares, as well as from nails and watches and clocks. N.F.R. Crafts (1985), *British Economic Growth during the Industrial Revolution*, Oxford: Oxford University Press, p. 23; E.B. Schumpeter (1960), *English Overseas Trade Statistics 1696–1808*, Oxford: Oxford University Press, p. 64; Ralph Davis (1979), *The Industrial Revolution and British Overseas Trade*, Leicester: Leicester University Press, Table 15, p. 27. See also S.D. Smith (1993), 'British Exports to Continental North America, 1690–1776', unpublished Ph.D. thesis, University of Cambridge; M. Overton (1992), 'The Prices of Commodities in Early Modern Britain', unpublished paper.
11. Mokyr, 'Technological Change, 1700–1830', p. 31; P.K. O'Brien, T. Griffiths and P. Hunt (1996), 'Technological Change during the First Industrial Revolution: the Paradigm Case of Textiles, 1688–1851', in Robert Fox (ed.), *Technological Change*, London: Harwood, pp. 155–76, p. 161.
12. K.L. Sokoloff (1988), 'Inventive Activity in Early Industrial America: Evidence from Patent Records, 1790–1846', *Journal of Economic History*, **48**, 813–50; K.L. Sokoloff and B.Z. Khan (1990), 'The Democratization of Invention during Early Industrialization: Evidence from the United States, 1790–1846', *Journal of Economic History*, **50**, 363–78; B.Z. Khan and K.L. Sokoloff (1993), '"Schemes of Practical Utility": Entrepreneurship and Innovation among "Great Inventors" in the United States, 1790–1865', *Journal of Economic History*, **53**, 289–307.
13. Christine MacLeod (1988), *Inventing the Industrial Revolution*, Cambridge: Cambridge University Press, p. 161; Christine MacLeod (1995), 'The Springs of Invention and British Industrialisation', *ReFresh*, **20**, 3.
14. T. Griffiths, P.A. Hunt and P.K. O'Brien (1992), 'Inventive Activity in the British Textile Industry, 1700–1800', *Journal of Economic History*, **52**, 881–906, p. 893.
15. Christine MacLeod (1992), 'Strategies for Innovation: the Diffusion of New Technology in Nineteenth-Century British Industry', *Economic History Review*, **45**, 285–307.
16. Griffiths, Hunt and O'Brien, 'Inventive Activity', p. 901; O'Brien, Griffiths and Hunt, 'Technological Change', pp. 167–8.
17. G.N. von Tunzelmann (1995), *Technology and Industrial Progress*, Cheltenham, U.K. and Brookfield, U.S.: Edward Elgar, pp. 107–8.
18. John Styles (1995), 'Product Innovation in Early Modern London', unpublished paper for Achievement Project on the Growth of the Skilled Workforce in London (can be consulted in Centre for Metropolitan History, University of London).

19. A.N. Link (1987), *Technological Change and Productivity Growth*, London: Harwood; B. Eichengreen (1986), 'What Have We Learned from Historical Comparisons of Income and Productivity?' in P.K. O'Brien (ed.), *International Productivity Comparisons and Problems of Measurement, 1750–1939*, Ninth Congress of the International Economic History Association, Berne: University of Berne, pp. 26–36.
20. Kelvin Lancaster (1971), *Consumer Demand, a New Approach,* New York: Columbia University Press, pp. 2–12. Recent attempts to include analysis of characteristics within a differentiated commodity have included a case study on Swedish wine consumers, choosing different qualities of wines, and taking prices as given. See M. Nerlove (1995), 'Hedonic Price Functions and the Measurement of Preferences: the Case of Swedish Wine Consumers', *European Economic Review*, **39**, 1697–1716.
21. D.S. Ironmonger (1972), *New Commodities and Consumer Behaviour*, Cambridge: Cambridge University Press, pp. 1–13.
22. R.J. Gordon (1990), *The Measurement of Durable Goods Prices*, Chicago and London: Chicago University Press, pp. 24, 77.
23. Gordon, ibidem, p. 96.
24. Gordon, ibidem, p. 24.
25. Gordon, ibidem, p. 560.
26. R.J. Sullivan (1990), 'The Revolution of Ideas: Widespread Patenting and Invention during the English Industrial Revolution', *Journal of Economic History*, **50**, 349–62.
27. Richard Prosser (1881), *Birmingham Inventors and Inventions*, Birmingham: privately printed; reprinted 1970, Wakefield: S.R. Publishers.
28. MacLeod, *Inventing the Industrial Revolution*, p. 76.
29. See *John Joseph Merlin. The Ingenious Mechanick* (1985), London: The Greater London Council. I am grateful to Dr. Helen Clifford for drawing my attention to this publication. The point is also made by O'Brien, Griffiths and Hunt, 'Technological Change', p. 175.
30. John Tuite, a silversmith, patented three land drainage schemes: in 1734 (no. 542, PRO C/210/3), in 1742 (no. 585, PRO C/73/4), and in 1752 (no. 671, PRO C/210/4). The latter two were for water engines for drainage and other purposes. Lewis Pantin patented a horsepower engine for raising ballast in 1743 (no. 598, PRO C/73/4). His invention appeared in J.T. Desaguliers (1744–45), *A Course of Experimental Philosophy*, 2nd edition, 2 vols, London: W. Innys.
31. Bennet Woodcroft (1854), *Titles of Patents of Invention, Chronologically Arranged from March 2, 1617 (14 James I) to October 1, 1852 (16 Victoriae)*, 2 vols, London: Patent Office, Great Britain, no. 3221, PRO C/210/122.
32. I am grateful to Dr. Helen Clifford for assistance with the collection of patent data, and to Dr. Frances White for assistance with computer analysis of these data.
33. For data on total patents grouped by 25-year periods, see O'Brien, Griffiths and Hunt, 'Technological Change', p. 167.
34. John Pickering, 7 March 1769, no. 920, in Woodcroft, *Titles of Patents of Invention.*
35. Woodcroft, *Titles of Patents of Invention.*
36. Bennet Woodcroft (1857), *Subject Matter Index of Patents of Invention*, London: Patent Office, no. 1568, PRO C/210/30. Rodgers described his ornamenting as 'more beautiful than either marble, wood or metal or composition', and 'less expensive than any of the present method of ornamenting as they will hold their lustre and colour to the last and not be liable to staining as other ornaments do'.
37. Hilary Young (1995),'From the Potteries to St. Petersburg: Wedgwood and the Making and Selling of Ceramics', in Hilary Young (ed.), *The Genius of Wedgwood*, London: Victoria and Albert Museum, pp. 9–20, p. 13.
38. Woodcroft, *Titles of Patents of Invention*, nos. 3083 and 3138.
39. Woodcroft, *Subject Matter Index*, nos. 297, 386, 461, 939, 1055 and 1299.
40. Woodcroft, *Titles of Patents of Invention*, nos. 1729 (1790) and 1148 (1777), PRO C/210/179.
41. For further details on japanning, see H. Huth (1971), *Lacquer of the West*, Chicago and London: University of Chicago Press; S.P. De Voc (1971) *English Papier Mâché of the Georgian and Victorian periods*, London: Barrie & Jenkins.
42. For example, see John Marston and Samuel Bellamy (no. 1165, 1777), stamping upon plated metal, hat & cloak pins; William Bell (no. 1242, 1779), die stamp with rolling cylinders; Isaac Mason (no. 3800, 1814), making stamped firefronts (Woodcroft, *Titles of Patents of Invention*).

43. See Gordon, *The Measurement of Durable Goods Prices*, pp. 420–23.
44. Nicholas Goodison (1975), 'The Victoria and Albert Museum's Collection of Metalwork Pattern Books', *Furniture History*, **11**, 1–30, p. 6; R. Timmins & Sons, Birmingham, *Tools for the Trades and Crafts: an Eighteenth-Century Pattern Book*, Birmingham: R. Timmins & Sons, reprinted 1976, Fitzwilliam, New Hampshire: K. Roberts Publishing Co.; Theodore R. Crom (1976), *Trade Catalogues 1542–1842*, Melrose, Florida: privately printed.
45. Charles F. Hummel (1964), 'Samuel Rowland Fisher's Catalogue of English Hardware', *Winterthur Portfolio*, **1**, 188–97; *Catalogue of English Hardware*, [1783–89], Birmingham (owned by Samuel Rowland Fisher), in the Winterthur Library; 'Samuel Rowland Fisher, 1767–68; 1783. Notes on English Manufactures and his Purchases from them', Microfilm 296 in the Joseph Downes Collection, Winterthur Library.
46. Goodison, 'The Victoria and Albert', p. 5.
47. See E. Richard McKinstry (1984), *Trade Catalogues at Winterthur: a Guide to the Literature of Merchandising 1750–1980*, New York: Garland Publishing Inc.; cf. K.D. Roberts, 'Introduction to English Pattern Books', in *Tools of the Trades*.
48. John Taylor [*c*. 1810], *Birmingham Catalog*, Birmingham, in the Winterthur Library.
49. *Catalogue of Tea Urns, Coffee Urns, Candlesticks, Trays, Goblets, Shaving Basins & other Metal Objects* [*c*. 1790], Sheffield, in the Winterthur Library.
50. *Catalogue of English Hardware*, [1783–89], Birmingham (owned by Samuel Rowland Fisher), in the Winterthur Library. See also *Catalogue of Candelabra, Escutcheons, etc.* [*c*. 1790], Birmingham, in the Winterthur Library.
51. *Catalogue of Benjamin Hadley* (1815), Birmingham, in the Winterthur Library. This last catalogue includes plated ware and glass goods such as candlesticks, bottles, pitchers, trays, salts, mustard-tankards, bottle-stands, tea and coffee pots and urns, egg cups, spoons, servers, sugar baskets.
52. James Whitehead (1798), *Designs for Sundry Articles of Earthen-ware: At the same manufactory may be had, a great variety of other articles, both useful and ornamental, as well printed, painted & enamelled; as likewise dry bodies, such as Egyptian, black, jasper etc.*, Birmingham (printed by Thomas Pearson), in the Winterthur Library.
53. Nathan Rosenberg (1976), *Perspectives on Technology*, Cambridge: Cambridge University Press; Nathan Rosenberg (1982), *Inside the Black Box: Technology and Economics*, Cambridge: Cambridge University Press, 1982; MacLeod, 'Strategies for Innovation'.
54. David S. Landes (1983), *Revolution in Time. Clocks and the Making of the Modern World*, Cambridge, MA: Harvard University Press, pp. 233–4.
55. E.H. Hundert (1994), *The Enlightenment's Fable: Bernard Mandeville and the Discovery of Society*, Cambridge: Cambridge University Press, p. 187.
56. Sonenscher, *Work and Wages*, pp. 210–43.
57. See H.W. Dickinson (1937), *Matthew Boulton*, Cambridge: Cambridge University Press.
58. Kenneth Quickenden (1995), 'Boulton and Fothergill's Silversmiths', *The Silver Society Journal*, **7**, 342–56.
59. Geoffrey de Bellaigue (1967), 'The Vulliamys and France', *Furniture History*, **3**, 45–53, pp. 45, 48.
60. Landes, *Revolution in Time*, pp. 233–4; Maxine Berg (1993), 'Small Producer Capitalism in Eighteenth-Century England', *Business History*, **35**, 17–39.
61. This defence of apprenticeship and guild structures is set out in S.R. Epstein (1995), 'Craft Guilds, Apprenticeship, Technological Change in Pre-Modern Europe', LSE Working Paper, London.

PART 4

Learning, Skills and Boundaries

9. Skills, Learning and the International Diffusion of Technology: a Perspective on Scandinavian Industrialization

Kristine Bruland

INTRODUCTION

This chapter is concerned with the process of technology transfer into Scandinavia (*c.* 1760–1870), focussing on the extent of transfer and on the learning processes through which such transfer was made possible. The underlying argument is that Scandinavian industrialization should be understood as a broad process of technological and structural change: it was not based simply on the rapid development of a few key sectors, but involved multi-sectoral learning and growth. These changes were made possible and supported by institutional and cultural changes, and by political initiatives which provided a general background to the acquisition and use of technologies from abroad.

The idea that technological change is a fundamental aspect of economic growth is widely accepted in applied economics and in economic history. However, the processes of creation, diffusion and transfer of technology through which industrialization spreads have been relatively neglected. Recent growth theories, both neo-classical and 'evolutionary', treat knowledge creation (that is to say, learning) as endogenous to the economy, and several papers have suggested that this has a major impact on quite basic issues in growth theory.[1] Yet there is little emphasis on what actually happens in learning, on how it is organized, financed and managed. We might expect to find significant inter-country differences in how learning occurs, in terms of institutions and organizations, and this is likely to have important implications for histories of economic growth.

In small economies, learning and industrial growth are in large part a matter of adaptation to foreign technological developments, since small economies are usually also 'follower' economies, seeking to catch up with technological leaders. Today, Sweden, Denmark and Norway are among the richest economies of the world, in terms of gross domestic product *per capita*. It was not always so, however, nor was it clear that they would become rich. Although these economies have certain resource advantages – in timber, fish stocks and iron ore, for example – they also have small populations, an extreme shortage of arable land, very formidable communication barriers, and unfavourable climatic conditions. A further obstacle to industrialization was the presence of powerful competitor economies near at hand, in a context of relatively open trading systems. Yet against these odds Scandinavia succeeded in industrializing, and in sustaining its industrial growth. As with other industrialization processes, the transition was a complex one, with no single determining factor. But what was it about these economies which made it possible for them to create high-income niches in the growing world economy?

The argument here is that access to and absorption of foreign technologies across a wide range of sectors was a key element in the story. Scandinavia has always been characterized by high levels of technology importation: very few of the technologies in use in Scandinavian industry have been developed within its boundaries. Many individual historical studies suggest an important rôle for foreign technology – in particular sectors, or with respect to particular technologies. But how can the overall process and effects be identified and understood? One issue concerns generality: how widespread was international technology acquisition across sectors during the period? A second issue concerns processes or modes of technology acquisition: what forms did they take, and did they vary across sectors and technologies? Finally, to what extent was the ability to acquire technology a general feature of these societies, shaped by social or cultural factors or policy actions? To put this in a different way, was there something about the so-called 'national innovation systems' of the Scandinavian area which promoted the ability to acquire and use foreign technology?[2]

INTERNATIONAL TECHNOLOGY ACQUISITION BY SCANDINAVIAN ECONOMIES

There is no general empirical or statistical evidence for technology transfer as a process. The attempt here is to map acquisition of foreign technology for various sectors in Norway, Denmark and Sweden by exploring available literature. Although a full literary survey has not been carried out, the studies which have been explored, and which deal either indirectly or directly with technology importation, permit some generalizations. I have constructed an overall summary of recipient sectors, the technologies transferred, periods, source and recipient countries, recipient enterprises and their locations; but this overview, even in summary form, is too extensive to include here.[3] To give a flavour of the overall summary, Table 9.1 shows the results for one sector, namely mining. Mining of both ferrous and non-ferrous metallic ores was (and remains) a major sector in both Sweden and Norway, and technology transfer into this sector began very early and continued over many years. Key elements of the technology were imported: pumps, explosives, and general mining equipment. In many cases, it was the ability to use these technologies which made mining possible in otherwise unfeasible conditions.

Table 9.2 goes beyond this, to the economy as a whole, summarizing the numbers of technology transfer cases which have been identified via empirical research, the broad periods involved, and the recipient countries.

For Denmark, Norway and Sweden, the process of technology importation began early and continued over a long period; since the table is based on cases which have been studied by historians for various purposes, it certainly underestimates – possibly to a very significant degree – the extent of the imports. Two broad inferences can be drawn from this kind of survey. The first is that technology transfer appears to have been integral to industrialization. Wherever detailed sectoral studies have been carried out, we find the presence of inward technology transfer, often on a very significant scale. Of course it is extremely difficult on current evidence to make the link between inward technology transfer and specific growth effects, and certainly not all transfer processes were economically successful. But the general prevalence of transfer, especially early in the industrialization process, suggests that it played a central rôle. The second point concerns the nature of the growth process. If the technological changes mapped by the transfer process were associated with industrialization and growth, then growth was multi-sectoral. It was not confined to a number of key industries, but occurred across a very wide range of activities and technologies. The sectoral studies are comprehensive enough to suggest that inward technology transfer was prevalent across the entire industrial economy of the Scandinavian area. This is the case for sectors such as building and construction, for all major manufacturing industries (including the largest in the Scandinavian economies, which were food processing, engineering and forest products), as well as for service industries (including transport) and public utilities. Understanding growth in the Scandinavian region therefore involves an understanding of the underlying learning process associated with multi-sectoral technology transfer.

Table 9.1. Technology Importation to Scandinavia – Mining Sector

Technology	Period	Source country	Products
Explosives	1630 – *c.* 1720s	Germany	Gunpowder
Power	1727–1736	U.K.	Newcomen steam engine
Equipment and explosives	1732–1816	Germany	Mining equipment, gunpowder
Power	1790s	U.K.	Newcomen steam engines
Steam technology	1804	U.K.	1 Watt steam engine
Power	1804 onwards	U.K.	4 steam engines
Pumps	1824	Germany	Pressure water machine
Elevation and suspension	1839	U.K.	Elevator techniques
Elevation and suspension	1837–40s	Sweden, U.K.	Wire
Power	1840s	Sweden	Steam engines
Explosives	*c.* 1840s	U.K.	Safety fuse
Transport	1850s	U.K.	Iron railways
Boring	1859–60s	Germany, U.K., Sweden	Steel bores
Pumps	1860	Germany	'Schwamkrug's' water turbine
Pumps	1869	Germany	Water lifting machine
Explosives	1863	Italy	Nitroglycerin
Explosives	1872	Sweden	Dynamite

Sources: B.I. Berg (1994), *Gruveteknikk ved Kongsberg Sølvverk 1623–1914*, Kongsberg: Norsk Bergverkmuseum; J. Hult, S. Lindqvist, W. Odelberg and S. Rydberg (1989), *Svensk Teknikhistoria*, Hedemora: Gidlunds Bokförlag; S. Lindqvist (1984), *Technology on Trial. The Introduction of Steam Power Technology into Sweden 1715–1736*, Uppsala: Almqvist & Wicksell.

Recipient country	Recipient enterprise and location	Reference
Sweden	Stora Kopperberget; Dannemora; others	Hult et al. (1989), pp. 141–2
Sweden	M. Triewald; Dannemora mines	Lindqvist (1984), passim
Norway	Kongsberg silver mines	Berg (1994), pp. 236, 321–30
Sweden	Höganäs, Ths. Stawford (British)	Hult et al. (1989), p. 180
Sweden	Dannemora mines	Hult et al. (1989), p. 203
Sweden	A.N. Edelcrantz	Hult et al. (1989), pp. 194, 203
Norway	Kongsberg silver mines	Berg (1994), p. 377
Norway	Kongsberg silver mines	Berg (1994), p. 390
Norway	Røros copper mines	Berg (1994), p. 405
Norway	Røros copper mines; Kongesberg silver mines; other mines	Berg (1994), p. 434
Norway	Røros copper mines; Kongsberg silver mines	Berg (1994), p. 363
Norway	Kongsberg silver mines	Berg (1994), p. 456
Norway	Kongsberg silver mines	Berg (1994), p. 359
Norway	Kongsberg silver mines	Berg (1994), p. 439
Norway	Kongsberg silver mines	Berg (1994), p. 439
Sweden	Åmmeberg	Hult et al. (1989), p. 189
Norway	Kongsberg silver mines	Berg (1994), p. 365

Table 9.2. Scandinavian Inward Technology Acquisition, by Sector

Sector	Period and number of cases	Technologies	Recipient countries studied
Agriculture	1830s–1890s; 6 cases	Reaping, dairy equipment, crops, etc.	All
Construction and building	1830s–1890s; 17 cases	Brick, tile and pipe machinery, steam engines, cement, kilns	Denmark
Engineering	1809–1898; 93 cases	Steam engines, all types of milling equipment, lathes, steam hammers, instruments	All
Fishing	1860s–70s	Boats, engines	Norway
Foundries	1769–1820; 3 cases	General foundry equipment	Sweden
Food and drink	1846–1895; 16 cases	Beer, flour milling, margarine, etc.	Norway, Denmark
Forestry	1730s–1850s; 3 cases	Saws and machinery	All
Glass techniques	1755–1850s; 8 cases	New glass types and furnace techniques	Norway
Iron and steel	1760–1890; 13 cases	Forging methods, castings, Bessemer, Siemens Martin and open hearth processes	Sweden
Mining	1727–1880s; 16 cases	Steam engines, blasting techniques, pumps, &c.	Norway, Sweden
Printing	1825–1829; 3 cases	High-speed presses	Denmark
Pulp and paper	1847–1899; 14 cases	Turbines, grinding machines. Integrated plants, paper-making machinery	All
Pottery and clay	1847–1851; 2 cases	Full plant and equipment	Norway, Denmark
Textiles	1780–1890s; at least 40 cases	All preparatory, spinning, weaving and finishing equipment	All
Transport and infra-structure	1809–1870s; 7 cases	Canal building equipment, vessels, marine power, dry docks, all rail techniques	All
Utilities	1847–1890s; 4 cases	Gas works; electricity	Denmark, Norway
Weapons	1867; 1 case	Remington rifles	Sweden

Source: K. Bruland (1997), 'Technology Transfer to Scandinavia: Sources and Periods', working paper.[4]

THE NATURE OF 'INNOVATION SYSTEMS', OF KNOWLEDGE AND LEARNING

What was happening in these transfer processes? The basic approach of this paper is to think in terms of *learning*. Whether we study industrialization in terms of a process which is endogenous to a particular economy, or whether we explore diffusion processes through which exogenously-developed technologies are absorbed and applied, it is clear that major learning processes are involved, because in either case important elements of novelty are introduced into the economy, and such novelty necessarily implies learning. For small economies, internal processes of industrial technology creation seem to have been limited in early industrialization, so this meant primarily learning from abroad. A key issue becomes how to conceptualize and map the learning processes involved. Economic actors had to learn new activities, but what and how? The argument developed here is that, given that technology transfer was so prevalent across the whole of the Scandinavian system, we need to think about learning and knowledge creation in a broad way, at the level of what would now be called the 'national innovation system'.[5]

Theories of the innovation system usually have two main components. The first, often implicit, is simply the view that economic growth does depend on innovation, and that innovation is widespread within high-growth and/or high-income economies. The second component is a view of the determinants of innovation capability or innovation performance. Here the argument is that the performance of innovating firms should be seen as determined by the wider socio-economic context within which they exist. A fundamental requirement is an institutional framework in the sense used by Douglass North: that is, a stable set of rules which regulates economic behaviour in such a way that innovation is promoted or at least not hindered. The relevant set of institutions includes a framework of law relating to contracts, property rights, entry/exit conditions and employment. Although this aspect of the Scandinavian systems will not be dealt with in this chapter, it does provide an important background to the analysis; as Hodne has remarked, 'Governments provided protection for private property and contracts, and jealously guarded this reputation.'[6] At a more detailed level, firms exist within more or less complex networks of suppliers and customers, sources of labour and skill, finance, and so on. These networks consist largely of inter-firm relations, but innovation performance is closely shaped by public-sector institutions such as technical organizations and universities. Taken together, this set of public and private institutions, regulations and policy constitute an integrated set of relationships, or networks, which persist over time and can therefore be thought of as systems. It seems clear, however, that such systems are constructed historically, either through specific political or policy decisions, or through some process of social evolution; and it is important to understand the ways in which they come into existence, and something of their empirical effects.

The empirical analysis in the following sections is mainly concerned with innovation, using foreign technologies, and this implies a broad process of learning about the existence, application and methods of novel techniques. But when learning occurs, what is it that is being learnt? Learning is a process through which knowledge is created, but what are the main characteristics of the kinds of knowledge which are relevant to production?

Lundvall and Johnson are amongst those who have argued that the ability to learn is central to economic processes. They define relevant economic knowledge along four main dimensions:[7]

- 'know what' – factual information about the technology concerned. As we shall see below, an important part of the learning-from-abroad process was simply access to information about what was happening. Learning that there is something to access is of course a precondition for undertaking other learning processes actually to absorb and to use it. This was by no means a trivial problem in the Scandinavian case, and we shall see a number of cases in which organizing 'know what' was central.

- 'know why' – knowledge of basic scientific and technological principles for the solution of problems.

- 'know who' – specific and selective social knowledge: learning and knowing who the relevant people are for the solution of problems. In the context of human-embodied knowledge and interactive learning and interactive problem solving, access to key personnel proved to be a key resource in Scandinavian inward technology acquisition.[8]

- 'know-how' – practical skills and capabilities; this covers skills and all aspects of production capabilities and marketing.

These types of learning and knowledge must cover at least three distinct areas: technological competences and capabilities, organizational capabilities, and management of links between different types of actors or institutions who are involved in the transfer process. I would argue that these categories – although somewhat abstract at this stage – are very relevant for understanding the historical processes of learning, transfer and subsequent knowledge creation in Scandinavia.

TWO PHASES OF LEARNING: CASE STUDIES

How were these learning processes carried out? In this section I argue that we can usefully think in terms of two phases of technology transfer. The first was a phase in which transfer involved individual workers or engineers as bearers of skills. The second was one in which suppliers of intermediate and capital goods, mainly in Britain and Germany, became the central source of transferred technology. Although these two phases should not be too sharply distinguished, they do represent different modes of transfer, and in this section I give examples of how this occurred.

Person-Embodied Skills: Steam Technology in Sweden

In the first phase of technology transfer, the flow of knowledge depended on the movement of individual people, either by immigration (a process which will be outlined in more detail below) or by individuals who travelled, acquired knowledge or skills, and returned. The most detailed case study of this early phase in Scandinavia is Svante Lindqvist's work on the transfer of Newcomen steam technology to Sweden.[9] After its development in the early eighteenth century, the Newcomen engine spread rapidly within Britain, and 70 engines were constructed during the period 1712–25; but it also travelled quite quickly to Europe, the first engine to be built on the Continent being constructed in Liège in 1721. It was introduced into Sweden less than fifteen years after its invention, after an official assessment by the Swedish Board of Mines. The machine was constructed in 1727–28, and operated intermittently at the Dannemora mines for about seven years. The machine was used for purposes similar to its most common application in Britain – that is, pumping. At the time iron bar production was Sweden's most important industry, accounting for 70 per cent of all exports, while mining was hampered by flooding problems similar to those found in Britain.[10]

What was involved in this transfer? First, there was the rapid transfer of information. By 1725, basic knowledge about the technology existed, an achievement generally due to a favourable social and institutional structure and scientific interests in Sweden. The Swedish Board of Mines played an especially important part, since the Swedish government increasingly controlled the mining industry and the Board was its policy instrument. The Board systematically collected and assimilated information about the new steam technology, and regularly sent people abroad to seek information. Thus the gathering of information about the Newcomen engine was 'a matter of routine by the Swedish civil service'.[11]

Lindqvist has argued that the rôle of an individual agent or entrepreneur was central in this transfer process, and has identified certain prerequisites which the agent had to possess: practical knowledge of how to apply the technology; the backing of a private group with the means to realize the project; 'high social status and a proven knowledge of science'.[12] The main agent, Mårten Triewald, was a Swede of German stock who, after his schooling, entered the service of the British merchant Samuel Worster, who was resident in Sweden. This connection eventually brought Triewald to Britain, where he gained practical engineering experience (in Newcastle) and detailed knowledge of the Newcomen engine. Triewald returned to Sweden in 1726, now fulfilling three of the above requirements, namely practical technological know-how, high social status and knowledge of science.[13] The Swedish Board of Mines gave him financial backing, and the new engine was constructed.

However, the engine was not a success. In Lindqvist's words, 'it worked sporadically – for anything from six minutes to three days at the time – punctuated with stoppages of up to three weeks',[14] and ceased operation altogether in 1736. A long-drawn-out court case between the mine owners and Mårten Triewald ensued, and the Newcomen technology languished. This episode is normally regarded, in Swedish historiography, as a failure for technical reasons. Lindqvist's study showed that this verdict must be revised: certainly there were technical problems, but these were overshadowed by much more complex social issues, which shaped the economics of the Newcomen machine in Sweden. In terms of ownership and control, the mines were decentralized and operated by a number of mine-owners during set alternating periods. The engine worked erratically, and this meant that the mines were kept water-free for some of the owners some of the time, while others were randomly penalized. Moreover, the technical adjustments necessary to make the engine reliable were equally prohibited by social constraints. While Triewald and his assistant possessed the expertise which could have overcome the shortage of trained mechanics, both in fact left the mining community. The assistant was in effect denied the social and technical standing which would have enabled him to do the job. Triewald's apparent lack of interest in the progress of the machine he had had constructed must be explained by reference to his social ambitions and aims. In many ways the technology was, for him, primarily a means of gaining membership of the Swedish Royal Academy of Sciences, which he did. While the transfer was a failure in the sense that the objectives of those introducing it, the mine-owners, were not realized, this view was not held by Mårten Triewald. Equipped with a view of science and technology founded on religious sentiments, he regarded the Newcomen engine as proof of God's wisdom and power, a 'revelation' which was introduced to

Sweden for the 'public good' (not necessarily profit), and such proof of an almighty and benevolent God was not to be questioned or criticized.[15]

Specialized Supplier Firms and Technology Transfer

What I call the second phase of technology transfer emerged around the mid-nineteenth century. During the first three decades of that century Britain developed new industries which significantly changed the character and pace of technology transfer. These were specialized industries, focussed on the development of technologies: capital goods industries, producing a wide variety of engineering products. The increasing use of large-scale transport – ships and railways, in particular – meant that suppliers of construction and maintenance equipment were increasingly capable of providing expertise to foreign users. At the same time, the mechanization of consumer goods production (notably, of course, textiles) led to the emergence of specialized machine-making firms. These firms constituted, for the first time, a mechanical engineering industry whose task was to produce machines and equipment for final-output producing industries.

This development shifted the burden of generating process innovations away from machinery-users onto specialized producers. This had important effects on the innovation process in British manufacturing industry: it affected the innovation rate, but also the diffusion environment. It had two significant implications. Firstly, technological change became increasingly complex and specialized: individual items of equipment became, not for the first time, but for the first time on a wide scale, components of a system which was integrated technically (rather than economically or managerially). This changed the diffusion problem, for it changed the nature of what it meant to acquire a technology. Secondly, it concentrated many of the technological skills necessary for the new production system, but it also incorporated a major incentive to spread those skills. Capital goods producers make their profits by promoting diffusion. This transition had very significant implications for the diffusion of industrialization as a whole, for the sale of equipment also involved, in one form or another, the diffusion of the adjunct skills which were necessary to industrial production.

It is clear from the British trade statistics that the pace of diffusion must have accelerated in the mid-1840s, since exports of 'machinery and millwork' to all European countries increased sharply at that time (see Table 9.3, below). This of course corresponds to the repeal of the prohibitions on the export of machinery after 1843.

Table 9.3. British Machinery and Millwork Exports, 1841–50 (Declared Value, £000)

Country	1841	1842	1843	1844	1845	1846	1847	1848	1849	1850
Russia	29.7	36.0	13.8	158.1	116.1	108.0	226.6	212.7	205.9	204.0
Sweden	4.8	6.5	1.8	2.4	5.2	10.1	30.2	10.5	29.6	22.0
Norway	0.8	4.2	1.4	2.5	9.4	15.5	5.3	5.7	4.2	12.2
Denmark	2.8	3.0	9.1	5.9	6.6	11.0	11.6	2.5	8.5	30.6
Germany	60.0	69.9	115.1	107.0	124.5	176.3	173.5	61.0	51.0	93.9
Holland	34.3	29.2	47.0	34.1	50.0	38.4	59.0	27.6	14.0	18.7
Belgium	20.5	11.4	19.4	27.8	36.6	20.2	20.0	5.0	16.4	22.6
France	96.6	106.4	90.6	84.3	103.1	167.2	78.8	35.2	26.9	59.1
Italy	58.8	64.0	96.0	96.3	60.8	60.3	108.9	83.6	45.1	117.4
Spain	13.5	27.8	36.2	54.6	65.3	109.2	97.5	97.7	35.8	73.2
Total	321.8	358.4	430.4	573.0	577.6	716.2	811.4	541.5	437.4	653.7

Source: British Parliamentary Papers, 1854–55, **LII**, 226.

An important point here is that British suppliers provided overall 'packages' of technology: information, equipment and machines, skilled labour, management expertise, and so on. The learning involved here was not that of learning a single new process, but rather a whole system. In the following I give two examples.[16]

A Package Example: Dry Dock Technology

The first example draws on Rasmussen's work on dock construction in Denmark.[17] In the mid-nineteenth century the Danish government decided to establish a dry dock in Copenhagen, mainly for naval purposes. This was to be a project on a completely new scale, and extensive enquiries were made in England, seeking relevant skills. A Danish naval officer, N.E. Tuxen, was chosen for the task. He already, in fact, had a track record in technology transfer: in the 1840s he had worked on problems related to the transition from sail to steam in warships, and had solved the maintenance problems associated with imported steam engines.

Tuxen's first strength lay in 'know who'. He contacted Swedish colleagues who had worked on the Gotha canal project in Sweden. They put him in touch with Thomas Telford, who had supervised the canal project. Tuxen also contacted British firms who were working on railway construction projects in Denmark. Tuxen then visited London (in June 1853), meeting Henry Maudslay and numerous other engineers.

During one of his visits to Britain, Tuxen discovered that a dock based on totally new principles was under construction in Southampton. He visited the site, met the engineer, Alfred Giles, and was impressed. Giles went across to Copenhagen, and a relationship of trust was established. The basic idea was to entrust the building of the dock entirely to a British contractor; but this fell through – only two serious bids were entered and neither was regarded as satisfactory. The decision was taken to build the dock with Danish resources, funded entirely by the state, but with Giles as a full-time technical advisor. The contract was signed in October 1854. Giles was to be fully responsible; he acted not only as technical advisor, but also as the main supplier of both skilled workers and materials. In the event, he spent very little time in Copenhagen, and the project has been described as 'the world's first correspondence course in dock construction'.[18]

This project has a number of important features. Significant numbers of British workers were involved, and often came with their families. There was extensive and successful co-operation between the British and the Danes in the construction process, and, from the Danish side, extensive learning by doing and imitation. But there was also extensive input from Alfred Giles in Southampton. He supplied

> both the technical knowledge, construction drawings, estimates, material, machines and workers. The development costs were paid in England and the system tested in Southampton. The skilled workers were experienced, and a network of trustworthy suppliers were able to deliver the required quality at the right time at the right prices.[19]

What we are seeing here is an example of something much more than the supply of a piece of equipment. We are seeing the supply of an integrated package of technological skills, covering the full dimensions of the complex elements which go to make up a technology.

Diffusion of the New Textile Technology to Norway

My second example is the Norwegian textile industry in the mid-nineteenth century. The discussion here is mainly based on my earlier study of the diffusion of British textile technologies to Norway.[20] The focus was on eleven Norwegian textile firms, with primary emphasis on one firm (the Hjula weavery) which is unique in that its nineteenth-century correspondence archives survive intact, giving a detailed picture of the process and the myriad problems of technological diffusion.

The fixed-capital stocks of the Norwegian textile firms rose rapidly from the mid-1840s, indicating a substantial programme of equipment acquisition. This was based on buying from Britain: the invoice records of two firms showed that out of the 557 specific items of machinery acquired over a fifteen-year period, 554 of them originated in Britain.[21] While general information on technical developments abroad, and on the availability of equipment, was transmitted through the Norwegian technical societies and the state-subsidized system of foreign travel, detailed technological information and skills were transmitted via the British textile engineering sector. More than 330 British firms of all types had some sort of contact with the small Norwegian textile industry during the period *c.* 1840–70; of these, 86 were textile engineers.[22] These firms keenly sought sales in the Norwegian market, and were prepared to provide a wide range of what might be called 'ancillary services' with the equipment they sold. They provided information: information relating to the technical aspects of setting up textile enterprises, and to the assessment and evaluation of new machines and equipment; information on the availability of equipment types, on the choice of technique, on prices and costs of production, and on the types and levels of output to be expected. The British engineering firms naturally also supplied machinery and equipment, as well as raw materials and shipping services. They also supplied labour: British textile engineers advertised for, interviewed, and recruited all types of skilled labour and supervisory staffs for Norwegian firms. They arranged contracts and journeys, and often made payments to the families of British workers while they were in Norway.

Among six of the Norwegian firms, a total of 96 British workers were traced for the period 1845–70, twenty of them women.[23] These workers performed important technical functions: the provision of specific skill inputs; technical information and advice; assistance and guidance in contacts with British firms; skill transmission (that is, instruction and training of Norwegian workers, and indeed of entrepreneurs); and supervision and management. From time to time the British workers advised on the employment of other British workers. They also travelled to Britain on behalf of their Norwegian employers to order equipment and parts. In general, this labour inflow appears to have been a central component of the overall process of diffusion.

The conclusion which emerged from my 1989 study was that British textile engineers were prepared to do a great deal more than simply sell machines. They were, in effect, selling 'packages' of technology which included a complex array of information, equipment and labour. The Norwegian textile entrepreneurs usually lacked technical experience; this potential obstacle to diffusion was overcome by the willingness of British machinery suppliers and agents to provide this wide range of technical services. It was this which solved the technology transfer problem noted above. The complex array of non-codified skills which were necessary to run modern textile equipment were provided by British textile engineers, who thus diffused not only equipment but technological capability. In other words, when we ask how

Norwegian firms acquired 'technological capability', the answer is that they bought it. The elements which made up such capability were provided, as an adjunct to the sale of hardware but also as something quite distinct from it, by British textile engineers and machinery-supplying agents.

The Continuing Importance of Skills: the Bessemer Process in Sweden

It is important to note that the distinction between these phases does not mean that the central rôle of individual skills lessened over time. Klaus Wohlert, in his study of the introduction of the Bessemer process to Sweden in 1857–58, emphasized that it is not enough simply to assume that the appropriate preconditions were in place; it is important 'to show *concretely* the workers' preconditions for being able to absorb the new techniques'.[24] In certain respects, the decision to introduce the Bessemer process was tied up with the well-known phenomenon of attempting to shift skills from craft workers to management. It was remarked at the time that the new method 'liberates steel preparation from dependence on the individual worker's hand skill and places the leadership of the process almost entirely within the sphere of intelligence'.[25] But as with other such projects, things appear to have turned out somewhat differently.

There was considerable interaction with Britain. Firstly, there was extensive travel by one of the key figures in the project, Sieurin: between July and October 1857 he visited Hull, London, Birmingham, Manchester, Newcastle, Liverpool and Wales. In his report he claimed that a large fee had been demanded when he asked actually to participate in the process. He had to make do with 'observation of the work methods'.[26] At the same time, people from Britain spent shorter or longer periods in Sweden: an instructor, Edward Scholey, and an engineer named Price, who set up a steam engine, while the Swedish engineer John Leffler 'who had worked at Bessemer's' led the installation and purchased necessary equipment from British firms.[27]

But behind these direct transfer operations lay two important aspects of learning which were domestic to Sweden. The first was the system of engineering training and sector-specific offices related to iron and steel. These included the Faluberg School, the Filipstad school and the Iron Office. The latter was funded 50 per cent by the state and 50 per cent by manufacturers. These organizations supported the process – not necessarily with existing skills and knowledge, but by supporting the process of learning. The Iron Office financed the experiments which led to the decision to establish the Bessemer process, and the schools developed instruction (and practical learning) in iron and related technologies (for example, steam engines).

But the second aspect, which made possible the learning involved in adapting the Bessemer process, was simply the accumulated skills of the Swedish workers. Existing skills are sometimes seen as a basis for resistance to new technology. Wohlert's argument is that these skills did not provide an obstacle for diffusing the Bessemer process; rather, they were a building block – 'The previously acquired skill of the Swedish workman was necessary for him to be able to absorb the new technology'.[28]

MECHANISMS, INSTITUTIONS AND ORGANIZATION

In Scandinavia a wide range of formal and informal mechanisms for monitoring foreign developments and for importing technologies emerged during the eighteenth and nineteenth centuries, and played a rather extensive part in acquiring basic factual information concerning the technological frontiers in the larger economies, and in the learning processes through which they were accessed and operated.

Industrial Espionage

Industrial espionage was a significant element in acquiring technological knowledge, particularly during the 'first phase' of transfer when many technologies were subject to formal or informal restrictions in terms of access. From the late eighteenth century, individuals, firms and governments in Scandinavia sponsored industrial espionage on a large scale. Espionage occurred in textiles, where the Danish and Swedish governments sent hand-picked agents to Britain;[29] in Norwegian glass production, where a large part of the development of this industry rested on the work of 'a young citizen ... Morten Waern, on an assignment of industrial espionage to Britain';[30] in metal working, where the Dane Jørgen Dalhoff bribed his way into plants in Britain and Germany (specifically paying skilled artisans for information);[31] and in iron production, where a Swedish production manager was advised to invite British mill foremen out, and ply them with drink, since 'that usually opens the Englishman's heart'.[32] Other fields were weaponry, pottery, instruments, general engineering machines and tools, canal building, coining and steam technology.[33] The internationally competitive strength of Danish and Norwegian salt, cobalt, glass and porcelain production, it has been argued, all depended on a continuous flow of information on foreign developments, gained from repeated spying in Britain and on the Continent. According to Christensen, it is 'difficult to see how Danish textile production could have been established without the espionage of Ljungberg and Nordberg'.[34] The information related not only to process innovations, but typically also included new products.[35]

A primary objective seems to have been to acquire knowledge of current developments – 'know what', as it were. However, there seems also to have been an important element of 'know who' in all this: at various times during the period 1750–1850, fifteen Norwegian and Danish spies operated in a wide range of fields, primarily in Britain – eight of these were members of Masonic lodges with extensive international links. It appears that London lodges significantly aided the access to technological information in Britain for three of the spies.[36] Furthermore, the Scandinavian spies were frequently members of key societies and associations in the 'target' countries, and their espionage efforts were aided by interaction with co-members both in Scandinavia and abroad. Christensen has argued that transfer to Denmark and Norway 'was undertaken by a small cosmopolitan élite': a majority of the spies were postgraduates or professors of natural philosophy, and were members of 'the Danish or Norwegian Societies of Sciences and Letters (and a few of the Royal Society in London), of the Society of Agriculture (and some of the Royal Society of Arts in London)'.[37] Lastly, important input came from contact with Scandinavian officials stationed abroad.

During his visit to England in 1807, the Danish state-supported spy Marstrand, for instance, contacted Scandinavian diplomats in London, who emphasized that he would get nowhere without British connections and their letters of introduction. In the event, his efforts were thwarted; he left when the British bombarded the Danish fleet.[38] Industrial espionage was regarded as a threat by the British, as David Jeremy has shown.[39] This concern was justified – and as a channel of general information such espionage was unquestionably important.

Technical Societies

By the mid-nineteenth century all the Scandinavian countries possessed a number of technical societies. The societies conducted regular meetings, published journals and facilitated discussion of technological issues; a central aspect of this activity was an outward-looking orientation. These societies were central to the dissemination of information on general foreign developments. In Norway, three technical societies were established between 1830 and 1840, plus a range of local and artisanal societies. The Norwegian Polytechnic Society averaged 25 meetings per year during the second half of the nineteenth century, addressing subjects such as 'Portraits of Brunel and Stephenson', 'Notes from a Journey to England' and 'Cotton and Cotton Spinning Mills'; by 1874 nearly 300 meetings had been held, with the proceedings published and widely distributed.[40] The Danish technical journal, the *Ursins Magazine*, first published in 1826, typically contained reports on foreign developments, drawing on *The Mechanic's Magazine* and *The Repertory of Arts, Manufacture & Agriculture*. At times, societies were started up by industrialists, or by governments, and most industrialists and many engineers were members of these societies, which also sponsored visits to foreign countries for the purpose of examining industry and acquiring information. The societies sometimes played direct rôles in transferring technology: in Denmark the Royal Danish Agricultural Society 'took on the transfer of agrarian technology, models of machinery and awarded prizes to pave the way for land reform'.[41] A Danish travel report on a visit to London, published in 1816, allegedly prompted the mechanic Winstrup to invent a paper machine (which he subsequently sought to have patented). This was nine years after the patent specification of a similar machine had been published in *The Repertory of Arts, Manufactures & Agriculture*: the journal was unavailable in Denmark at the time because of the war.[42]

In Sweden, the Royal Academy of Sciences and various technical societies played important rôles, according to Lindqvist and others, in the overall development of technological capability:

> The technical changes during the eighteenth century can to a large extent be characterised as qualitative, and these changes were to be important for the industrialisation process which started during the nineteenth century. Generally, we can say that the largest technical change during the eighteenth century was that 'technique became technology'. This was because unified and systematised knowledge about technique was developed ... This knowledge was sustained, developed and communicated through new institutions such as the Iron office, the Academy of Sciences and various agricultural societies. These institutions raised technical problems which earlier had just been local questions to a general level and made them national concerns. Through journals and books which were published by these institutions the knowledge was spread to new groups. This institutional development was perhaps the most important technical change in Sweden during the eighteenth century.[43]

In the early years, these societies were focussed on the need to monitor foreign developments, simply reporting on technological developments abroad. But during what I have referred to above as the 'second phase' of transfer, technological knowledge in general became more codified and accessible in books and journals, and the Scandinavian societies acquired substantial libraries of books and journals. Such libraries did not in themselves make 'know-how' accessible, but the availability of specific operating manuals, guides to machine construction, and descriptions of production procedures were increasingly important in the education and training of engineers and managers throughout the industrializing world; and to the extent that they were accessible and usable, they constituted a mode of learning and a transfer mechanism.

Travel and Foreign Work Experience

Turning to the question of 'know-how', the main channels of transfer were travel abroad by Scandinavian engineers, and immigration by foreigners. Such travel, giving a kind of 'hands-on' experience, was important throughout the period.

Like many European governments, Scandinavian legislators were anxious for close contact with foreign industrial developments and promoted foreign travel; and many of the state-supported espionage cases discussed above involved prolonged stays abroad. During the second half of the nineteenth century, the Norwegian government granted a total of 1006 travel stipends, of which 187 went to 'mechanics'. Furthermore, most industrialists spent considerable periods of time abroad, many gaining work experience (or 'know how'), and at times firms financed employees' going abroad.[44] Jörberg points to the importance of 'the many journeys that Swedes made to British enterprises' for the introduction of innovations to Sweden in general.[45] Ljungberg, who was Swedish-born but was hired as a spy by the Danish government, spent altogether 15 years in England; the Dane Bidstrup was there for 6 to 7 years in the late eighteenth century, including a period as apprentice (he then ran his own engineering workshop in Leicester Square); while Nordberg, who was Swedish-born but grew up in England, worked in the British textile industry.[46]

Travelling and working abroad were important mechanisms for gathering specific information about technical developments, and for building up individual technological capacity. To keep abreast of the developments required frequent visits, as did the demands of the learning involved in mastering the new techniques. For example, the Dane Nordberg's travels to England were undertaken for the sake of keeping up to date, and of solving persistent problems linked with reconstructing and running textile equipment he had smuggled to Denmark. Bidstrup entered a course of what can be thought of as British continuous education in instrument making, which may have involved learning some basic scientific and technological principles, or 'know why'. At the same time his many long visits were undertaken because 'he realised that he had to master all phases of the division of labour himself, since the various skills were absent in Copenhagen'.[47] Presumably he succeeded in this, and went on to establish modern Danish instrument making.

In some cases, work experience and training from abroad made the hiring of foreign workers in Scandinavia redundant: in Danish instrument making at the end of the eighteenth century, and in the 1790s Swedish iron industry, when the Huntsman method was successfully introduced by two Swedes with work experience from Sheffield, who 'after their study tours ... needed no further help from Britain'. Two decades earlier the introduction of new production methods had involved Britons 'in important functions ... as owners, technical supervisors and skilled workers'.[48]

Foreign travel and work experience was pervasive in the Danish woollen industry, sometimes preceding the establishment of firms. The dyer M.K. Brandt spent several years in Germany while an apprentice in the 1820s. His son travelled in Germany and worked in a textile firm in 1865. On his return he managed the installation of machinery at the new textile factory which the family established, and which he later directed together with his brother. Before the establishment of the Bruunshåb factory in 1821, the owner-manager sent his son to Hamburg, Berlin and Bohemia to study modern wool weaving. His son worked two years in a factory in Belgium and became a manager of the family firm on his return. J.L. Binder, who established a woollen factory in Denmark in 1883, had also visited Germany, and his son travelled in Germany for three years in the early 1890s, gaining work experience from several German factories.[49] A number of foreign visits were financially supported by the Danish Reiersens Foundation.

Immigration

During the first phases of technology transfer, the development of Scandinavian industry relied significantly on input from foreign workers. Workers were recruited in the fields of glass making, production of leather, salt, steam, cobalt, porcelain, textiles, iron and steel, and mining. A Norwegian ironwork, Bærum, successfully operated steel-making equipment smuggled from Sheffield and the Carron Iron works, supplying Sheffield customers, with the aid of two British workers.[50] Dan Christensen, in his study of technology transfer to Denmark and Norway during this earlier period, concluded that 'Expertise from abroad made up the foundation on which the development of modern Danish and Norwegian industries could be built'. Yet, he goes on, 'In view of the significant British emigration to France and the U.S.A., it is remarkable how few came to Denmark–Norway'.[51] Nevertheless, immigrant workers are a persistent feature in sectoral studies of Scandinavian industrialization. Some of the relevant cases are summarized in Table 9.4, which, like Table 9.2 above, is drawn from a survey (by no means complete) of the literature on history of technology in Scandinavia. It should be noted, however, that immigration was not purely a matter of 'first-phase' transfer – it continued into the later nineteenth century, but the evidence seems to suggest that whereas earlier immigrants were highly-skilled artisans, by the second half of the nineteenth century they were primarily supervisors and overseers; the relevant skills may have shifted from technical knowledge to organizational or managerial skills.

Table 9.4. Some Immigrant Workers in Scandinavia

From	To	Sector	Personnel	Date	Function	Source
U.K.	Sweden	Iron	2 'British workers' (smiths). Samuel Houlder Sr.	1829	Taught British forging methods	Adamson p. 113
U.K.	Sweden	Iron	'Smiths from Monmouthshire'	1830	Continued 1829 activity; refining; steel	Adamson p. 114
U.K.	Sweden	Iron	2 workers	1830s	Taught blister steel production	Adamson p. 97
U.K.	Sweden	Engineering	Several workers	1830s on	Managers; teachers	Jörberg p. 191
U.K., Germany	Sweden	Textiles	Several workers	1830s on	Foremen; various	Schön pp. 122, 142, 225
U.K.	Norway	Engineering	John Wilson	*c.* 1840s	Taught drawing and mill construction	Hodne p. 166
U.K.	Norway	Engineering	At least 33 workers	1843–90		Bruland 2 pp. 259–60
U.K.	Norway	Utilities	James Malam	1847–48	Constructed first gasworks in Oslo	Hodne p. 167
U.K.	Norway	Textiles	86 workers	1845–71	Skilled workers and foremen	Bruland 1 pp. 173–81
U.K.	Norway	Utilities	James Small	1851–53	Constructed first gasworks in Trondheim	Hodne p. 167
Germany	Norway	Food & drink	German engineer	1866	Set up machinery for flour milling	Hodne p. 169
U.K.	Denmark	Construction	Construction engineer	1857	Brick and tile work	Hyldtoft 1 p. 202
Germany, Duchies	Denmark	Construction	Many migrant seasonal workers	1830s on	Often brick burners	Hyldtoft 1 pp. 203, 208
Duchies	Denmark	Construction	'Manager'	1830s	Manager	Hyldtoft 1 p. 208
Germany	Denmark	Glass	21 workers	1833		Christensen p. 407
U.K.	Denmark	Textiles	Hugh Francis Rennie	1840s	'British engineer', manager	Hyldtoft 3 p. 185
Germany, U.K.	Denmark	Textiles	Several workers	1840s on		Hyldtoft 3 pp. 171, 179, 185–6, 193
U.K.	Denmark	Utilities	At least 14 workers	1850s–1890s	Workers and managers	Hyldtoft 2 pp. 37, 41–2, 79
U.K.	Denmark	Construction	British master	1860	Organized production of cement	Hyldtoft 3 p. 127
Sweden	Denmark	Construction (Nivågård)	Swedish workers	1872	Seasonal work, brick and tiles	Hyldtoft 3 p. 132
Sweden	Denmark	Construction (Hakkemose)	80 Swedish workers	early 1880s	Seasonal work, brick and tiles; every summer	Hyldtoft 3 p. 132
Sweden	Denmark	Construction	1200 Swedish workers (16% of labour force)	1897	Seasonal work, summer	Hyldtoft 3 p. 132

Sources to Table 9.4:

Adamson	R. Adamson (1991), 'Borrowing and Adaptation of British Technology by the Swedish Iron Industry in the Early Nineteenth Century', in K. Bruland (ed.), *Technology Transfer and Scandinavian Industrialisation*, New York and Oxford: Berg Publishers.
Jörberg	L. Jörberg (1991), 'The Diffusion of Technology and Industrial Change in Sweden during the Nineteenth Century', in Bruland, *Technology Transfer*.
Schön	L. Schön (1979), *Frä Hantverk till Fabriksindustri. Svensk Textiltilverkning 1820–1870*, Lund: Arkiv.
Hodne	F. Hodne (1991), 'Transfer of Technology: Theory and Evidence', in Bruland, *Technology Transfer*.
Bruland 2	K. Bruland (1991), 'The Norwegian Mechanical Engineering Industry and the Transfer of Technology, 1800–1900', in Bruland, *Technology Transfer*.
Bruland 1	K. Bruland (1989), *British Technology and European Industrialization: the Norwegian Textile Industry in the Mid-Nineteenth Century*, Cambridge: Cambridge University Press.
Hyldtoft 1	O. Hyldtoft (1991), 'Foreign Technology and the Danish Brick and Tile Industry, 1830–1870', in Bruland, *Technology Transfer*.
Christensen	D.C. Christensen (1996), *Det Moderne Projekt. Teknik & Kultur i Danmark-Norge 1750–1850*, (The Modern Project), Viborg: Gyldendal.
Hyldtoft 3	O. Hyldtoft (1996), *Teknologiske Forandringer i dansk Industri 1870–1896*, Odense: Odense universitetsforlag.
Hyldtoft 2	O. Hyldtoft (1994), *Den Lysende Gas. Etableringen af det danske Gasssystem 1880–1890*, Herning: Systime.

International Industrial Exhibitions

An important aspect of the emergence of Britain and France as industrial economies was the gradual development of formal exhibitions, which presented, in effect, a review of the state of the art in mechanical and industrial technique. During the nineteenth century, exhibitions were held in a number of major European cities, including Prague, Lyons, Mulhouse and Birmingham – but they were not confined to major cities: during the years 1823–49, no fewer than 16 exhibitions were held in Sweden, for example.

The combined Kingdom of Norway and Sweden made major efforts to participate in international exhibitions. Ahlström, who has produced the most detailed study of such exhibitions, sees the usefulness and potential gains of international industrial exhibitions as 'yardsticks for measuring the relative technological positions of different countries', stressing their 'importance to the diffusion of technology and knowledge'.[52] His work has given an overall picture of Swedish participation at eight industrial exhibitions (London 1851, Paris 1855, London 1862, Paris 1867, Vienna 1873, Philadelphia 1876,

Chicago 1893, Paris 1900), and he has analysed Sweden's technological and industrial development during the nineteenth century against the background of that participation. Participation in the exhibitions was not a marginal matter; the Swedish contribution to the 1862 exhibition in London, for example, was organized by a prestigious committee, chaired by Prince Oscar (the Swedish King's brother) with members drawn from the academic élite. The committee held approximately 50 meetings, and local and regional sub-committees were established all over Sweden and in Norway. In the event Sweden had 511 exhibitors and Norway 216. The results indicated continuing technological backwardness and the jury for metal products remarked that

> The collections from Denmark, Sweden and Norway are small, and contain little that requires notice. ... although Sweden has long produced the most valuable iron, as the raw material of the finest steel and the most finished cutlery, it does not appear that the manufacture itself has made any great advance.[53]

Subsequently, the judgements on Scandinavian products were more favourable, as product quality improved sharply in many fields.

These exhibitions were clearly linked to transfer activities. For example, the Norwegian parliament funded travel by two mechanics to visit the 1851 exhibition in London: this led to the importing of reaping equipment which was diffused in Norway via further local farm fairs, at which the new technology was demonstrated.[54] In general, Ahlström sees the exhibitions firmly in the context of international diffusion, particularly of radical technologies: his analytical approach is based on the long-wave and neo-Schumpeterian analysis associated with Freeman, Perez and others. In this framework, exhibitions were seen as crucial in publicizing the content of technological revolutions, in locating the technological frontiers, and in providing information about diffusion possibilities from the frontiers.

ELEMENTARY ACADEMIC COMPETENCE AS AN ESSENTIAL PRECONDITION FOR 'TECHNOLOGICAL LEARNING'

In the sections above, I have outlined some of the specific mechanisms through which technology transfer was promoted. But there are also more general factors, particularly to do with literacy and the value placed upon it, which may well be important in this context. Certainly human capital, for which literacy and education are often used as proxies, has become a major part of recent growth theory. How did Scandinavia perform in this respect?

Lars Sandberg, in two well-known papers,[55] has argued for dividing 'backwardness' in the early phase of industrialization into two components: firstly, low levels of income relative to leader economies, and secondly, 'ignorance', meaning low stocks of human capital. Sandberg's argument is that the industrialization of Britain and France provided major opportunities for growth in follower economies. But to seize these opportunities depended critically on human capabilities and the human capital stock.

Sandberg treats basic literacy rates as 'a good index of per capita human stocks in the middle of the XIXth century'.[56] He divides the European economies into three groups: those with literacy rates above 70%, those between 50% and 70%, and those below 50%. Of the eight countries in the first group, four were the Nordic states; the others were Germany, Scotland, Switzerland and Holland. Sweden, Denmark and Norway were the top three in the ranking. These literacy rankings, argues Sandberg, correspond to elementary school attendance, and also to longer-term *per capita* income for countries in all three groups: 'Of the top literacy countries of 1850, only Scotland is not one of the top per capita income countries today'.[57] This literacy achievement was mirrored in Sweden by high levels of elementary school provision and attendance: 'by 1859, only about 7,000 of more than 500,000 children of school age were receiving no formal education'.[58]

Sandberg's work has recently been criticized by Hortshøj, O'Rourke and Williamson, who argue that although human capital factors were important in Scandinavian industrialization, they in fact account for relatively little of the overall growth performance.[59] Their work essentially consists of tests of the contributions of various factors to Scandinavian convergence with the leader economies after 1850, extending Sandberg's arguments to the cases of Norway and Denmark as well as Sweden. Their argument is that inward capital mobility and trade growth account for most of the 'catch up'; but they nonetheless show that education made a significant contribution to residual growth, and argue that education 'certainly helped make the nineteenth century Scandinavian catch up possible'.[60]

In Sweden, elementary formal education was paralleled by developments in higher education – especially in engineering training. From the early nineteenth century Sweden had two technical universities, one in Stockholm and one in Gothenburg. The Stockholm school, according to its 1846 statutes, aimed to provide 'education for those young men who pursue some kind of industrial occupation which cannot be properly performed without knowledge of nature, chemical and mechanical technical knowledge especially'.[61] Although absolute numbers were small, rates of growth were high. Thus high basic literacy and school attendance rates and strong engineering education, particularly in Sweden, seem to have been important prerequisites for learning and knowledge creation.

CONCLUSION

The focus of this chapter has been on the nature and workings of the learning processes through which foreign technology was accessed. The main intention here has been to draw attention to the prevalence of technology transfer, and associated learning activities, across the industrializing Scandinavian economies. It would not be appropriate to draw over-strong conclusions from the evidence which has been examined here; except for a few cases, we lack analysis of the economic significance and effectiveness of the technology transfer processes, and of how they integrated with the development of indigenous capabilities. However, there seems to be strong *prima facie* evidence for two conclusions. Firstly, inward technology transfer was central to the growth of these economies; these were follower countries, 'catching up' with the technological leaders of the era, and this process appears to have been linked with the direct diffusion of technologies from leaders to followers. Secondly, there was no specific sector, or mode of learning, or technological change which underlay Scandinavian industrialization. The process of change may have been revolutionary, but it was broadly based across many sectors: competence was acquired and developed at many levels and in many ways – it was the overall impact rather than specific activities which mattered. This in turn suggests that there were wider processes – perhaps economy-wide – shaping the capacity to import technologies. The various mechanisms of technology or knowledge importation which have been described above were not narrow in their field of application, and behind them lay wider social patterns of education, literacy and general culture. It should also be noted that behind such organizations or processes as technical and scientific societies, travel funding, participation in exhibitions, and so on, stood the governments and social élites of the countries concerned. It does seem to be relevant to think in terms of a social or cultural system, with strong international links and an associated infrastructure, generating the complex capacity for technological learning and absorption which characterized industrialization in this region.

NOTES

1. J. Fagerberg (1994), 'Technology and International Differences in Growth Rates', *Journal of Economic Literature*, **32**, 1147–75.
2. During the period under discussion, Norway was part of Denmark until 1814, when Norway entered into a union with Sweden in accordance with the settlements following the Napoleonic wars. This Swedish-Norwegian union, which lasted until 1905 when Norway became independent, left Norway with increased economic independence, and with its own government, finances and legislation. See, for example, T.K. Derry (1973), *A History of Modern Norway*, Oxford: Clarendon Press.

3. The overall summary is available on request from the present author.
4. Unpublished, available on request from the present author.
5. See B.A. Lundvall (ed.) (1992), *National Systems of Innovation*, London: Frances Pinter.
6. F. Hodne (1991), 'Transfer Patterns of Technology: Theory and Evidence', in K. Bruland (ed.), *Technology Transfer and Scandinavian Industrialisation*, New York and Oxford: Berg Publishers, p. 177.
7. B. Lundvall and B. Johnson (1994), 'The Learning Economy', *Journal of Industry Studies*, **1**, 23–42.
8. See especially E. von Hippel (1988), *Sources of Innovation*, Oxford: Oxford University Press.
9. S. Lindqvist (1984), *Technology on Trial. The Introduction of Steam Power Technology Into Sweden 1715–1736*, Uppsala: Almqvist & Wicksell.
10. Lindqvist, ibidem, p. 34.
11. Lindqvist, ibidem, p. 134.
12. Lindqvist, ibidem, p. 180.
13. Lindqvist, ibidem, p. 214.
14. Lindqvist, ibidem, p. 270.
15. Lindqvist, ibidem, pp. 299–307.
16. A number of interesting case histories exist, notably O. Hyldtoft (1994), *Den Lysende Gas. Etableringen af det danske Gasssystem 1880–1890*, Herning: Systime, which identifies a transition in the development of Danish gas production and distribution: from import of technology 'packages' to the establishment of a 'national system' of innovation towards the end of the nineteenth century.
17. F.A. Rasmussen (1995), 'From Southampton to Copenhagen: Transfer of Dockyard Technology to Holmen (The Danish Naval Dockyard), 1852–58', *Fabrik og Bolig*, **1**, 5–39.
18. Rasmussen, ibidem, p. 19.
19. Rasmussen, ibidem, p. 36.
20. K. Bruland (1989), *British Technology and European Industrialization: the Norwegian Textile Industry in the Mid-Nineteenth Century*, Cambridge: Cambridge University Press.
21. Bruland, ibidem, pp. 51–55.
22. Bruland, ibidem, pp. 69–75.
23. Bruland, ibidem, p. 112.
24. K. Wohlert (1979), 'Svenskt Yrkeskunnande och Teknologi under 1800-tallet: en Fallstudie av Forutsätningar för Kunskapstransfer (Swedish Know-how and Technology during the Nineteenth Century: Prerequisites for the Transfer of Knowledge – a Case Study)', *Historisk Tidskrift*, **4**, 398–421, p. 404. All translations from this article are by the present author except where specified otherwise.
25. Wohlert, ibidem, p. 416.
26. Wohlert, ibidem, p. 414, n. 5.
27. Wohlert, ibidem, p. 410.
28. Wohlert, ibidem, p. 421 (original summary in English).
29. See, for instance, D.C. Christensen (1996), *Det Moderne Projekt. Teknik & Kultur i Danmark-Norge 1750–1850*, (The Modern Project), Viborg: Gyldendal, pp. 268–84. All translations from this book are by the present author except where specified otherwise.
30. R.P. Amdam (1991), 'Industrial Espionage and the Transfer of Technology to the Early Norwegian Glass Industry', in Bruland, *Technology Transfer*, p. 75.
31. P. Strömstad (1991), 'Artisan Travel and Technology Transfer to Denmark, 1750–1900', in Bruland, *Technology Transfer*, pp. 144–7.
32. L. Jörberg (1991), 'The Diffusion of Technology and Industrial Change in Sweden during the Nineteenth Century', in Bruland, *Technology Transfer*, p. 194.
33. See Christensen, *Det Moderne Projekt*, pp. 507–8.
34. Christensen, ibidem, pp. 426, 517.
35. Christensen, ibidem, pp. 407–21.
36. Christensen, ibidem, pp. 99–105, 516–22.
37. Christensen, ibidem, pp. 818–19 (original summary in English).
38. Christensen, ibidem, p. 518.
39. D. Jeremy (1977), 'Damming the Flood: British Government Efforts to Check the Outflow of Technicians and Machinery, 1780–1843', *Business History Review*, **51**, 1–34.
40. Bruland, *British Technology*, pp. 58–9.
41. Christensen, *Det Moderne Projekt*, p. 801 (original summary in English).

42. Christensen, ibidem, pp. 500, 460.
43. J. Hult, S. Lindqvist, W. Odelberg, S. Rydberg (1989), *Svensk Teknikhistoria*, Värnamo: Gidlunds Bokförlag, p. 121.
44. For instance, see Bruland, 'The Norwegian Mechanical Engineering Industry and the Transfer of Technology, 1800–1900', in *Technology Transfer*, p. 245, for the Norwegian engineering sector; for textiles, see Bruland, *British Technology*, pp. 61–7.
45. Jörberg, 'The Diffusion of Technology', p. 188. See L. Schön (1979), *Frä Hantverk till Fabriksindustri. Svensk Textiltilverkning 1820–1870*, Lund: Arkiv.
46. Christensen, *Det Moderne Projekt*, pp. 510–22.
47. Christensen, ibidem, pp. 510–22, 780.
48. R. Adamson (1991), 'Borrowing and Adaptation of British Technology by the Swedish Iron Industry in the Early Nineteenth Century', in Bruland, *Technology Transfer*, p. 96.
49. O. Hyldtoft (1996), *Teknologiske Forandringer i dansk Industri 1870–1896*, Odense: Odense universitetsforlag, p. 171.
50. Christensen, *Det Moderne Projekt*, p. 517.
51. Christensen, ibidem, pp. 818 (original summary in English), 519.
52. G. Ahlström (1996), *Technological Development and Industrial Exhibitions 1850–1914. Sweden In An International Perspective,* Lund: Lund University, p. 251.
53. Ahlström, ibidem, p. 37.
54. Hodne, 'Transfer Patterns', p. 171.
55. L.G. Sandberg (1979), 'The Case of the Impoverished Sophisticate: Human Capital and Swedish Economic Growth before World War I', *Journal of Economic History*, **39**, 225–41; L.G. Sandberg (1982), 'Ignorance, Poverty and Economic Backwardness in the Early Stages of European Industrialization: Variations on Alexander Gerschenkron's Great Theme', *Journal of European Economic History*, **2**, 675–97, p. 686.
56. Sandberg, 'Ignorance, Poverty and Economic Backwardness', p. 686.
57. Sandberg, ibidem, p. 688.
58. Sandberg, 'The Case of the Impoverished Sophisticate', p. 230.
59. Kevin Hortshøj, O. Rourke and J.G. Williamson (1995), 'Education, Globalization and Catch-Up: Scandinavia in the Swedish Mirror', *Scandinavian Economic History Review*, **43**, 287–309.
60. Hortshøj et al., ibidem, p. 309; residual growth here is 'the residual left over after controlling for initial wage levels', which the authors test against a variable representing how much Scandinavian schooling exceeded the international average. Taking school enrolment as the education variable, higher rates of schooling in Scandinavia accounted for 31% of residual growth in Sweden, 40% in Denmark and 57% in Norway (p. 300).
61. G. Ahlström (1982), *Engineers and Industrial Growth*, London: Croom Helm, p. 36.

10. Kinship and the Transmission of Skills: Bar Iron Production in Britain and Sweden, 1500–1860

Chris Evans and Göran Rydén

INTRODUCTION

In the early industrial world the success of production rested very largely on the skills of workers. Technological change was generated from within the workshop rather than the laboratory. Indeed, 'technology' cannot be detached from the procedures and habits of the working day. For this reason it is useful to explore technological change from the perspective of the workplace, taking account of the possibility that innovation could involve changes in organization as well as changes in plant or equipment. This can be demonstrated very clearly with regard to the iron industry of northern Europe in the early industrial era. Despite the considerable capital resources devoted to bar iron making, a pivotal rôle was played by small groups of workmen amongst whom links of kinship provided a framework for the transmission of technological knowledge as well as a set of practices which governed the day-to-day conduct of work and life. This had implications for technological development, which will be examined in two different national contexts: Sweden and Britain. Here, kinship networks operated in differing social and cultural environments, and with rather different results.

The international iron trade in the seventeenth and eighteenth centuries hinged upon the relationship between Sweden and Britain.[1] Between the early seventeenth century and the middle of the eighteenth, Sweden was the single most important exporter of iron in Europe. During the same period Britain emerged as the most significant market for iron in the global economy, becoming thereby the major recipient of Swedish iron exports. Despite this level of import penetration, Britain had a busy iron industry of its own, albeit one that was constrained by limited energy resources before the advent of coal-fired ironmaking techniques in the mid-eighteenth century.

The ability to exploit mineral energy reversed the relationship between the iron industries of Sweden (rich in charcoal, but poor in coal) and Britain (poor in charcoal, but rich in coal). In particular, the introduction of iron puddling enabled British ironmasters to sweep their Swedish rivals from all but the most specialized of niche markets in Britain. This precipitated a crisis in the Swedish iron trade, one which was only resolved in the mid-nineteenth century, not by embracing coal technology as such, but by adopting new forms of labour organization which permitted the bulk production of bar iron whilst still employing charcoal technology.

Yet, before the spread of iron puddling, the technologies employed in early modern Sweden and Britain for the production of malleable bar iron were essentially the same. And in both countries the organization of labour within the forge took on a strongly dynastic flavour, with family groups playing a key rôle in the transmission of skill. However, the institutional conditions under which bar iron production took place were divergent, Britain being precociously capitalist, Sweden as yet imperfectly commercialized. This difference was reflected in the conduct of work and in the nature of technological change in the two countries, the subject of this chapter.

Let us begin by noting that iron production in north-west Europe in the early modern period was dominated by the so-called indirect method – a two-fold operation involving the smelting of ore at a blast furnace and then the refining of pig iron into malleable bars at a forge. Whereas the direct-reduction techniques of the medieval world had produced malleable iron straight from the ore, they were slow and yielded only small quantities of metal. The blast furnace was enormously more productive, yielding several tons of pig iron per week, but this pig iron was brittle and had to undergo additional processing at a forge. The indirect technique therefore introduced a new division of labour into iron production and a new element of skilled labour: the forgeman.

The forgeman's task was not easily mastered. He worked at a charcoal-fired hearth in which he melted pig iron, oxidizing most of the carbon impurities in the iron and leaving behind a pasty mass of iron and slags. This mass was then subjected to the blows of a large water-driven hammer which expelled the slags and consolidated the iron into a squared-off lump known as a bloom. Further reheating and hammering eventually gave a thin bar of iron some two to three metres in length, the form which refined iron took as an international commodity. This fundamental refining technique had two variants: Walloon forging and German forging. Walloon forging employed two separate hearths – the finery, in which the actual refining of the metal took place, and the chafery, which was used to reheat the iron between bouts of hammering. In the German method the same hearth was used for carrying out both tasks. In both, the success of the process depended completely upon the experience and dexterity of the forgeman.

The transfer of technologies in the early modern world often took the form of transplanting skilled workers wholesale into a new environment, and it was thus that the indirect technique was introduced to England: abruptly, under the patronage of the Tudor state which was alert to the significance of the blast furnace in the production of heavy ordnance. At least 500 French furnace and forge workers arrived in the Weald of Kent and Sussex between the 1490s and the 1540s, almost all of them from the same, quite small area of Normandy.[2] They introduced Walloon forging to the British Isles and, as will be seen, their descendants continued to play a central rôle in the dissemination of ironmaking techniques for the next two centuries. In Sweden the indirect technique developed spontan-eously in the late medieval period and was initially in the hands of *bergsmän*, peasant proprietors who ran furnaces and forges on a part-time, co-operative basis, each working up their own ore and charcoal. At first they used a fairly elementary forging technique, producing so-called *osmund* iron: refined iron cut into small pieces. This iron was exported in the main to the Hanseatic towns where it was welded into bars by local hammermen. During the sixteenth century, however, skilled German forgemen were brought to Sweden under the Vasa monarchy, implanting the more developed German forging method. During the great expansion of the iron industry in the early seventeenth century there was again an influx of expert foreign workers: Walloons from the bishopric of Liège and the Spanish Netherlands. Some 2,000 men, women and children moved to Sweden, most of them to the county of Uppland where the Swedish state was promoting the growth of forges devoted to the export of high-quality bar iron made the Walloon way. Here they remained as forgeworkers over many generations, assimilated into Swedish society, but their origins announced by their family names: Gille, Bonnivier, Pjerrou.

The iron industries of Britain and Sweden shared, then, some essential features. There was a fundamental uniformity of technology, and in both countries this technology came to be deployed by skilled labour organized upon a dynastic basis. But there the similarity ends, for the social and institutional context of ironmaking differed radically between the two countries. In Britain the rôle of the state was muted after its short-lived patronage of the indirect system in the early sixteenth century. The spread of the indirect technique from its Wealden heartland to all areas of the British archipelago was largely at the spontaneous initiative of aristocratic landowners who were anxious to put reserves of ore and wood on their estates to profitable use. By the later seventeenth century most landowners had decided that this was most easily done by leasing out their furnaces and forges to professional ironmasters. But regardless of who exercised managerial control, it is clear that throughout the early modern period the iron industry in Britain responded to an essentially capitalist dynamic, and did so within a society that was heavily commercialized and in which labour was a commodity.

The Swedish iron industry knew a quite different context. There, the state played a far more prominent rôle. Indeed, at the start of the seventeenth century the Swedish state attempted a far-reaching socio-spatial reconfiguration of iron production. Peasants in *Bergslagen* – the traditional mining zone of central Sweden – were to be restricted to producing pig iron. Henceforth, the production of bar iron – the high-value-added sector of the iron trade – was to be monopolized by politically privileged forge-owners drawn from the ranks of the nobility or from mercantile circles. The making of bar iron became centred upon the *bruk* – the ironworks estate – at which a specialized industrial workforce was quartered and from which the fuel and food needed to sustain the ironmaking community was to be drawn. Every *bruk* was presided over by its *brukspatron* (ironmaster), but overall the system of production was policed by the *Bergskollegium*, a state agency established in 1637, which was endowed with wide-ranging powers. It alone could license the construction of new plant, and most importantly it could regulate the market for charcoal by compelling peasants to supply charcoal to a particular *bruk* in lieu of taxes. Taken together, the *bruk* and the regulatory system of the state defined the environment in which Swedish forgemen laboured.[3]

FORGEMEN IN SWEDEN

The introduction of the Walloon forging method into Sweden was the result of co-operation between the Vasa state and Amsterdam-based merchants, often of Walloon origin. The 'Walloon' ironmasters took over a number of former royal ironworks in Uppland, in the vicinity of the Dannemora iron ore mine, and looked to their native region for skilled workers to staff their new industrial undertakings. Most of those who left the iron-producing regions of the southern Netherlands for Sweden moved between the 1620s and 1650s. Not all were forgemen. Walloon carters, charcoal-makers and forest workers arrived in Sweden, as did skilled furnace-keepers and artisans in the metal trades. Nevertheless, it was as forgemen that these Walloon immigrants won renown and it is as forgemen that they are remembered in Sweden.[4] This is not so paradoxical, as the Walloons did introduce a new way of making bar iron, distinct from the German forging method that had been established in the preceding century. And within the *bruk* the forgemen enjoyed a special eminence among their fellow artisans, reflecting their strategic importance in bar iron making. The iron they produced was of exceptional quality, due largely to the purity of the ore from Dannemora. As a result, Walloon iron from Sweden found its most important market in the steelmaking centres of Britain, where it was known as *oregrund* after the port of Öregrund through which it was exported.

By the end of the seventeenth century there were approximately twenty-five forges using the Walloon method in Uppland, distributed between fourteen ironworks, the so-called *Vallonbruk*. (Outside Uppland only a limited number of works employed Walloon forging.) At each forge a nine-man team undertook the task of refining pig iron: a master finer, a finer hand and two apprentices at the finery; and a master hammerman, two apprentices and two *goujards* – fetching charcoal – at the chafery. All in all, approximately 250 forgemen were occupied in making *oregrund* iron in Uppland, and almost all of them were of Walloon origin until at least the first decades of the eighteenth century. Very few Swedes indeed gained positions of eminence at Walloon forges during the seventeenth century.

Not only were Walloon forgemen numerically dominant in bar iron making in the *Vallonbruk*, but the organization of work was structured along family lines. One or two families tended to dominate bar iron making at particular ironworks, or particular forges. At Leufsta – together with Österby and Forsmark, the most important of the *Vallonbruk* – the dominant families were called Tillman, Guilliuem, Boivy and Martinell. In 1677 twenty out of thirty-five forgemen belonged to these families. At Österby the forgemen were drawn from the Gouffeng, Bouveng and Gille families, while those at Forsmark bore the names Godeau, Gjers and Sporran. In his study of Walloon immigrants in the seventeenth century, Bernt Douhan has established that sons followed in their fathers' footsteps to a remarkable degree. The sons of forgemen became forgemen in their turn. Not only that, but the sons of finers became finers, whilst hammermen brought up hammermen. There was thus a clear division between families at fineries and those at chaferies.[5]

Only in the eighteenth century did Swedish forgemen assume some significance in the organization of bar iron making at the *Vallonbruk*, and they usually remained a minority. In 1801 thirty-six persons were employed at the three forges in Österby. Only five of them were of Swedish origin. Twelve of them belonged to the Gille family, eight were called Goufeng, and four Bouveng.[6] At Forsmark Swedish forgemen made a greater impact. During the 1760s more than half of the forgemen bore Swedish names, but forgemen with the names Gjers and Sporran were still to be found. In 1765, for instance, eight of the nineteen forgemen at the two Forsmark forges were of Walloon origin, and Walloons tended to monopolize the rôle of master forgeman. Only in the middle decades of the nineteenth century did Swedes begin to appear as masters with any consistency. By 1870 both master finers were Swedes, but there were still forgemen with the familiar Walloon names at the *Vallonbruk*.[7] Indeed, the Gille family at Österby provides one of the most striking examples of workplace continuity, being present at the forge from the seventeenth century until its closure in 1941.

The *Vallonbruk* accounted for only 10 to 15 per cent of Swedish bar iron output in the seventeenth and eighteenth centuries. Most iron was made according to the German method. Far less is known about the organization of work in the smaller and less sophisticated German forges, although there were in fact some Walloons engaged in German forging as well. As early as the seventeenth century there was some migration of Walloon forgemen from Uppland to other iron-producing areas. The two Walloon families of Goude and Herou, for example, arrived at Horndal in Dalarna in the 1650s and were to dominate work at the two forges there for two centuries. Their presence at the heart of forge crews can be documented throughout the course of the eighteenth century.[8] And at Horndal, as at the *Vallonbruk*, kinship links remained a salient feature of production into the nineteenth century. Of the eight master forgemen at Horndal between 1830 and 1850, five answered to the name Goude or Herou. It may be that Horndal was exceptional in this respect, since no other detailed studies of forges employing the German method in the seventeenth and eighteenth centuries are available, but in the second quarter of the nineteenth century much the same kind of kinship structure can be found in ironworks in the county of Gästrikland. Here, at the works of Tore Petré, almost half the forgemen belonged to three different kin groups. The Skog family dominated on the eastern fringe of the complex while the Pjerrou family was strongest in the northern works. In Hammarby, the centre of the enterprise, bar iron manufacture was controlled by the Ekengren family – whose pre-eminence dated back to the seventeenth century and reached a new intensity in the post-Napoleonic period.[9] This is an under-researched area, but since many company histories make brief mention of so-called 'dynasties of forgemen', it seems likely that kin-based structures prevailed at other ironworks too.[10]

Kinship provided a medium for the transmission of skill and know-how in the Swedish iron industry. Ironmasters recognized that an intergenerational continuity of skill could best be achieved through stable family structures at the *bruk*. As Carl Rinman, the leading metallurgist of his day, wrote in his *Handbook of Heavy Iron and Steel Making* (1829):

> At ironworks to which a clan of experienced forgemen and workers moved a long time ago, and where, under a good and unbroken rule, they have remained both within their trade and kin and within the simple and good customs of their ancestors,

> here is commonly found more skill in work, more happiness and more order than at works where workers from different places with different customs are recruited.[11]

Thus kinship and skill were inextricably linked in the eyes of ironmasters. As a consequence, the masters were concerned to immobilize families of forgemen, to pin them down at their own *bruk*. In this they were assisted by *Bergskollegium* decrees which capped forgemen's wages (thereby hampering the growth of an active labour market) or which placed restrictions upon the movement of forgemen who were indebted to their masters. As a consequence, migration was limited and age-specific: it was the lot of young men who were surplus to requirements at their native *bruk* or who were as yet free of debt.

For their part, forgemen had good reasons of their own for adopting a dynastic approach to the organization of production. These concerned the 'internal economy' of a forgeman's household. Decrees concerning forgemen stated that payment for production was to be split among the members of the hammer crews in the following fashion: the master received 5/12, the forgehand 4/12 and the apprentice 3/12. If, however, the apprentice was to be included in the master's household and 'clothed and fed' by him, then the master received the apprentice's 3/12 share. As the costs of feeding and clothing an apprentice were much lower than 3/12 of the whole payment it made economic sense for a master forgeman to have someone from his own household as an apprentice.[12]

A master forgemen could not, however, always include a son of his own in his hammer crew, for very obvious demographic reasons. Boys started work in the forge as apprentices in their late teens – seventeen being the mean age – and they remained in that position for around seven to eight years until they were fully trained and could take the step up to become forgehands. They were then in their mid-twenties. They could also now get married. After another eight to ten years they became masters in their own right. At that time the eldest son could not be more than ten years old, too young to be an apprentice, and for eight to ten years the master forgeman had to recruit apprentices from outside his own household. Evidence from the county of Gästrikland suggests that master forgemen would whenever possible pick their own sons as soon as they were of the right age. Before that they were forced to look beyond their own households for apprentices, but only rarely did they look beyond their own kin. Nephews or relatives of their wives were picked before other boys.

It ought to be clear that forgemen as well as ironmasters had something to gain from the dynastic organization of labour. The masters supported a system that would in the long run guarantee the presence of skilled forgemen and hence the production of high-quality bar iron on their *bruk*. For forgemen the dynastic organization of labour offered the opportunity of making the privileged position of master forgeman accessible to their kin, a position that was all the more lucrative if apprentices were drawn from within the household. This remained the case until the middle of the nineteenth century when the development of large-scale processing techniques removed from the forgemen their ability to organize production on the basis of kinship.

FINERS AND HAMMERMEN IN BRITAIN

The organization of forge labour in Britain[13] was in one sense more straightforward than in Sweden. There was no split between forges using the German method and those using the Walloon. From the outset, bar iron appears to have been made using the Walloon method only, whereby a team of finers effected a chemical change in the iron at the finery hearth and a team of hammermen shaped the decarburized iron into bars at a chafery. When the Swedish traveller Reinhold Angerstein visited Britain in the 1750s he illustrated the typical arrangement of an English forge by the example of Bromford near Birmingham: two finery hearths linked to one chafery and a helve hammer, each hearth being attended by a three-man forge crew. Angerstein's sketch (see overleaf) shows Walloon forging in Britain at its most developed: in the generation following Angerstein's tour the nature of bar iron manufacture changed dramatically with the successful development of refining methods based upon coal. Thereafter, the use of mineral energy allowed British ironmasters to produce bar iron in unprecedentedly large volumes at a low cost.

Broadly calculated, the trajectory of bar iron production in Britain involved a sharp rise in the sixteenth century when the finery hearth was first introduced, a period of steady but unspectacular growth in the seventeenth and early eighteenth centuries, and a period of relative stagnation in the mid-eighteenth century when the forge trade was constrained by fuel shortages and was unable to climb above the production plateau it had attained. A critical break in this pattern came only with the introduction of coal technology between the 1760s and the 1790s, allowing British bar iron output to soar. Given this pattern of growth, it can be suggested that the forge trade could recruit a workforce very largely from its own internal resources for much of the seventeenth and eighteenth centuries. Indeed, the direct labour force in the British forge trade was never very large. Contemporary surveys of the industry suggest that the charcoal forge trade, even at its greatest extent in the mid-eighteenth century, on the eve of the transition to mineral fuel, never needed many more than a thousand forgemen.[14]

Figure 10.1. Reinhold Angerstein's Sketch of Bromford Forge, near Birmingham, early 1750s

Within this small, compact workforce the descendants of those who had introduced Walloon forging into the British Isles seem to have played a prominent rôle. The importance of Francophone immigrants for the growth of the British iron industry has already been noted. The furnaces and forges that were established in the Weald in the early sixteenth century were staffed to a significant degree by Frenchmen from Normandy, who were themselves the descendants of Walloon furnacemen and forgemen who had settled in the *pays de Bray* between Dieppe and Beauvais a generation or two earlier.[15] Several hundred crossed the English Channel and once in England they continued to be mobile. The spread of the indirect technique across the British Isles was spearheaded by ironworkers of French descent, so that by the end of the seventeenth century forgemen with suitably anglicized names – Leonard, Jarrett, Vinton – were present in every ironmaking district of Britain. Moreover, this was not a once-and-for-all movement. They moved repeatedly. There was a circuit of labour, flowing back and forth between a network of sites, and this circuit was national in scope.[16] Unfortunately, this very mobility makes them very difficult to study. 'Stayers' become visible in the written historical record; at best, only fleeting glances can be had of 'movers'. In contrast to Sweden, the presence of forgemen's dynasties in Britain often has to be inferred: it is not easily demonstrated. Nevertheless, some family names occur within such regularity that it seems safe to suggest that their bearers were a continuous element in the British iron trade.

Members of the Lavender family may serve as an example. Of French origin, they had settled in the Weald in the early sixteenth century. By the early seventeenth century members of the family had found their way to a variety of ironworking districts (Yorkshire, Shropshire, Staffordshire). A century later, finers and hammermen named Lavender were at work in the far north-west and north-east of England, but the principal branch of the family was based in the heartland of the British iron industry in the West Midlands. Family members worked at Wilden and neighbouring forges in Worcestershire over several generations, although this cannot be plotted with complete accuracy. The survival of an early nineteenth-century poor law examination does, however, allow the reconstruction of the working life of one of the Lavender forgemen. James Lavender was born at Wilden in the early 1760s. At the age of fifteen he was apprenticed to a carpet weaver in nearby Kidderminster. Yet this evidently did not suit, and he returned to the trade of his ancestors within a couple of years. By 1787 he had settled in the Staffordshire parish of Bradley, presumably working in the Bradley works of the great ironmaster John Wilkinson. From there he moved into South Wales. He worked for three years at Llangrwyney forge in the Usk valley, then for five years at the Beaufort ironworks in the Monmouthshire uplands, and then at the forges at Pentyrch and Machen further south. At this point James Lavender is lost to view, but it is clear that he was not the only one of his clan to come to South Wales. His descendants and kin were ubiquitous in the vestigial charcoal iron industry of South Wales, far into the late nineteenth century, over three hundred years after their forgemen ancestors had come to Britain.[17]

The evidence we have of forgemen's migration is scrappy but it is consistent. Whereas the movement of forge workers in Sweden was intentionally restricted and in practice it was only apprentices who moved in large numbers, the movement of forgemen in Britain was regular and appears to have been undertaken by forgemen of all ages. Father-and-son teams, part of a wider cousinage that extended over several generations, were as characteristic of the British forge trade as they were of the Swedish. But in Britain father and son were more apt to move together. They shifted from site to site within the iron trade, moving on when charcoal supplies were interrupted, when the lease of a forge fell in, when plant had to undergo serious refurbishment, when disagreements arose between forgemen and an ironmaster, or when disagreements broke out among forgemen themselves ('Leonard had notice to give up his Finery ... and says he will never work where Coslett is any more').[18] Sometimes workmen simply wanted to be united with other family members: 'Tim Jenkins Father and Brother who work'd at Caerleon as master finer and Bloom maker are here', the manager of the Melingriffith forge reported in 1774, 'and say they are Entirely off from that Place and wish to be imploy'd here so that they might be together...'[19]

Nevertheless, the dynamics of family labour in the forge trade are largely mysterious. Nothing is known about household structure in ironmaking communities. Nothing is known about the economic activity of forgemen's wives and children. Even empirical evidence about recruitment and payment practices is scant in comparison with Sweden. However, some hypotheses can be ventured. It is clear, for example, that the dynasticism of the iron trade facilitated the mobility of labour. Ironworkers in widely separated areas of the country were known to one another; often they were related to one another; and they were immersed together in a common culture of work. A life punctuated by regular movement, often inter-regional, held no terrors. The dynasticism of the iron trade had a further positive attribute: it provided a framework for the transmission of information and rumour about the efficacy of new methods of working. Dynasticism supplied an environment in which incremental shifts in technique could be quickly communicated and generalized as forgemen moved between the iron districts of the British Isles. This may do much to explain the technological fertility of the British iron industry in the eighteenth century. Certainly, it may help explain the contrast between British and Swedish attitudes to the dissemination of technical knowledge. In Sweden the gathering and dissemination of knowledge tended to be centralized under the auspices of the *Bergskollegium* and the Ironmasters Association (*Jernkontoret*). Both organizations sponsored technical experts to travel to other important iron-producing regions of Europe and report upon new developments. It is for this reason that a succession of Swedish observers toured Britain between the late seventeenth and mid-nineteenth centuries. Equally, the *Jernkontoret* was responsible for a technical literature – scientific monographs on problems of metallurgy – for which there is no British equivalent. Indeed, the absence of a technical literature in English is quite striking. What descriptions there are of forge practice occur haphazardly in works of topography such as Robert Plot's *The Natural History of Staffordshire* (1686) or in miscellanies such as John Ray's *A Compleat Collection of English Proverbs* (1737). The best early descriptions of the puddling process are in Swedish and French.[20] Overall, there is little of substance before David Mushet's *Papers on Iron and Steel* (1841). All attempts to remedy this absence before the mid-nineteenth century were conspicuous failures. One such attempt was made by Joseph Dawson, managing partner of the Low Moor ironworks in Yorkshire in the last years of the eighteenth century. Dawson was an true Enlightenment figure, educated at Daventry Dissenting Academy and Glasgow University, a life-long friend of Dr. Joseph Priestley, and a scientific enthusiast. Accordingly, in 1800 Dawson tried to interest his fellow ironmasters in forming a technical discussion group as part of their regular deliberations on pricing policy. But Dawson was to be disillusioned:

> Some years ago, from a desire to have the Nature of making & Manufacturing of Iron better understood, I proposed to the Ironmasters of Yorkshire and Derbyshire at one of their Meetings that every person in rotation should read a short Essay upon some operation or upon the property of the Materials ... The proposal was accepted & I was requested to deliver the 1st Essay. I did & after that a 2nd but I could not prevail upon any other Person to contribute to our Instruction & Entertainment[21]

This was symptomatic. The collective organizations of British ironmasters focused resolutely on matters of pricing and marketing. In Sweden the *Jernkontoret* was far more active in fostering the take-up of new techniques. It had to be: the *Jernkontoret*'s mission was to counteract the isolation of the *bruk*. Forgemen who were immobilized on a particular *bruk* had, after all, only a limited opportunity to expand their repertoire of skills. By comparison with their British counterparts, they had less experience of shifting from region to region, accumulating in the course of a standard working life various forms of expertise which could be synthesized into qualitatively new techniques.

In Britain technological advancement was directed first and foremost by skilled forgemen, not by ironmasters. If skill was the 'property' of the workman during the period of manufacture, then forgemen were rich in property, a property from which ironmasters were largely excluded. Ironmasters were alert to this and the more daring of them dreamt of downgrading the skill and self-assurance of 'that proud Rogueish & Ignorant Sett of men Calld forgemen'. The problem with existing forge techniques, wrote one ironmaster in the 1780s, was that they would 'admit no other man to work at the Refinery but what have been Bred up to it from Their Cradles'.[22] His solution was twofold: firstly, to attempt a wholesale redivision of labour within the forge so that no one man could understand 'the whole Branch of Refining as anciently', and secondly to recruit common labourers unversed in the ways of iron and from districts with no tradition of metal working. This was no mere pipe-dream. Not the least of the reasons why Richard Crawshay, the greatest ironmaster of his day, sponsored the new technique of iron puddling in the 1780s and 1790s was the opportunity it seemed to offer of breaking down the craft control of long-established forgemen's dynasties. Crawshay told his managers at the Cyfarthfa works in Merthyr Tydfil to have the new methods taught to complete novices, recruited locally: 'teach the Welch your mode of making Iron ... the *created servants* of the vicinity will be the best security for peace and performance'.[23] One of the attractions of coal-fired technology for ironmasters was the promise it held of breaking with the dynastic organization of bar iron production. For that very reason it was viewed with mixed feelings by many forgemen who feared that 'it would entirely do away with the difficult process that had before been practised by them ... and thereby very much reduce their consequence, and wages also ...'[24]

INDUSTRIALIZATION AND THE DECLINE OF KINSHIP

Iron puddling and rolling transformed the production of bar iron in Britain in the generation after 1790. Puddling exploited cheap mineral energy, and rolling dispensed with the need to shape bars under the forge hammer. In all, this was a cheaper and far more expeditious mode of producing bars than had hitherto been used. The triumph of iron puddling is, of course, one of the staples of textbook history, but in the telling of that history nothing is said of the fate of the finers and hammermen whose skills were apparently superseded. Nor is anything said about where the great numbers of puddlers and rollers came from. Research on the British iron industry during this momentous period has concentrated on the history of technology rather than the social history of work, but it is beyond doubt that this dramatic departure in technology was accompanied by some sort of rupture in the social organization of iron production.

Having said that, the consequences of puddling within the forge trade in Britain are far from clear. It is plain that puddling, like earlier forge techniques, was and remained a highly skilled operation. The expulsion of the skilled workforce from the forge which certain ironmasters had hankered after came to nought. Puddling was assimilated into the repertoire of refining techniques of which forgemen were masters. The forgeman retained sovereign control of the forge until bar iron was superseded by mass-produced steel in the decades after 1860. On the other hand, the epochal expansion of the iron trade which followed the introduction of coal-based technologies required a massive growth in the numbers of forgemen. The 32,000 tons of bar iron produced in Britain in 1788, half by charcoal-fired methods, half by coal-fired methods, had grown to 100,000 tons by 1805, produced largely by coal technology. By 1827 national output had reached 354,000 tons.[25] In other words, bar iron production had increased more than tenfold in less than forty years. As has been stated, labour organization in the early coal-fuelled sector remains frustratingly opaque. One thing is certain, though: revolutionary though puddling was in many respects, it rested upon human muscle-power. The great increase in British bar iron production in the early nineteenth century depended upon an endless multiplication of puddling furnaces and hence an endless multiplication of puddlers.

In the present state of research we can point out three important areas where this new organization of iron production differed from the previous century: firstly, in the geographical distribution of the trade; secondly, in the creation of integrated production units, with blast furnaces, puddling forges and rolling mills combined on a single site, and on a much grander scale than formerly; thirdly, in a re-organization of production within these large integrated ironworks, one characterized by a greatly elaborated division of labour.

The geographical nature of the iron industry altered. Production was now centred on the coalfields, principally those of Staffordshire, South Wales, Shropshire and Scotland. For the first time iron was produced on a centralized, not a decentralized, basis. Forgemen may well have experienced regular changes of job, but these changes were now more likely to occur within regions rather than between regions. Occupational migration between regions still took place – how else could new centres of iron production, such as that in Cleveland in the 1850s, have emerged? – but this was increasingly an occasional rather than a systemic feature of the industry.

The spatial redistribution of the iron industry was accompanied by a spatial re-organization of production at plant level. The small and scattered production units of the old charcoal iron industry gave way to larger integrated works. The scale of employment changed in parallel. Whereas the largest forges of the mid-eighteenth century had employed no more than fifteen workers, the furnace–forge–mill complexes of the 1830s employed many hundreds of workers. Not only did the number of forgemen increase, they also became more differentiated in function. Puddling and rolling, from a very early stage, was marked by a far more complex division of labour than the relatively straightforward distinction between finers and hammermen which had characterized Walloon forging. The making of bar iron now followed an elaborate sequence of operations, involving the collective labour of refiners, puddlers, shinglers, heaters, cutters, bundlers and two teams of rollers, all with an array of underhands and boys.[26]

Where did all these iron workers come from? No research has been carried out in this area. Suffice it to say that the pre-eminence which the old forge dynasties had exercised since the sixteenth century could not – on a common-sense calculation – have survived the influx of hundreds of new forge workers in the early nineteenth century. This is not to say that kin groups were expelled from the iron trade. In any large industrial undertaking family links are likely to be found among the workforce. The situation in two integrated ironworks in Shropshire can be used to emphasize this point. In the 1830s the Old Park and Stirchley ironworks, established by the Botfield family in the late eighteenth century, consisted of seven blast furnaces, two forges with a total of thirty puddling furnaces, a number of different rolling mills, and even two charcoal-fired fineries. Whilst not a gigantic undertaking on the scale of the largest works in South Wales, these works were substantial. Here, clusters of interrelated forgemen could still be found. Of the twenty-three puddlers working at Stirchley in November 1834, five were called Parker, while four of the thirteen puddlers at Old Park were named Ellis.[27] Yet this was kinship as a *presence*, no longer as an *organizing principle*.

In Sweden the take-up of coal technology in Britain also had profound consequences. Puddled iron swept Swedish imports from the British market, or rather it restricted them to niche markets where high-quality charcoal iron, forged in the Walloon manner, was still esteemed. But whereas in Britain coal technology undermined kinship as the agency through which skill was nurtured and transmitted, in Sweden the impact of puddling and rolling had at first the reverse effect. There was a renewed attention to the importance of kin organization as a guarantee of quality. For quality was now the key to Swedish survival on the international market. German-forged bar (which accounted for nearly 90 per cent of Swedish exports) could no longer compete with British puddled iron in terms of price; therefore its quality had to be enhanced. How this was to be done was a matter of active debate among Swedish ironmasters during the crisis into which their industry was plunged in the post-Napoleonic period. A consensus arose that too many forgemen could not produce good bar iron. Everyone, it was stated in 1823, 'seems to agree that knowledge of bar iron making is in decay'. How, asked A.G. Tamm of the Ironmasters' Association six years later, was the 'the present scarcity of forgemen' to be remedied?[28]

The fullest analysis of the problem came from Carl Rinman. Rinman identified the resources which ironmasters needed if they were to remain viable competitors in the international market: pig iron of better quality, abundant supplies of charcoal, technically advanced forges, but above all skilled forgemen. And skilled workmen should ideally be knit together by ties of kinship. Rinman's diagnosis from 1829, quoted above, stressed the links between skills and kinship. The renewal of these links was urgently needed if the quality of Swedish iron was to be improved. Rinman therefore urged ironmasters to introduce a paternalistic social régime that would restore the stability of kinship within the *bruk*. Ironmasters should monitor the social and mental well-being of their workspeople, providing for the children's schooling and 'decent leisure activities'; they should encourage good work and an eagerness to learn, seeking to improve their workers' skills and dexterity. In economic terms, it was perfectly clear to Rinman that skilled forgemen were a valuable asset worth paying for and that they should be well rewarded.[29]

The redoubled emphasis on kin was part of a general refurbishment of the German forging technique. The lay-out of forges was altered, hot blast was introduced to hearths, and the organization of production was transformed with enlarged forge crews providing opportunities for a more rational pattern of bar iron making. This was done to effect: productivity increased, production costs fell and the quality of bar iron rose. In the 1830s and 1840s iron made by the German method began to be sold on the Sheffield market alongside Walloon-forged iron for the first time. German-made iron was never among the top brands, but it had gained a share of a market that had formerly been closed.

Thus, the Swedish iron industry clawed its way back into the international market by upgrading its traditional techniques and actually intensifying the rôle of kinship, rather than by embracing coal technology. The full rehabilitation of Swedish iron came in the 1850s and 1860s with the adaptation of new organizational arrangements, modelled on British lines but using charcoal fuel. The new technique, known as Lancashire forging, was, like puddling, characterized by a multiplication of refining hearths, a more elaborate division of labour (presided over by a designated supervisor for the first time), and an accelerated rate of production through the use of rolls rather than hammers for the shaping of bars. As with Britain, two or three generations earlier, production was centralized in integrated plants. These new forge-and-mill units were very much larger than the older forges which had used either the Walloon or the German methods. Admittedly, they remained small by British standards, but where eight to twelve forgemen had once worked, there were now more than 50, sometimes as many as 100 workers.[30]

In these circumstances kinship lost its earlier meaning as an institution for recruitment and the transmission of skill. The sheer number of forgemen in the Lancashire forges meant that one kin group was too small to support it with new apprentices and workers. It is also likely that the introduction of foremen meant that the forgemen lost some of the power they had previously exercised in matters concerning recruitment. In spite of this, old families and kin groups of forgemen remained in the forges. Kinship continued to be of importance, but as in the British case, in a diluted form. The Ekengren family continued to work at Hammarby in Gästrikland, as they had done when it was a German forge. But after the plant had converted to a Lancashire forge with eight hearths in the 1870s, they could barely control one hearth, let alone the whole forge as they had done half a century earlier.

CONCLUSION

Kinship or a dynastic organization of labour is by no means the only way of recruiting labour or of transmitting skill, but for three centuries it played a central rôle in both the Swedish and British iron industries. In part the clannish self-identity of many ironmaking families may have reflected their foreign origins. It is more likely, however, that dynasticism endured because of its value as a means of retaining and transmitting skill. Forgemen recognized the advantage of being able to confine knowledge of ironmaking techniques to their own kin. It was sure to enhance their bargaining position *vis-à-vis* their masters and rivals in the labour market.[31] This arrangement was not without its advantages for ironmasters. Dynasticism could underpin the intergenerational transmission of skill. Moreover, hiring a forgeman from a well-known family offered some sort of guarantee of competence. This was of particular value in the relatively fluid and nation-wide market for forge labour that operated in Britain. Yet dynasties of forgemen had somewhat different destinies in Sweden and Britain. As long as work in the forge remained a matter for the discretion of forgemen rather than of ironmasters, it was likely that the skills which forgemen exercised would be viewed ambivalently by their masters. Craft expertise from which ironmasters were excluded was as likely to be damned as the 'Ignorance and vile wickedness of forgemen'[32] as it was to be lauded. The old-established forge workforce in Britain, productive though it had been of many incremental technological advances, was not viewed with wholehearted enthusiasm by ironmasters. When coal technology provided the opportunity to dispense with, or at least thoroughly dilute, the workplace authority of charcoal forgemen, that opportunity was taken. In Sweden things were very different. The threat posed by cheap puddled iron on the crucial British market meant an intensification of the dynastic basis of production, as ironmasters and the authorities sought to enhance the skill levels in the Swedish iron industry and thereby to maintain some sort of position in the international market. It was only after two or three generations, with the full blooming of the Lancashire method, that the importance of kinship declined in the Swedish iron industry – with the adoption not of coal technology itself but of the principles of large-scale industrial organization.

NOTES

1. C. Evans and G. Rydén (1995-96), 'Iron in Sweden and Britain 1600–1850: Interdependence and Difference', in G. Magnusson (ed.), *The Importance of Ironmaking: Technical Innovation and Social Change* (2 vols), Stockholm: Jernkontoret, vol. I, pp. 407–14.
2. B.G. Awty (1981), 'The Continental Origins of Wealden Ironworkers, 1451–1544', *Economic History Review*, **34**, 524–39.
3. K.-G. Hildebrand (1992), *Swedish Iron in the Seventeenth and Eighteenth Centuries: Export Industry Before the Industrialisation*, Stockholm: Jernkontoret; G. Rydén and M. Ågren (eds) (1993), *Ironmaking in Sweden and Russia: a Survey of the Social Organization of Production Before 1900* (Opuscula Historica Upsaliensia, 12), Uppsala: Uppsala University Press.
4. A. Florén (1994), 'Le "Wallon" comme Métaphore: Quelques Réflexions Suédoises', in L. Courtois and J. Pirotte (eds), *L'Imaginaire Wallon: Jalons pour une Identité qui se Construit*, Louvain-la-Neuve: Fondation Wallonne Pierre-Marie et Jean-François Humblet, pp. 109–29.
5. B. Douhan (1985), *Arbete, Kapital och Migration. Valloninvandringen till Sverige under 1600-talet* (Acta Universitatis Upsaliensis), Uppsala: Uppsala University Press, chapter v.
6. K. Kilbom (1958), *Vallonerna. Valloninvandringen och den Svenska Järnhanteringen*, Stockholm: Tiden, pp. 374 ff.; S. Montelius (1959), *Fagerstabrukens Historia. Band V. Arbetare och Arbetsförhållanden*, Uppsala: Almqvist and Wiksell, p. 54.
7. Uppsala Landsarkiv, Forsmark Bruksarkiv, Avräkningsböcker, 1765–70, 1820–40, 1860–80.
8. Montelius, *Fagerstabruken*, pp. 53–4.
9. G. Rydén (1991), *Hammarlag och Hushåll. Om Relationen Mellan Smidesarbetet och Smedshushållen vid Tore Petrés Brukskomplex 1830–1850*, Stockholm: Jernkontoret. For a condensed version of this argument in English, see G. Rydén (1995), 'Iron Production and the Household as a Production Unit in Nineteenth-Century Sweden', *Continuity and Change*, **10**, 69–104.
10. See S. Montelius (1962), *Säfsnäsbrukens Arbetskraft och Försörjning 1600–1865*, Falun: Falu nya Boktryckeri AB, pp. 210–11, for forge dynasties at Säfsnäsbruken in Dalarna.
11. C. Rinman (1829), *Handbok uti den gröfre Jern och Stål Förädlingen*, Falun: Roselli, p. 247.
12. Rydén, *Hammarlag och Hushåll*, pp. 180–87.
13. For a general treatment of the British industry, see J.R. Harris (1988), *The British Iron Industry 1700–1850*, London: Macmillan. The literature on technological change in Britain is extensive: see C. Hyde (1977), *Technological Change and the British Iron Industry, 1700–1870*, Princeton: Princeton University Press. The social history of ironmaking in

Britain has yet to be written, although see for an overview C. Evans (1997), 'The Social Foundations of Iron Smelting and Manufacture in the British Isles 1500–1800', in D. Ebeling and W. Mager (eds), *Protoindustrie in der Region. Europäische Gewerbelandschaften vom 16. bis zum 19. Jahrhundert* (Studien zur Regionalgeschichte), Bielefeld, and C. Evans (1993), *The Labyrinth of Flames: Work and Social Conflict in Early Industrial Merthyr Tydfil*, Cardiff: University of Wales Press, for a study of one important centre.

14. Compare the surveys of the industry made in 1749 and 1788 (Science Museum Library [hereafter S.M.L.], MS 371/1, ff. 86–7) and in 1794 (Birmingham Central Library, Archives Division, Boulton and Watt MSS, MII/5/10).
15. J.-F. Belhoste, Y. Lecherbonnier, M. Arnoux, D. Arribet, B.G. Awty and M. Rioult (1991), *La Métallurgie Normande, XIIe–XVIIe Siècle. La Révolution du Haut Fourneau*, Paris: Inventaire Générale, 'Liste des Ouvriers du Fer Normands ayant Emigré en Angleterre', pp. 297–9.
16. In this, British forgemen differed from their French counterparts, who occupied two entirely discrete circuits, one in the east and one in the west of the country: J.-F. Belhoste (1994), 'Dynasties de Maîtres de Forges et d'Ouvriers du Fer (1600–1800). Origines, Renouvellements et Rôles dans le Transfert des Techniques', paper presented to Session C32 of the 11th International Economic History Congress, Milan, 1994.
17. Gwent Record Office, D314.13, examination of James Lavender, 3 July 1817. We are indebted to Grahame and Janet Lavender of Theydon Bois, Essex, and to Colin Lavender of Wolverhampton, for much additional genealogical information on this family.
18. Glamorgan Archives, D/D Mg 2/64, John Vaughan to Joseph Vaughan, 6 November 1783.
19. Glamorgan Archives, D/D Mg 1/14, Joseph Vaughan to the Melingriffith Company, 20 September 1774.
20. C. Rinman (1789), *Bergsverks Lexicon*, **II**, 413; A.H. de Bonnard (1805 and 1806), 'Sur les Procédés Employés en Angleterre pour le Traitement du Fer par le Moyen de la Houille', *Annales des Arts et Manufactures*, **23**, 113–51, 225–54, and **24**, 44–62.
21. S.M.L., MS 371/2, f. 525, Joseph Dawson to James Weale, 4 November 1809.
22. National Library of Wales, Bedford MSS (among the I.A. Williams MSS), 'Forge Rule Settled to Employ Carefull Labourers for forgemen', 26 March 1787.
23. S.M.L., MS 371/3, f. 192, Richard Crawshay to Henry Cort, 3 November 1787. Our emphasis.
24. S.M.L., MS 371/3, f. 214, Robert Thompson to William Cort, 19 February 1812. For a wider discussion of this see C. Evans (1994), 'Iron Puddling: the Quest for a New Technology in Eighteenth-Century Industry', *Llafur*, **6**, 44–7.
25. S. Pollard and R.S.W. Davies (1988), 'The Iron Industry 1750–1850', in C.H. Feinstein and S. Pollard (eds), *Studies in Capital Investment in the United Kingdom 1750–1920*, Oxford: Clarendon Press, p. 87.
26. The division of labour, particularly where it related to juvenile labour, is described in the report of the 1842 Royal Commission on Children's Employment: British Parliamentary Papers 1842 (381), **XVI**, 478–9.
27. John Rylands University Library, Manchester, BOT2/8, wage accounts for Old Park and Stirchley forges.
28. *Jernkontorets Annaler*, 1823, 138; A.G. Tamm (1829), *Försök till Besvarande af den af Herrar Bruks-ägare i Wermland och Dahl år 1827 Framställde Prisfråga*, Karlstad: Wallencrona, p. 4.
29. Rinman, *Handbok*, pp. 247 ff.
30. G. Rydén (1994), 'Gustaf Ekman, Jernkontoret och Lancashiresmidet – Ett Inlägg I Synen på Teknisk Utveckling', *Polhem. Tidskrift för Teknikhistoria*, **12**, 132–64.
31. D. Woronoff (1984), *L'Industrie Sidérurgique en France pendant la Révolution et l'Empire*, Paris: Editions de l'Ecole des Hautes Etudes, pp. 159–65.
32. National Library of Wales, Bedford MSS, 'Forge Rule...'

11. Why Change to Steam Power? Institutional Constraints, Risk-Aversion and Path-Dependence in the Swedish Sawmill Industry, 1850-1900

Alf O. Johansson

The diffusion of steam-powered technologies in the industrializing world of the nineteenth century was an uneven process.[1] The uneven development of industrial manufacturing, the improved efficiency of competing technologies, particularly water wheels and turbines, the different economic conditions governing various industries, technological path-dependence and a broad range of institutional and social factors all contribute to an explanation.[2]

This chapter presents the case of water versus steam power in the nineteenth-century Swedish sawmill industry. This industry was developed in response to foreign large-scale demand by industrializing countries and has been considered among the prime movers of nineteenth-century industrialization in Sweden.[3] From 1850 to 1900 the number of Swedish export sawmills tripled and their export volume increased more than twelve times (see Tables 11.1 and 11.2). As early as 1850 Swedish sawmill owners and technicians were familiar with complete production systems of steam engines, frame saws and edging machines, and the first steam engines were introduced at that time.[4] As late as the mid-1880s, however, water-powered technology was preferred by most producers (see Table 11.1).

The first aim of this chapter is to situate this late introduction of steam-powered technology in Swedish sawmills in its broad political and institutional context. Steam-powered technology made little sense until restrictions to free trade were lifted. Resistance to this technology also came from established industrial interests and the various institutional obstacles they represented.[5]

Secondly, it will be demonstrated that demand, conceptualized as a steadily growing willingness by foreign European markets to buy Swedish timber products, was not enough to pave the way for the introduction of the steam engine. To understand this, one has to analyse the multifarious and risky nature of the whole process of production and sale.[6]

We shall show that even when these factors were being brought under control, it took time to develop the technical system and the new division of labour which were needed for the steam-powered technology to realize its potential. The process of introducing steam was also delayed by the competition from successively improved water-powered technologies. The notion of path-dependence seems to cover these phenomena.[7]

The protracted process of substituting steam for water power was the overall outcome of a large number of individual trajectories. Particular sets of factors made particular groups of sawmill owners change technology at different points in time and to different degrees. The individual selection of technology was also the outcome of cultural and social values, power relations and entrepreneurial initiatives in various local contexts.[8]

Our discussion will be limited to the northern half of Sweden (the region north of the Dalecarlia river), called 'Norrland', which was the main export area for wood products.

THE NUMBER OF SAWMILLS AND THE GROWTH OF EXPORTS

Table 11.1 gives the growth in the number of water- and steam-powered saws from 1850 to 1900. These numbers indicate that the latter did not outnumber the former until the late 1880s.

Table 11.1. Number of Sawmills in 'Norrland' Grouped According to Source of Power, 1850–1900

Year	Water-powered saws	Steam-powered saws	Total number of saws
1850	106	1	107
1862	231	22	253
1871	211	61	272
1880	203	121	324
1890	151	185	336
1900	54	230	284

Source: Wik, *Norra Sveriges Sågverksindustri*, pp. 85, 98, 115, 131, 314. (See n. 11 below.)

It is quite easy to demonstrate the rough correlation in time between the massive introduction of steam-powered saws and the rise in the volume of export of timber products (see Table 11.2). Export of sawn products and handcut beams increased steadily but modestly from 1750 to 1830. From 1830 to 1846 the sale of deals and boards then tripled while that of handcut beams increased no less than seven and a half times, achieving a 46 per cent share of total exports. After the economically turbulent 1850s, the upward trend for sawn products continued at a slower pace, while the share of handcut timber began to decrease. The international boom years of 1869 to 1874 then brought an impressive growth also to Swedish timber exports. In that period it grew by 60 per cent; numerous new sawmills were established and many old ones were modernized and enlarged.[9]

Table 11.2. Export Volume of Swedish Sawmills 1830–1900 (1000s of standards)[10]

1830	1840	1850	1860	1870	1880	1890	1900
30	65	88	200	470	620	860	1000

Source: Söderlund, *Svensk Trävaruexport*, pp. 36, 166-7.

In the mid-1870s sawmills were hit hard by falling foreign demand caused by the Great Depression. Stiffened competition from Finnish and Russian exporters lowered prices and profitability during the 1880s. Bankruptcies were numerous and the whole industry was restructured. The survivors were large-scale producing companies controlling large forest areas and floating systems which guaranteed timber supplies and reasonable timber prices. The mills of these companies increasingly began to use steam-powered technologies and became dominant from the early 1890s onwards. From the early twentieth century we can divide sawmills into two groups: large-scale export producers located along the coast, and a reduced number of predominantly small mills situated in the interior of the country. Competition forced the small mills to switch from export to more flexible production strategies for domestic markets.[11]

Apparently there was a rough connection between the export pattern and the large producers' introduction of steam technology. But why did Swedish exporters lag behind those in Norway, who had responded to the demand for wood products in the industrializing countries of Europe much earlier,[12] and why did a majority of sawmills refuse to invest in steam technology as late as the mid-1880s?

THE SLOW DISMANTLING OF INSTITUTIONAL BARRIERS

Institutional factors go some way to explaining this late response to international demand. For centuries, the extraction of timber had been limited by state regulations protecting the ironworks' demand for charcoal. Restrictions on production volumes for sawmills were not lifted until 1842, and those against the founding of new mills not until 1863. It has been the unanimous view of Swedish historians that these regulations significantly retarded exports until after the mid-1860s.[13]

Another important barrier was the control by the state over vast forest areas, especially in the north of Sweden. The general view is that restrictions against private use of state forests did not cease to be a problem until the 1850s. After decades of parliamentary debate, vast areas of virgin and cheap forest lands were then generally opened to private exploitation.

A basic prerequisite for large-scale export was the lifting of trade barriers, most notably the successive reductions of duties on imported timber products in Britain during the years 1842–1866. Not until the mid-1850s were these duties low enough to make Swedish sawmills competitive with Canadian producers, and Britain became the most important export market for the Swedes. The second-largest buyer, France, also lowered its duties and allowed free importation of timber products in 1865. Not until the late 1860s did the combined effect of these deregulations and of economic growth create an atmosphere in which a considerable number of Swedish merchants and sawmill owners began to consider large-scale exports and the introduction of the steam engine. Yet, for reasons discussed below, for another decade only a few dared to take the step.[14]

Lastly, another important obstacle was the ironworks' control over waterways and forests. Since the seventeenth century the ironworks had been located at the mouths of rivers and streams and at their inland waterfalls, which made it difficult for independent sawmill owners to operate. Even where forests were owned by peasants, permission from, or co-operation with, the ironwork owners was needed for the floating of logs and the running of water wheels. The ironwork owners had their own networks for trade and finance, including harbour constructions, merchant houses and sale and export agents. Socially and economically, these were powerful counterforces to newcomers. In certain districts ironwork owners played an important rôle by establishing their own sawmills, but these were mostly of secondary interest to them as complements to iron production. Through networks and partnerships with traders and merchants in the iron trade, they often acted to limit the scope and influence of new sawmill interests in their forest areas.[15]

In Finland and Norway, too, a similar and contemporary chain of events took place.[16] As demonstrated in Francis Sejersted's contribution to this book, political obstacles were important to delay the whole industrialization process in mid-nineteenth-century Norway. The political process of the day reflects the contradictory interests of new entrepreneurs and established interests in the timber trade, with different outlooks and different attitudes towards social and economic development in general, environmental considerations, and the full play of free-trade capitalism and modernization.[17]

ECONOMIC UNCERTAINTIES

Large-scale response to foreign demand required the buying of forest stands or of the rights to cut trees in forest areas belonging to others, as well as the extremely costly cleaning of rivers and construction of floating systems deep into the country. The mere size of such projects, and the general lack of investment capital in society, made most sawmill owners hesitate. Not until 1864 was the first Swedish business bank established, and only to a minor extent did foreign capital enter this risky trade. The traditional merchant houses favoured the iron industry and were reluctant to go into sawing. Although joint-stock companies were allowed by law in 1848, there were only 13 such sawmill companies in 1870. Their main purpose was not to attract capital but to protect owners from financial losses. The normal way for ordinary sawmills to get capital was to accumulate their own surpluses.[18]

Another important aspect to consider is the composition of total costs. Table 11.3 illustrates the main elements of total production costs for a medium-sized export producer. The share of costs for cutting the logs at the mill was only a minor part of the total, while the main part was for raw materials, including the buying of forests, the cutting of trees and the floating and cleaning of rivers. This means that technical improvements in the mill could lower the total costs of the producer only to a minor extent.

Table 11.3. Percentage Shares of Main Costs of Production at the Stocka Sawmill, 1868–1893

Year	Timber logs	Sawing	Yard work & loading	Miscellaneous
1868	70%	16%	------------14%------------	
1878	74%	11%	9%	6%
1888	72%	12%	10%	6%

Source: Johansson, *Arbetets Delning*, p. 102.

In addition, dramatic price fluctuations were recurrent. To take some examples, in 1854 average prices fell by 30 per cent. A slow rise from 1859 was followed in 1866 by an abrupt fall in the range of 20 to 25 per cent. From 1869 to 1874 prices increased by 50 per cent. The Great Depression suddenly reduced prices by 20 per cent in 1874–75 and by another 20 per cent before 1880. During the 1880s, average prices were more stable, although in single years falls or rises were considerable on specific dimensions and qualities. On top of this came fluctuating interest rates and freight costs. It was no surprise that bankruptcies were frequent in this business.[19]

The supply of labour was not a major problem before the 1880s and 1890s. Large numbers of workers were needed during the spring and summer months, when rivers allowed floating and ships could sail the Baltic Sea and be loaded at the ports. Due to a very strong population growth, the dissolution of the old villages, new production methods and inheritance rights which favoured a more labour-saving agriculture, large numbers of landless seasonal workers were pushed onto the labour market after the 1850s.[20] Yet, also, labour could be a recurrent cost problem for employers. Costs rose when new rivers and streams had to be cleared to new forest stands. Regular production, too, frequently required costly extra undertakings when weather and temperature made work extremely difficult in the yard, in forests and along rivers. Costs rose dramatically when expansion could not be achieved without massive reconstruction work along rivers and lakes. In periods of rapid growth of production, the construction of new dwellings and the establishment of a local infrastructure for the additional labour force also became necessary, especially in remote areas. To become a large-scale producer in this trade was no doubt both a costly and a risky undertaking.[21]

For the individual employer, the growing markets of the 1850s and 1860s did not automatically signal what to produce, and for whom. It was not even clear that sawn products were to be preferred, when, in the 1860s, markets for unsawn 'Dutch timber' in the Netherlands, for 'pitprops' in Belgian mines and for handcut beams in Britain grew faster than markets for sawn products. But even if the share of handcut beams for construction and building in European urban centres went down from the early 1860s, its volume continued to increase into the 1880s. As 'rough-sawn deals' remained the most important item for joiners and dealers in Britain and the Netherlands, a strategy of large-scale mechanization and specialization was not undisputed. The economic problems and uncertainties of this trade meant that most sawmill owners continued to meet the shifting opportunities of markets by risk-aversive short-term behaviour as seasonal producers of both sawn and handcut products.

In the late 1860s, however, the relative decline in prices of beams, and the demand from German, French and British large-scale buyers for more diversified qualities of sawn products and splitwood,[22] indicated that changes were under way.[23] To illustrate, while the number of sawn dimensions in stock for one producer was 10 in 1862, it was 30 in 1875. For the same producer the number of qualities were 2 in 1864, 4 in 1878 and 7 in 1891. The number of various sawn products (dimension × quality × fir or pine) was 48 in 1864, 120 in 1878 and over 200 in 1898. The share of splitwood increased from 4 per cent in the early 1860s to 12 per cent in the late 1870s.[24]

To concentrate on producing this broad variety of fine sawn products was the new strategy, and in the 1860s it was still quite hazardous. It was, however, slowly and gradually facilitated by the growing need among large-scale buyers to have more stable relations with producers with a variety of products in stock and who could keep agreed delivery dates.[25] From the mid-1870s, such relations became organized by specialized timber merchant companies working through networks of agents abroad. In 1875–76 the National Society of Sawmill Owners and Timber Traders was created, with the stated aim of having producers agree collectively upon output restrictions and standardized rules for sale.

To conclude: the basic conditions for large-scale mechanized production and the introduction of steam-powered technology were not in place until the late 1870s, when the institutional barriers had been eliminated and the economic uncertainties reduced.[26]

TECHNICAL ADVANTAGES OF STEAM ENGINES

Contemporary handbooks repeatedly emphasize the technical advantages of steam power for sawing. Steam would free sawmills from the harsh constraints of nature, in particular ice and the seasonal variations in water levels. Ice made water-powered sawing impossible from December to April, and the summer and early autumn drought did so from August to November. The actual sawing period normally lasted for 3 to 4 weeks along the smaller streams and about 3 months along the main rivers. Traditional water-powered sawing also involved costly inland investments, as deals and boards had to be piled up twice: first at the sawmill located in the inland and later at the port before loading onto the ships. Timber yards had to be constructed and yard workers to be paid at both ends of the flow. Costs were high for the handling of sawdust and of huge amounts of waste around the mill, and for long-distance transport of finished products. Damage to finished products was also a recurrent problem when they were floated on the streams.

Considerable economic losses resulted when floated goods were lost, or when the wood darkened and had to be sold at a reduced price. To avoid reclamation, boards and deals often had to be hand-washed. These problems increased dramatically in the 1860s and 1870s when markets demanded a wider variety of dimensions and qualities.

Technically, steam-powered sawing could solve the problems of season, water, ice and inferior qualities. Sawdust and waste could be used as fuels. Sawing could be done during most of the year and costly transports of finished goods could be avoided. Sawing, drying and loading of ships could be concentrated at the ports and systematically organized. In addition, by varying the size and technical construction of steam engines, the speed limits of water wheels, set by the amount and velocity of the running water, could be overcome.[27]

Yet for a long time these technical advantages did not convince sawmill owners to change motive power. According to the technical handbooks, the fuel needed by the steam engines was no problem, but in practice it was. To use wood waste and sawdust was not an obvious choice for the individual sawmill, because, until the late 1880s, this waste offered the traditional ironworks an inexpensive raw material for making charcoal. Sawdust was even used to cover the kilns. When the Stocka sawmill, controlled by the nearby ironworks of Ströms Bruk, was built in 1856 and modernized in 1871–73, such considerations were behind its choice of a water turbine instead of a steam engine. Increasing market demand for smaller sawn dimensions and the need in the depressed 1880s 'to use everything' of raw materials made competition for fuel still more acute. Also, from the late 1880s the growing pulp industry needed waste and smaller dimensions of trees. The Stocka sawmill installed a steam engine as a complement to its turbine in 1874, but removed it altogether in 1900, when the company expanded its pulp production. Competition for fuel was less acute in districts rich in forests and where ironworks were few, but even there the individual sawmill often found it profitable to sell its waste rather than to invest in a steam engine.[28]

The steam engines of the 1870s and 1880s were more solid and powerful than the earlier ones, which could drive only one or two frame saws and their edging machines. To achieve the full benefit from their potential, however, a range of complementary equipment was required. In the 1870s, the saw blades were still being fitted into wooden frames, which made sawing uncertain. Stable iron constructions belonged to the 1880s. Higher throughput, in combination with smaller dimensions and increased demands on accuracy, also required thinner, stronger and more accurately ground saw blades. In the late 1870s, hand-grinding gave way to grinding machines, and the use of petroleum for lubrication and of kerosene for lighting (both instead of vegetable oils) made breakdowns less frequent and allowed accurate and rapid processing.[29]

In the late 1880s to 1890s one single steam engine could run timber conveyors, ten to twelve frame saws, five to six edging machines, splitting machines, cutting machines, lath saws, sawdust conveyor belts, and so on. By the use of wire systems, deals, battens and boards could be pulled on rails to the timber yard, where the steam engine also powered cutting saws and conveyor belts through various mechanical devices. Steam could be used for heating and drying, and electric power for lighting could be generated by a dynamo.[30] Such large-scale mechanical systems were, however, the outcome of a protracted adaptation to oscillating demand and prices and of the piecemeal innovation of complementary machinery and equipment. The extreme ups and downs of the yearly sales figures of the leading producer of various wood-working machines in Sweden illustrates the precarious nature of this process (see Figure 11.1).

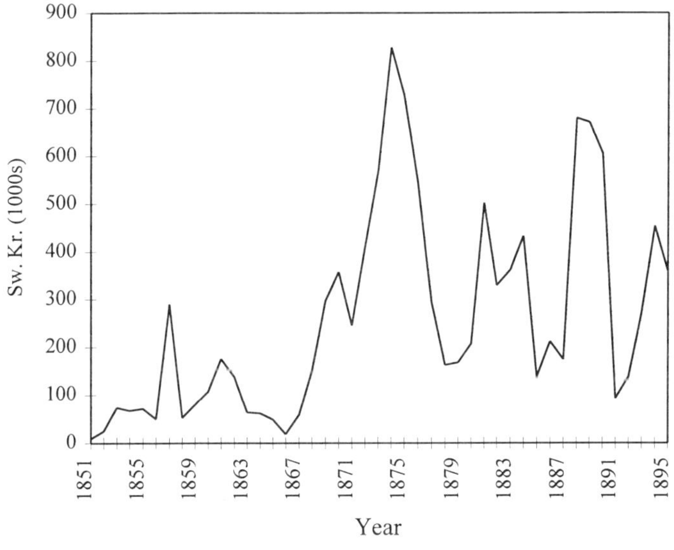

Source: Gårdlund, *Bolinders*, pp. 255–6.

Figure 11.1. Sales of Wood-Working Machinery (including Steam Engines) of the Bolinder Mechanical Workshop 1851–95

Large-scale mechanized production also depended on the emergence, after the 1870s, of a more detailed division of labour and specialized skills to handle the new work tasks and to meet the more specified demands of the buyers. The permanent access to skilled master sawyers, edgers, oilers, grinders, cutters, and so on, throughout the year meant permanent employment of these workers and the costly construction of a village around the mill. Furthermore, to run large-scale sawing profitably required the floating during the concentrated flood period in March and April of an uninterrupted flow of timber into the sawmill, and the handling of large amounts of dimensions and qualities in the yard, which required hundreds of workers. Uncontrollable problems associated with rain, snow, cold, flood, lack of water, and so on, were frequent complaints. Frozen water basins at the mill and ice in the harbour recurrently made both sawing and the loading of ships impossible. When deals and boards could not be covered and piled up for drying, considerable losses in quality and price ensued. To a certain extent, these problems were aggravated by the higher throughput of more advanced sawing systems, which created new bottlenecks at the various stages of the production flow. Artificial drying and heating systems, and mechanical devices for the handling of both logs and sawn products, were not generally introduced before the 1920s. Before that time interruptions in the production flow were frequent and the theoretical sawing period of 12 months was in practice generally reduced to 9 to 10 months.[31]

Lastly, large-scale production based on steam power involved risks of explosion and fire. From figures presented by Hellström, it can be estimated that out of 62 steam-powered saws 23 were totally devastated by fire on one or more occasion before 1920. Most fires started around the steam engine and many of them spread to the timber yards.[32] Before the official Steam Commission began to control steam engines in the mid-1880s, many sawmills suffered from considerable losses due to fires, and others hesitated to install the dangerous machine.

Thus, in practice, continuous large-scale production in general, and the introduction of steam engines in particular, were less obvious choices than the technical arguments of handbooks might lead us to believe. Even if demand became more predictable after the mid-1870s, competition for fuel and raw materials and the sudden shifts in the pattern of costs made the average sawmill owner first of all consider flexible and low-cost alternatives to expensive and more risky large-scale systems. These created new technical and economic bottlenecks and required the development of complementary equipment and new skills which could only be achieved over time. Was there a third way, an alternative both to traditional small-scale production and to large-scale introduction of steam-powered technology?

THE WATER TURBINE

Scandinavian historians have almost totally neglected the water turbine as an alternative to water wheel and steam technology. According to a recent study by Nerheim, however, from the mid-nineteenth century

> water turbines increasingly began to supplant water wheels in central Europe, North America and Scandinavia in regions rich in water resources. The turbine, running faster than the water wheel, required less transmission and was easier to regulate. Above all, the turbine had a higher efficiency, and produced more power from the water available ... Power needs combined with abundant water resources led many entrepreneurs to install water turbines instead of steam engines ... Around the mid-nineteenth century it seems that no special skills or expertise were required to build turbines other than those that most machine shops had at their disposal.[33]

Unfortunately, lack of sources makes Nerheim's empirical evidence meagre for the period before the wood-grinding industries emerged in the 1870s.[34] Yet it is possible to claim that there was a strong presence of turbine technology in Swedish industry and that it contributed to the delay in the introduction of steam engines. We know that Swedish and Norwegian mechanical workshops were producing turbines for various industries in Scandinavia by the late 1840s. Among the buyers were several sawmills. From the late 1860s to the mid-1880s turbine-powered technology made great progress, and sawmill owners could choose between various domestic and foreign constructions. In particular, turbines of the Jonval and Girard types seem to have been extensively used in Scandinavian industries of this period. Successive improvements in the pressure-regulating mechanisms made them particularly suitable for rivers characterized by seasonally varying water levels.[35]

There is also empirical evidence of combinations of turbine and steam-powered technology. As already mentioned, the Stocka sawmill in Hälsingland was constructed with a turbine-powered saw in 1856, but changed to a combined turbine-and steam-powered saw between 1874 and 1900; the turbine was used for three of the frame saws and the steam engine for the additional two. From 1900 Stocka relied only on the turbine to run all the frames and from 1920 new turbines were installed to produce electricity as the new motive power. A similar case was the larger Ljusne sawmill in the same district. It was run by a turbine from the early 1850s, subsequently combined turbines and steam, with 12 frames run by a turbine and 8 by a steam engine, switched to steam-powered sawing in 1881 and finally, in 1917, converted to electricity.[36] According to a handbook on Nordic wood industries, such combinations of technologies were common in several medium-sized sawmills as late as 1913.[37]

The presence of turbines can also be verified indirectly. In the literature, the construction, as late as the 1880s, of large 'modern water-powered saws', which could operate for six to eight months of the year, is explained by their ability to use modern frames and sawing equipment, by specific and fortunate geographical conditions, tradition and culture, the need not to waste capital already invested, etc. It is also said, however, that these 'modern' water-powered saws were situated near the coast in areas without natural waterfalls.[38] The conclusion to be drawn is that turbines made these sawmills competitive in such areas. An important factor making them competitive was the use of traditional skills in regulating and collecting running water through damming constructions in lakes and along rivers. The above-mentioned Stocka and Ljusne turbine-driven sawmills were situated along the coast, and used a system of canals and dams to collect the water before it was led through an iron pipe to the turbine under the sawmill.[39]

A number of examples demonstrate that best-practice turbine saws could be technically and economically competitive with steam-powered saws during the 1860s to 1880s. In 1856, Stocka's turbine yielded 75 hp, while 45 to 50 hp was the norm of early steam engines. The steam engine installed at Stocka in 1874 yielded 40 hp compared to the 75 hp of the new turbine installed in 1888.[40] Another case is the Bergvik sawmill, which was constructed in 1872–3 with three turbines capable of running the most modern frame saws at the speed recommended by the producer as their maximum: 170 strokes per minute.[41] In terms of power, speed and length of yearly operation, the turbines of the 1860s to 1880s seem to have been competitive with steam engines. In 1863–4 the turbine at the Stocka sawmill was kept running for 5½ months, and in 1868–70 this period was extended to no less than 9 months. Not until the 1870s did such long sawing periods become general for steam-powered saws.[42] Sällström-Nygren has made a comparative study of the accumulated costs of producing one standard of finished goods by several large water- and steam-powered saws, which shows that the former were competitive as late as the mid-1880s.[43]

Thus, two of the basic arguments in favour of the steam-powered saws, their longer periods of operation and their independence of river transport for finished goods, seem to be much more relative and conditional than the technical handbooks admitted. Empirical evidence does not invalidate the impression of a general victory for steam engines in export sawmills around 1885–90, but it strongly supports the view that turbines could be competitive until that time. What, then, made them lose out to steam? In most cases the supply of water seems to have caused trouble as production was extended after the 1870s. Only some of the large water-powered sawmills along the coast were in the fortunate geographical position of having enough water throughout the whole year. Turbines were still sensitive to extreme flooding during the spring and often had to be repaired. This was the case at the Stocka sawmill, where the flood 'destroyed' the lower sections of the sawmill on several occasions.[44] More fundamentally, beyond a certain point investments in large turbines became too expensive. To use the potential of larger and stronger turbines meant the large-scale collection of water and the extremely costly construction of large dams and high waterfalls at the sawmill. Often it also meant expensive damming constructions far inland. In short, although technically the steam engine was not necessarily superior, beyond a certain size and duration of production water-powered technology in most cases became economically disadvantageous.

EXPLAINING GEOGRAPHICAL DIFFUSION

The geographical diffusion of steam- and water-powered saws respectively illustrates the impact of the economic, institutional and technical factors discussed above.[45] During the second half of the nineteenth century the sawmill industry moved from the interior to the coast. To a large extent this was the result of the closing down of the smaller water wheel-powered saws in the interior and of the construction of sawmills run by steam engines at the river mouths. However, the maps also suggest that the remaining and technically improved water-powered saws moved the same way (see Figures 11.2 – 11.7). The main explanation of this is the presence of competitive turbine technology.

A special case is the region of Jämtland. For obvious geographical reasons and due to its long winter, this was a district without a market-oriented sawmill industry until the 1880s, when the railway between the Swedish east coast and the Norwegian city of Trondheim offered steam-powered sawmills new conditions. In the interior of the Hälsingland region in the south, the railway offered similar opportunities.

In the most northern district, Norrbotten, the steam engine came relatively late, to a few large coastal mills, and traditional inland saws remained predominant for a long time. Relatively short sawing and loading periods, due to climate, and specific institutional conditions, notably the small number of ironworks, largely explain this structure. When the radical shift came in the late 1880s, it was mainly due to the change from using sailing ships to using large steamers able to travel in icy waters in spring and autumn. This considerably lengthened the sawing period.

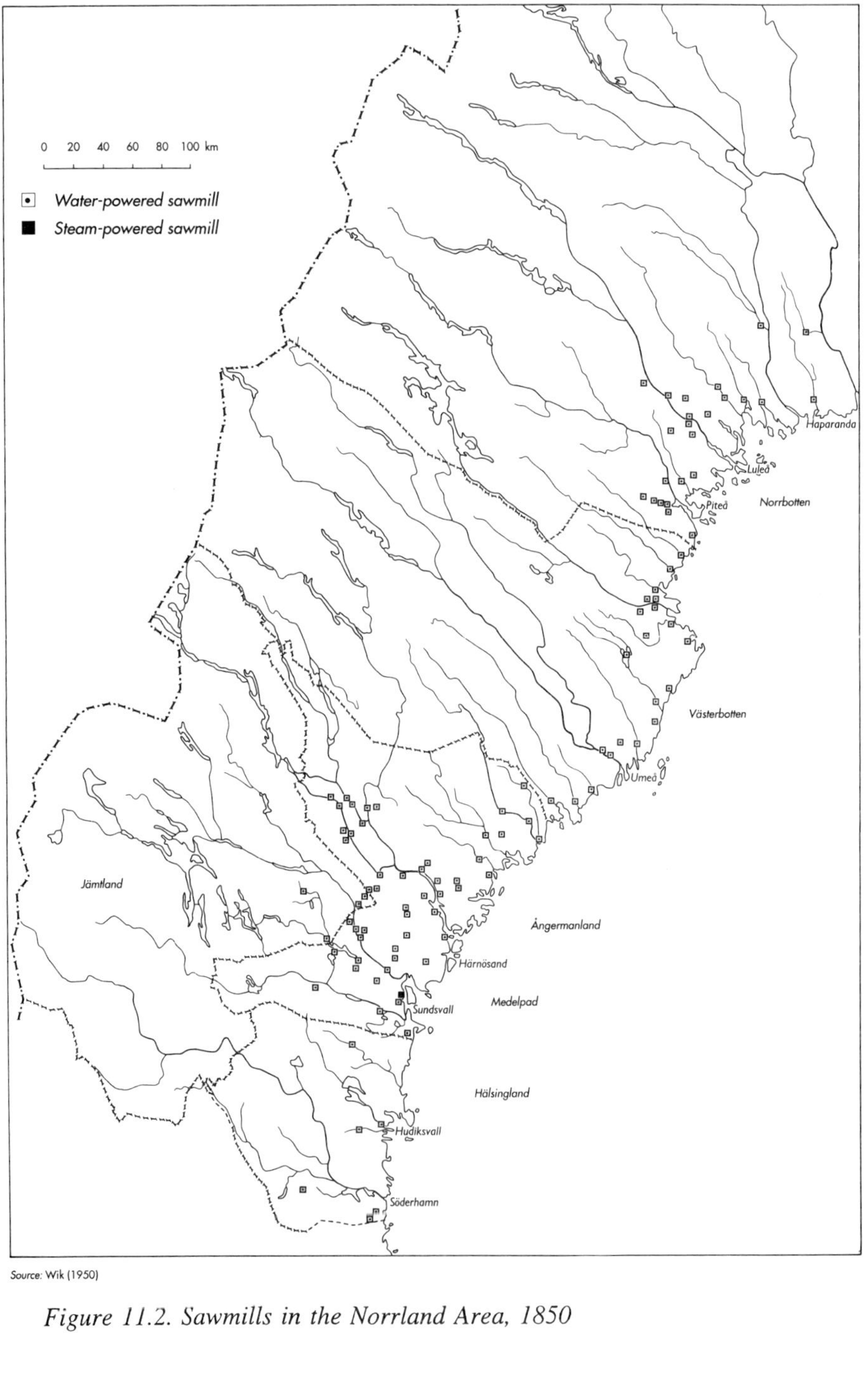

Source: Wik (1950)

Figure 11.2. Sawmills in the Norrland Area, 1850

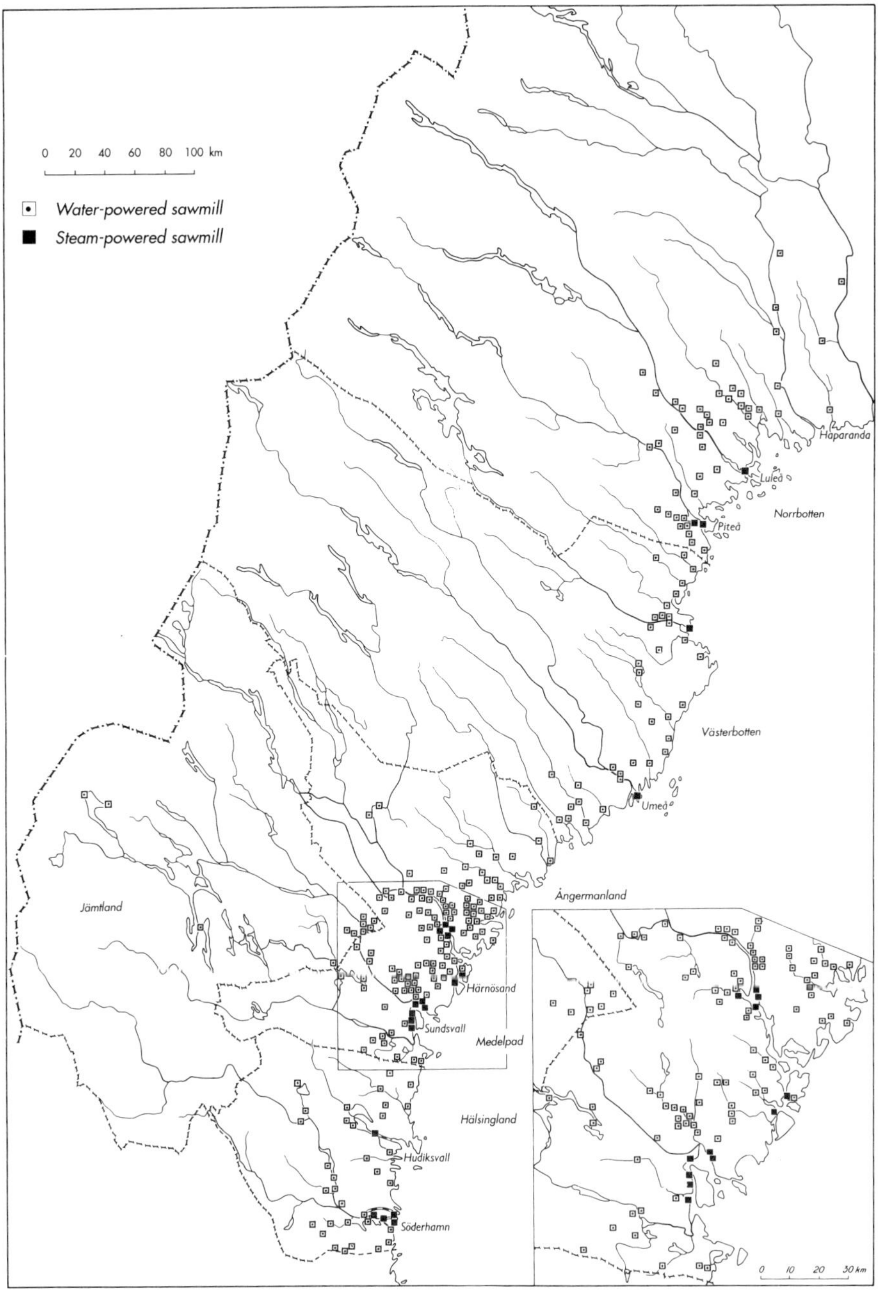

Source: Wik (1950)

Figure 11.3. Sawmills in the Norrland Area, 1862

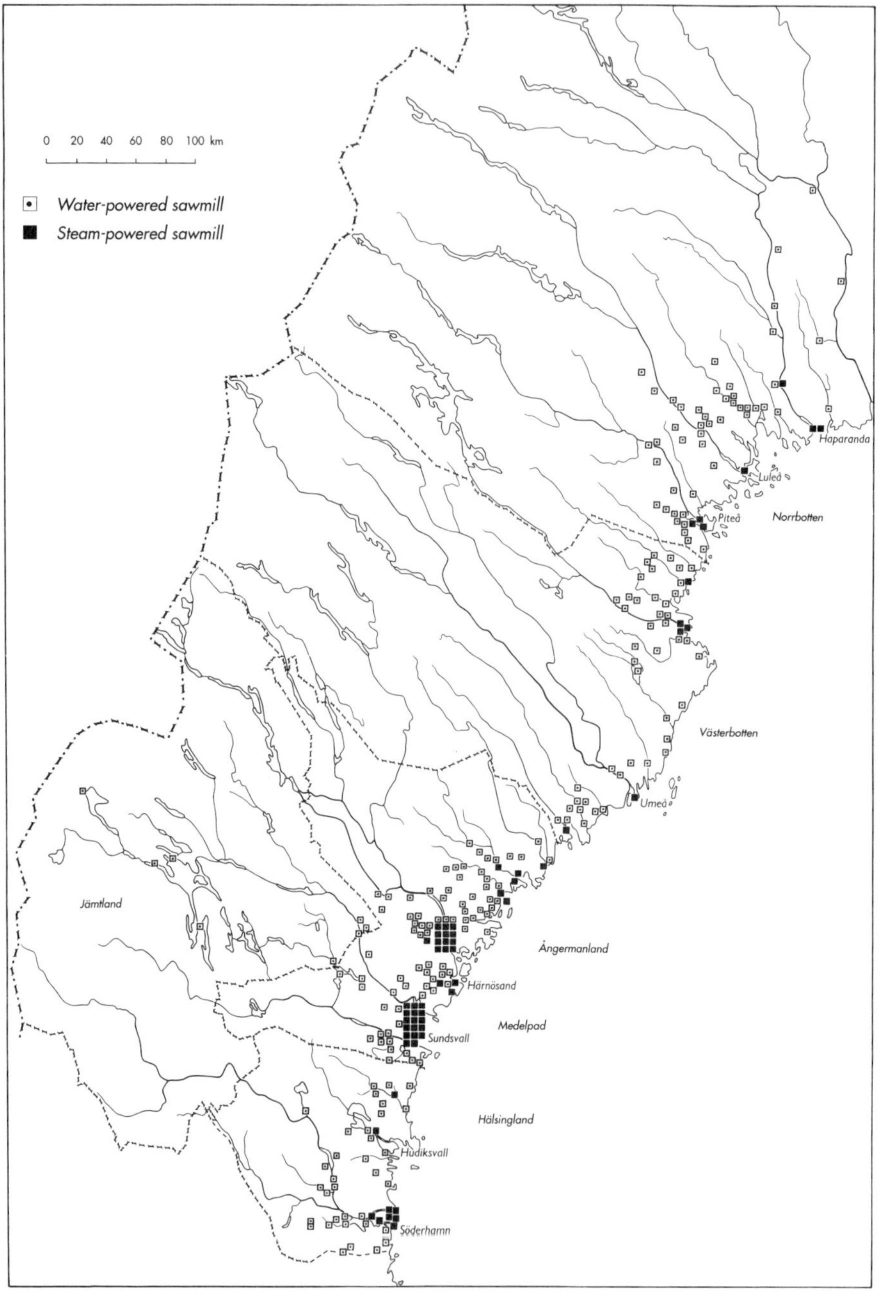

Source: Wik (1950)

Figure 11.4. Sawmills in the Norrland Area, 1871

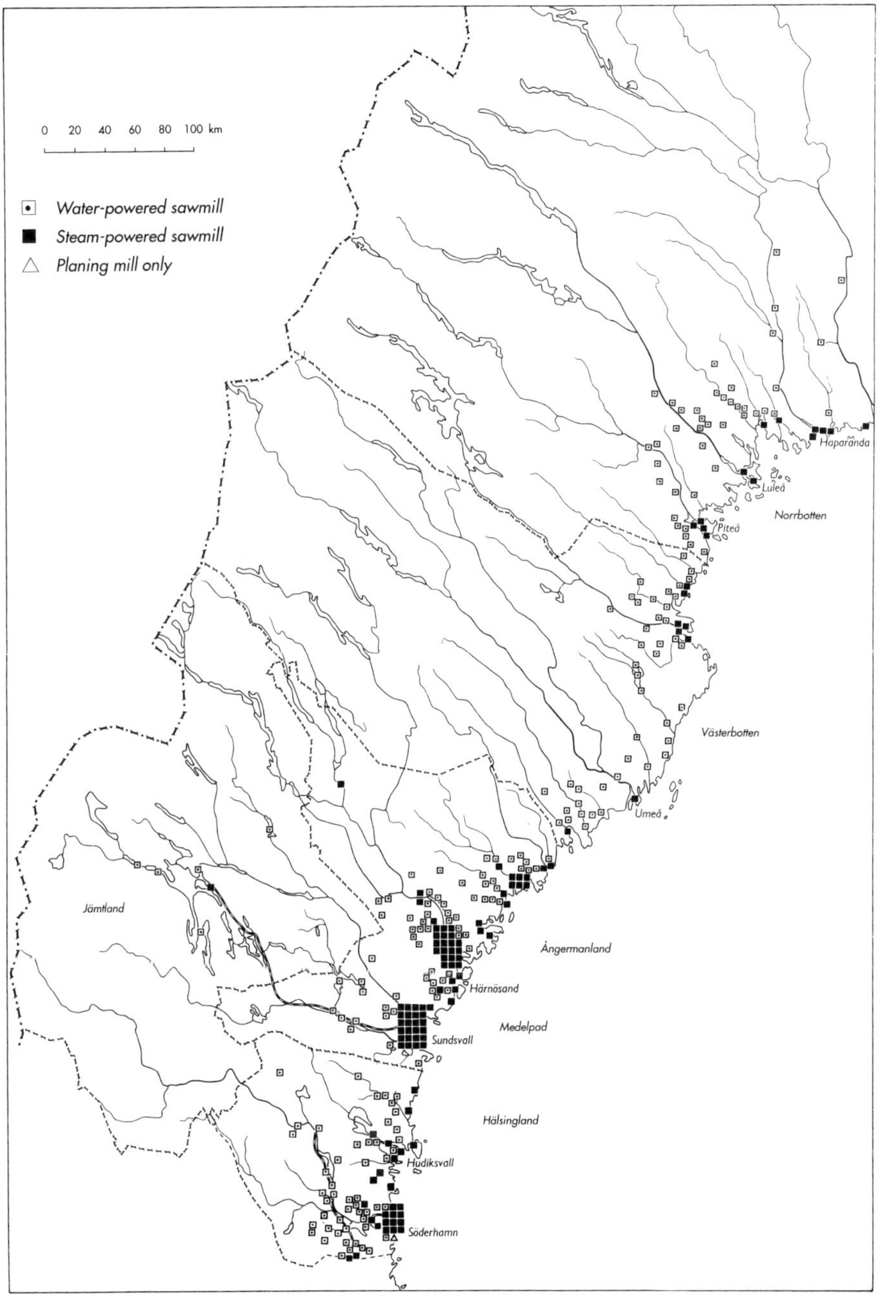

Source: Wik (1950)

Figure 11.5. Sawmills in the Norrland Area, 1880

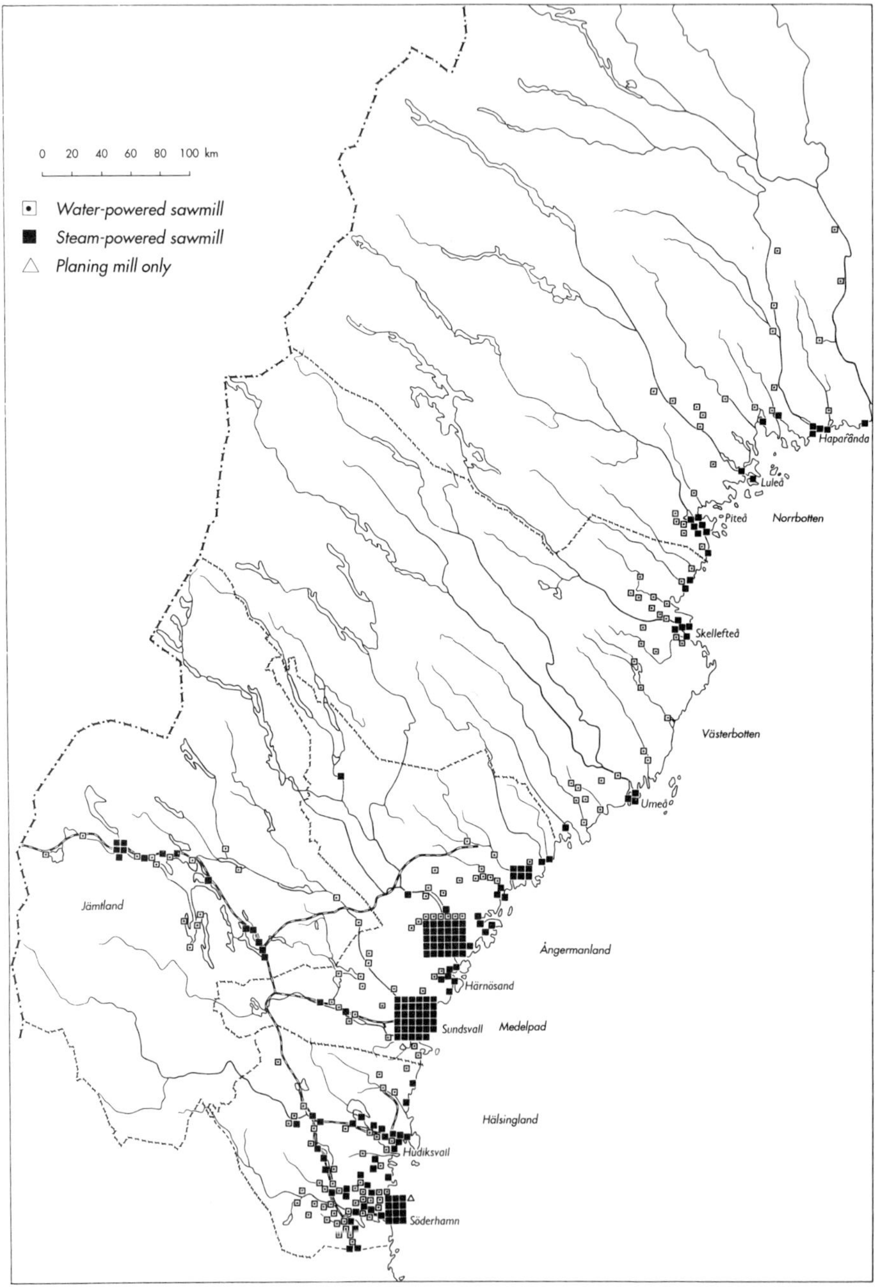

Source: Wik (1950)

Figure 11.6. Sawmills in the Norrland Area, 1890

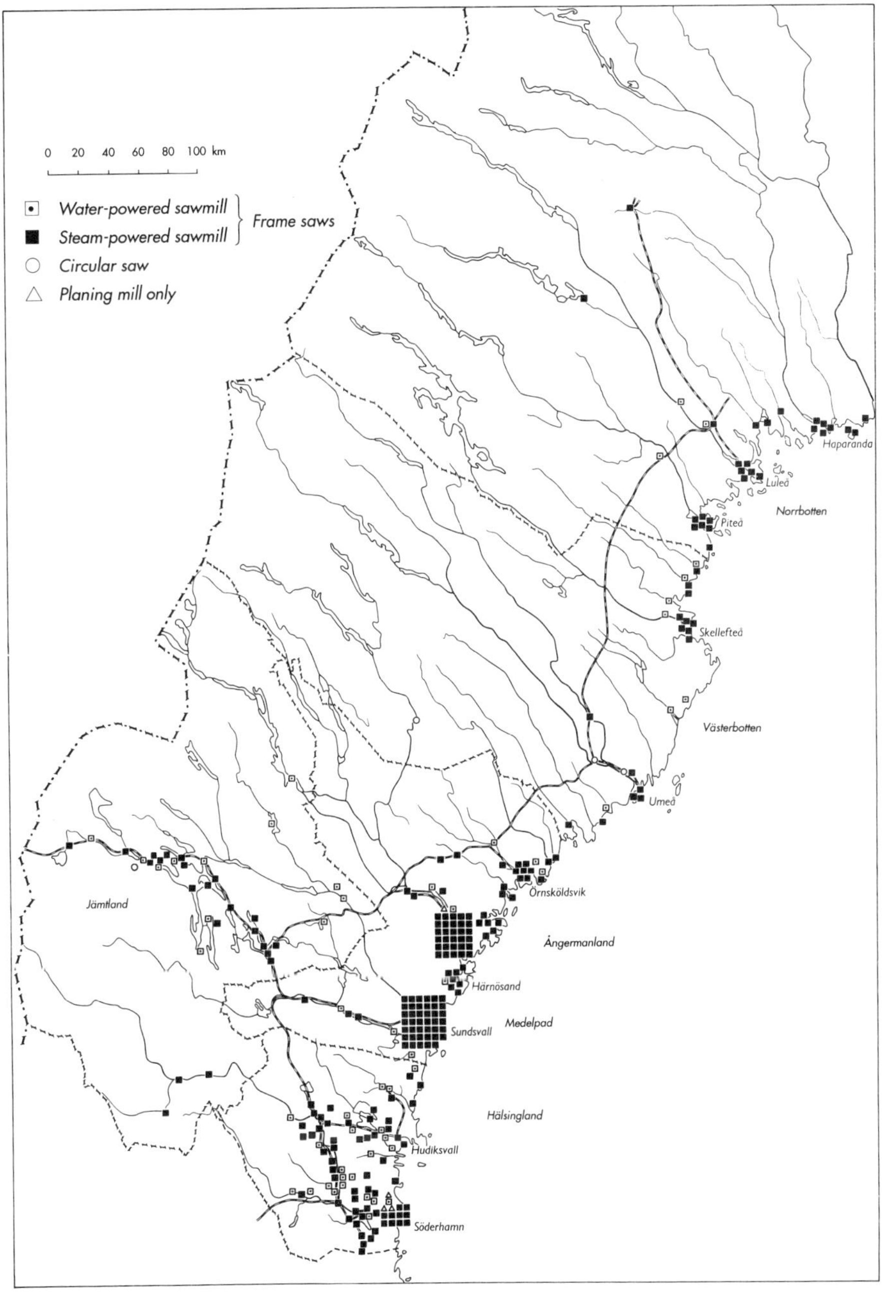

Source: Wik (1950)

Figure 11.7. Sawmills in the Norrland Area, 1900

The Västerbotten district was also special in several respects. This district had no old traditions of sawing or of iron production. In the 1860s to 1880s it developed a specific structure with quite large water-powered saws along the coastline. The location of several waterfalls close to the coast has been the traditional explanation. More important, however, was the increased use of improved turbines capable of running large sawing systems during considerably longer sawing periods, even in areas without natural waterfalls.

In terms of the number of mills, their size and the volume of their exports, the most important regions were those of Ångermanland and Medelpad. As early as the 1860s, a considerable number of traditional saws had to close down, and the change to steam-powered technology was more definite than elsewhere. The specific combination of large, deep rivers, rich forests, and the scattered structure of a limited number of ironworks made the co-existence of various groups of owners and interests less problematic than elsewhere. This district became the leader in the large-scale application of steam-powered technology.

The impact of conflicting interests is best illustrated in the case of Hälsingland in the south. Here, the ironworks were numerous and of an early origin. The maps show the slow diffusion of the steam engine as well as its co-existence with water-powered alternatives, in a tenacious structure of both small and medium-sized sawmills. The owners of both ironworks and sawmills tried to solve conflicts about resources either by using only water power or by combining the two technologies. Sawmills not directly controlled by ironworks had to operate on a small-scale and seasonal basis and on terms largely set by their owners.

WHY A WATER-POWER REVOLUTION?

In his book on energy and the environment in industrializing America, Martin Melosi coined the expression 'the turbine revolution' for the transformation and wide application of water-powered technologies and the late introduction of steam engines. In districts where running water was abundant, access to it was cheaper than coal, and since skills in taming and using water power had long been developed, steam power did not become predominant in these districts until the late nineteenth century.[46] This chapter has demonstrated a similar pattern, but a slightly different causation, in the Swedish sawmill industry. Here pit-coal played no part and the institutional context was different. On the other hand, the development and refinement of traditional skills in water-powered technologies and rich access to running water also made the mechanization of production methods of Swedish sawmills a 'turbine revolution'.

This chapter has argued that powerful Swedish institutions operated at different levels to retard steam power: through laws that did not permit free trade and the establishment and free operation of sawmills, and through the specific institutions of the traditional ironworks industry which permeated business and social behaviour in the forest industry at large. The hope and aspiration, not necessarily economically well founded, of a future for traditional iron production, in combination with short-term competition for material resources, apparently also retarded the innovation of new sawmill technology.

Secondly, path-dependence, rooted in local cultures, traditional skills and experience, also favoured water power. Far into the 1870s, such factors were decisive in the labour-intensive operations of cleaning rivers and constructing dams, canals and water systems. The art of controlling running water had been improved for centuries, and the improvement of water-powered technologies, and especially the introduction of water turbines, was a logical and successful means of defence against the new steam engine. This is a demonstration of a protracted process of change involving both the transformation and successive improvement of old technologies and the introduction of new ones. The process was characterized both by path-dependence and by innovation.[47]

The third important factor was economic risk. The immense problem of calculating future costs and incomes for a long time made mechanized large-scale production, and consequently also investment in new steam-powered technologies, a highly uncertain strategy. Even when costs and incomes could be reasonably well predicted, the ever-present risks of sudden shifts in markets and demand, and of large economic losses, seem to have been strong restrictions on innovation.

Lastly, the chapter has argued that economic factors, institutions and path-dependence were, to a large extent, interrelated and interdependent. Prohibitive legislation protected the interests and influence of established industrial groups. Technological path-dependence was related to the interests and social relations of established companies and institutions, as were access to basic resources: capital, running water, forest areas and commercial networks.[48]

NOTES

1. In countries rich in mountains and rivers – Scotland, parts of France, northern Italy, central Europe, Scandinavia and parts of the United States – water-powered technologies were competitive with steam in several industries until the 1870s to 1880s. See G. Basalla (1988), *The Evolution of Technology*, Cambridge: Cambridge University Press, pp. 144–51; J. Mokyr (1990), *The Lever of Riches. Technological Creativity and Economic Progress*, Oxford: Oxford University Press, pp. 84–90; G. Nerheim (1991), 'The Development and Diffusion of European Water Turbines', in K. Bruland (ed.), *Technology Transfer and Scandinavian Industrialisation*, Oxford and New York: Berg, pp. 333–62; and J.C. Williams (1997), *Energy and the Making of Modern California*, Akron: The University of Akron Press, pp. 90–114.
2. See G.N. von Tunzelmann (1978), *Steam Power and British Industrialization to 1860*, Oxford: Clarendon Press; L.C. Hunter (1980), *A History of Industrial Power in the United States, 1780–1930: Water Power in the Century of the Steam Engine*, Charlottesville: University Press of Virginia, vol. I; J. Atack, F. Bateman and T. Weiss (1980), 'The Regional Diffusion and Adoption of the Steam Engine in American Manufacturing', *Journal of Economic History*, **40**, 281–308.
3. For a short overview, see L. Jörberg (1971), 'The Industrial Revolution in Scandinavia 1850–1914', *Fontana Economic History of Europa*, vol. IV, chapter 8.
4. See T. Gårdlund (1945), *Bolinders. En Svensk Verkstad*, Stockholm: AB Bolinder-Munktell, pp. 80–9.
5. For a theoretical reflection on the notion of resistance to new technology, see K. Bruland (1995), 'Patterns of Resistance to New Technologies in Scandinavia: an Historical Perspective', in M. Bauer (ed.), *Resistance to New Technology*, Cambridge: Cambridge University Press, pp. 125–45.
6. For an elaboration of this argument, see Mokyr, *The Lever of Riches*, pp. 151–5.
7. For empirical evidence of early industrial path-dependence and organizational adaptation to new technology, see D. Landes (1977), *The Unbound Prometheus. Technological Change and Industrial Development in Western Europe from 1750 to the Present*, Cambridge: Cambridge University Press, pp. 293–326.
8. For a systematic treatment of this perspective, see D. MacKenzie and J. Wajcman (eds) (1987), *The Social Shaping of Technology*, Milton Keynes: Open University Press.
9. See E. Söderlund (1951), *Svensk Trävaruexport under Hundra År*, Uppsala: Svenska Trävaruexportföreningen, pp. 21–40, 50–6, 166–81.
10. The most frequently used 'St. Petersburg Standard' contained 4,672 cubic metres: see A. Johansson (1988), *Arbetets Delning. Stocka Sågverk under Omvandling 1856–1900*, Lund: Arkiv Förlag, p. 456.
11. See H. Wik (1950), *Norra Sveriges Sågverksindustri*, Uppsala: Uppsala Universitets Geografiska Institution, pp. 257–64; and Söderlund, *Svensk Trävaruexport*, pp. 50–76. The share of sawn products of the total export value was 15% in 1850, an impressive 50% in 1870 and remained stable at around 40% from 1880 to 1900. See Söderlund, ibidem, p. 3.
12. See F. Sejersted (1978), *Den Vanskelige Frihet 1814–1851* (*Norges Historie*, Vol. 10), Oslo: J.W. Cappelens Forlag, pp. 182–7.
13. See Söderlund, *Svensk Trävaruexport*, pp. 5–6.
14. See Wik, *Norra Sveriges Sågverksindustri*, pp. 81–3; and Söderlund, *Svensk Trävaruexport*, pp. 10–17.
15. See G. Utterström (1985), *Iggesunds Bruks Historia 1685–1985, II*, Iggesund: AB Iggesunds Bruk, pp. 14–15.
16. See N. Meinander (1945), *En Krönika om Vattensågen*, Helsingfors: Finska Sågägareföreningen, pp. 220–42.
17. See also Bruland, 'Patterns of resistance'.
18. See Söderlund, *Svensk Trävaruexport*, pp. 182–97.
19. See Söderlund, ibidem, pp. 77–121 and passim.
20. See G. Utterström (1957), *Jordbrukets Arbetare, I*, Stockholm: Tiden, pp. 727–73.
21. See Johansson, *Arbetets Delning*, chapters 1 and 2.
22. Splitwood was used mainly as firewood or to construct packing boxes.
23. See Johansson, *Arbetets Delning*, pp. 53–56. On the mix of production and exports, see also Söderlund, *Svensk Trävaruexport*, pp. 57–76, 139–55, 164–81.

24. See Johansson, *Arbetets Delning*, pp. 47–59.
25. Even as late as the 1860s, it was not uncommon for producers to rely on the ships' captains personally to find buyers in the foreign country. See Söderlund, *Svensk Trävaruexport*, p. 91.
26. For a detailed study of how new market conditions affected the production process, see Johansson, *Arbetets Delning*. References to this are also found in Söderlund, *Svensk Trävaruexport*, pp. 51–6, 100–102, 123–38, 155–7.
27. See O. Hellström (1924), *Den Norrländska Sågverksindustrien*, Uppsala: Almqvist & Wiksell, pp. 274–5.
28. See G. Arpi (1955), 'Sågverken', in F. Hjulström, et al. (eds), *Sundsvallsdistriktet 1850–1950*, Uppsala: Uppsala Universitets Geografiska Institution, pp. 283–325, p. 315; and Johansson, *Arbetets Delning*, pp. 113–20, 184–8.
29. See Johansson, *Arbetets Delning*, pp. 203–26.
30. See Hellström, *Den Norrländska Sågverksindustrien*, pp. 287–8.
31. See Johansson, *Arbetets Delning*, chapter 4.
32. See Hellström, *Den Norrländska Sågverksindustrien*, pp. 207–72; Arpi, 'Sågverken', p. 297; Wik, *Norra Sveriges Sågverksindustri*, p. 120, n. 3; and Gårdlund, *Bolinders*, pp. 164–5.
33. See Nerheim, 'The development and diffusion', pp. 335–6.
34. See Nerheim, ibidem, pp. 344–7.
35. For empirical evidence, see for Finland, Meinander, *En Krönika om Vattensågen*, p. 257; for Sweden, Y. Bergström (1980), *Industriorten Bergvik – Gamla Dokument Berättar*, Söderhamn: Bergvik och Ala Aktiebolag, pp. 56–61, 72; U. Hamilton (1982), 'Den Blygsamma Ångmaskinen. En Studie av Arbetsresurser i Sverige, 1800–1850', *Fataburen 1982* (Yearbook of the Nordiska Museet and Skansen Folklore and Zoological Park), p. 31; Johansson, *Arbetets Delning*, p. 184; and J.E. Pettersson (1990–91), 'Nydqvist, Johan Henrik Antenor', in G. Nilzén (ed.), *Svenskt Biografiskt Lexikon*, Vol. 27, Stockholm: Svenskt Biografiskt Lexikon, pp. 690–91; and for Norway, H. Thoresen (1942), 'En oversikt over vannturbinens utvikling', *Kvalitet*, **23**, 78–87; C. Gierloff (1948), *Et Bruk ved Akerselven. Myrens Verksteds Hundre Års Minne*, Oslo: A.S. Myrens Verksted, pp. 102–9; K. Anker Olsen (1953), *Kvaerner Brug gjennom 199 År*, Oslo: A.S. Kvaerner Brug, pp. 54–86; E.B. Schieldrop (1961), *Christiania Spigerverk 1853–1961*, Oslo: Grondahl & Son, p. 68; and Nerheim, 'The Development and Diffusion'.
36. See Hellström, *Den Norrländska Sågverksindustrien*, p. 264; Wik, *Norra Sveriges Sågverksindustri*, p. 132; and Johansson, *Arbetets Delning*, pp. 186–92.
37. See Th.A. Askengren (ed.) (1913) *Handbok för Nordisk Trävaruindustri*, Stockholm: Svenska Trävaruexportföreningen, passim.
38. See Wik, *Norra Sveriges Sågverksindustri*, pp. 98–143; Söderlund, *Svensk Trävaruexport*, pp. 109–15; and Arpi, 'Sågverken', pp. 291–3, 310–25.
39. See Johansson, *Arbetets Delning*, p. 192.
40. See Johansson, ibidem, pp. 184, 188, 191.
41. See Bergström, *Industriorten Bergvik*, p. 58; and Johansson, *Arbetets Delning*, p. 207.
42. See Johansson, *Arbetets Delning*, p. 126.
43. See C. Sällström–Nygren (1967), *Vattensågar och Ångsågar i Norrland under 1800-talet*, Göteborg: Meddelanden från Ekonomisk-Historiska Institutionen vid Göteborgs Universitet, pp. 50–7.
44. See Johansson, *Arbetets Delning*, pp. 186–92.
45. For the following section, see Wik, *Norra Sveriges Sågverksindustri*, pp. 97–154; Söderlund, *Svensk Trävaruexport*, pp. 50–121, 139–97; F. Hjulström (1955), 'Sundsvallsdistriktet vid 1800-talets Mitt', in F. Hjulström (ed), *Sundsvallsdistriktet 1850–1950*, Uppsala: Uppsala Universitets Geografiska Institution, pp. 97–120; and Arpi, 'Sågverken', pp. 283–325.
46. See M. Melosi (1985), *Coping with Abundance. Energy and Environment in Industrial America*, Philadelphia: Temple University Press, p. 23.
47. See Johansson, *Arbetets Delning*, pp. 33–44, 72–79, 316–34.
48. On path-dependence, see Mokyr, *The Lever of Riches*, pp. 162–5. On the social shaping of technology and access to resources, see MacKenzie and Wajcman, *The Social Shaping of Technology*.

12. An Old Production Method Mobilizes for Self-Defence

Francis Sejersted

A TECHNOLOGICAL REVOLUTION

Joel Mokyr has stated in his paper that 'the cases in which the owners of physical capital have fought against the introduction of new techniques are comparatively few'. He refers explicitly to the mill owners in Lancashire who might have resisted the replacement of water power with steam technology, but did not, because they could 'make up for the losses by buying into the steam technology themselves'.[1] I shall present a case which in many aspects is parallel to the case mentioned by Mokyr.[2] It concerns the political process around the middle of the nineteenth century by which the way was opened up for the replacement of water power by steam technology in the Norwegian sawmill industry. The outcome of the case proves that it was indeed the owners of old capital who most successfully bought into the new technology. But that was only after they had mobilized to resist the transition, and had succeeded in postponing it. This apparent paradox, that those who resisted a change were the ones to benefit most by that same change, calls for discussion of the social and cultural basis for the utility maximization assumption, of the different conceptions of utility, and of who were the agents of change.

Norwegian exporting of deals and battens, mainly to England, had grown substantially during the seventeenth century and was a well-established trade in the eighteenth. It was heavily regulated by privileges, the main clause of which said that timber for export could only be sawn on specially privileged sawmills. Deregulation started in the late eighteenth century, but the main clause lived on for many decades even after the liberal revolution in 1814. At that time new privileges were usually not granted, whereas the old privileges were connected to the old water-powered sawmills, deciding their value on the market. A paradoxical situation emerged in so far as, in this period of liberalization, the old system of privileges was no longer actively used, so that what remained of the old system was 'frozen': this had the originally unintentional consequence that the old system came to serve as an effective barrier to new mills with new technologies. In the 1840s the conflict arose over whether the system should be abolished altogether. The final outcome was the passing of a law in 1854 saying that the trade should be completely free from 1860.

In Norwegian industrial history the almost explosive growth of the steam-powered sawmill industry from 1860 onwards is a central event. Within a short time a revolutionizing technological change took place in Norway's dominant export industry. The change was more comprehensive than just the substitution of water power for steam power. Most of the old water-powered saws had, right up to the time of this conversion, been constructed and built by local builders and blacksmiths, who represented a long tradition of technical insight. The steam-powered sawmills, however, got not only their steam engines, but also their saw machinery, delivered from the new machine shops at home, or from abroad. For instance, Myren's engineering workshop in Christiania (Oslo) quickly managed to acquire a technical insight within this field which was on a level with that of foreign competitors.

This technological revolution also involved a relocation of the traditional industry. The steam-powered sawmills did not need to be situated by waterfalls. They were built in connection with ports, or by railways. Thus Fredrikstad, which was a sleepy little garrison town at the mouth of the Glomma, Norway's biggest river, quickly became the new dominant sawmill centre.

It is interesting to note that we also can trace a well-known phenomenon in technological history, namely that a new method of production becomes the direct cause of a technological development within the old production method it competes against. In 1861, one of the leading timber companies, Tostrup & Mathiesen, established a large new steam-powered sawmill at Hurdalssjøen, 60 kilometres north of Christiania; this mill was connected to the first railway in Norway from Christiania to Eidsvold, opened in 1854. Part of the machinery was delivered by Myren's engineering workshop. When this mill started production, Tostrup & Mathiesen closed down one of its two large old mills based on traditional water-powered saws, but let Myren's workshop modernize the other. In other words, the old master-built saws were replaced by saw machinery delivered by an engineering workshop.[3] Thus the change in 1860 also involved, to some degree, a technological renewal of the water-powered saws, which were kept on for some while.

The attempts to modernize the old water-powered saws can be looked upon as an attempt at self-defence on the part of the old production method and pattern of location. No study has been made of how extensive these attempts were. However, it is clear that the era of the water-powered saws was over. There was no adequate answer within water-powered sawing technology to the challenge represented by the steam-powered saws after 1860. The relatively quick and complete technological transition after that year demonstrates how the new technology had matured and indeed that the transition was overdue.

TRANSITION DELAYED

The problem is, why was the transition delayed until 1860? We have already seen that the old system of privileges had been 'frozen' after 1814 and that it then served as a barrier to the new technology. This could happen partly because the new liberal régime was faced with a dilemma as to how to deal with the old-established rights from the mercantilist period. On the one hand, the old privileges represented an alien element in the new order; on the other hand, established rights, and not least privileges bought on the free market, should not be expropriated. In addition to this ideological dilemma, the pressure put on the Government to change the system was not extensive, as there was a general and long-drawn-out crisis in the trade, because the British timber duties in effect cut the Norwegians off from what had been their main timber market.

In the 1840s things changed. The Norwegians had gained a foothold in the French market, the British started to lower their duties, times in general became more prosperous, and pressure was put on the old system from different quarters, among others from potential new sawmill entrepreneurs and from farmers who wanted a higher price for their timber. In 1849 a commission was appointed whose task, among other things, was to pronounce on whether the privileges should be abolished. One of the members of the commission was the important timber merchant and sawmill owner, Christopher Tostrup, who represented the established group within the trade. In May 1850 the commission made its report, recommending unanimously that the system of privileges should be opened up.[4]

The main argument for lifting the restrictions was 'the free competition, which the present spirit of the age and our notions of freedom of trade require in all branches'. The report also implied that there was an economic reason behind the spirit of the age: 'In general any ... interference in the conditions of trade on the part of the Government will never be healthy. Commerce itself will find the best way'. The commission used nothing more profound than catchword arguments on this point. It was sufficient to mention the spirit of the age; there were few who as a matter of general principle would have opposed these basic standpoints.

The old principal argument for privileges had been that a completely free trade could lead to ruthless exploitation of forest resources. But this does not seem to have caused the commission much worry. It is in fact quite striking that the commission regarded the question of the abolition of sawmill privileges as a case of little practical importance. Or to put it in another way: the commission seems to have taken it for granted that the established and privileged sawmill business would prove to be the strongest even if the privileges were removed.

In this connection it was pointed out that all the best waterfalls had already been taken by the privileged sawmills. The advantage which some felt they could obtain by establishing sawmills after the abolition of the system of privileges was 'more illusion than reality'. The abolition was concerned, the commission stated in an interesting formulation, 'more with the principle than with the neccessity'. Notwithstanding this natural strength of the traditional system, the commission proposed additional help. Firstly, it proposed a ten-year period before the privileges were abolished. The old-established trade should have ten years to prepare for a possible increase in competition. In addition, the commission maintained that the holders of the privileges were entitled to compensation, arguing that privileges could be considered as property acquired on the free market.

There was one cloud on the horizon. The commission remarked on the possibility that 'a dangerous rivalry in the steam-powered sawmills' could develop. It even mentioned the possibility that a steam-powered sawmill industry might grow up at the mouth of the Glomma. From hindsight, we know that this was exactly what happened, but the commission did not stick to its prophecy – it stated that improved communications would make the old sawmills competitive, so that the steam-powered saws would only 'upset the existing state of affairs to a certain extent'.

Steam-powered sawmills had become a possibility in the 1840s, and in 1843 the Ministry of Finance had proposed their compulsory licensing. Two applications for a licence had been sent to the Government before the commission was appointed. In 1845 A. Blehr, a merchant in Skien, and in 1847 B. Wessman, a merchant in Larvik, applied for licences to run steam-powered sawmills. In both cases Frederik Stang, Minister of Home Affairs, recommended that the licence should be granted. Both times he was opposed by Severin Løvenskiold, the governor of Norway, and the end result was that the licences were refused.[5]

Stang's argumentation was, first and foremost, liberalist in principle. He wanted to open the doors to competition on the free market. He also deduced that there was a need for steam-powered saws from the fact that serious plans had been proposed. His attitude was passive, as far as it went, because it followed the path of automatically granting licences from the applications at hand. The interesting point is that this way of using the old system, by automatically granting licences, could have served as an effective way of abolishing the old system, or indeed of directing the old trade onto a new technological path. The barrier to the new technology did not lie in the old system as such, but in the way in which it was used, or rather not used, by the government.

Stang used a line of argument which we have recognized in the commission's report of 1850. He maintained that a steam-powered saw was expensive both in construction and in running costs. He based this on the advice of the timber merchants in Skien, who had asserted that there was every reason to regard sawing by steam sawmills as a transitory enterprise which could only under exceptional circumstances be employed advantageously. In other words, the steam-powered sawmills represented no real threat to the established water-powered sawmills. That, too, seemed to Stang a reason why licences could be given. But, as we have seen, the governor Løvenskiold refused the licences.

The drift of the governor's argument is noteworthy, coming at the time it did. He held that the steam-powered saws had completely altered the premises behind existing regulations and practice. For steam-powered saws there were no natural limitations, as there were for water-powered sawmills. Therefore new legislation would be necessary to take this situation into account. The reasoning here reflects a clear understanding of the new technological possibilities. Løvenskiold, who is generally reckoned to have been conservative, seems in fact to have been one of the first to see the threat from the steam-powered saws. He was himself a sawmill owner and was accused of feathering his own nest by his refusal. However, his argumentation assumed that the question of privileges had to be taken up, and this position was what ultimately led to the appointment of the commission in 1849.

The commission's report, submitted in 1850, concluded, as we have seen, that the system of privileges should be abolished, the negative argument being that it was of little significance. The Ministry of Finance fastened onto this point and suggested, typically, that since the system had so little significance it was unnecessary to burden the public purse with large payments of compensation. This way of reasoning led the Ministry to the conclusion indicated earlier by Frederik Stang, that one could achieve the same result just by automatically granting all licence applications.

Faced with this problem of compensation, the Government decided that, before any bill was submitted, a report should be made to find out how many compensation cases there would be and what the total costs would amount to. So the matter was postponed and a new commission appointed. This included three out of the five members of the previous commission, one of whom was Christopher Tostrup. The commission was intended to be a 'Valuation Commission', but it was also part of its mandate to consider again the question of the abolition of privileges.[6]

In December 1853 the report of the second commission was ready.[7] The remarkable aspect of their proposals is that, on the very question of the abolition of privileges, the commission reached a different conclusion from the earlier commission. It unanimously recommended that the system of privileges should be maintained. So what had made the three members – particularly Tostrup, the representative of the group of privileged sawmill owners – change their minds? It is immediately clear that their reasoning was based on a new appraisal of the actual situation.

It is at once striking that the first commission had carried out its task against a background of bad times. The boom of the 1840s had been superseded by crisis by its time. By contrast, the second commission had worked during prosperous times. These good times, as before, had increased the pressure from those without privileges who wished to saw for export. Whilst there had been three applications for licenses for this in 1849, there were 39 in 1853. The Government had also started the practice of granting new licences, although not for steam-powered mills.[8] There was strongly increased pressure from the forest owners, who considered that the privileges led to an exorbitant part of the profits from the boom ending up in the hands of the small group of sawmill owners. However, more dramatic events had taken place than a change in the trade outlook.

The three members of the commission had formed a completely new opinion of the steam-powered saw's competitive ability. The advantages of the steam-driven saws 'are so overwhelming', they wrote, 'that within a short time they will be established and will then replace and eliminate the water-powered saws'. They could also refer to 'several applications for permission to set up steam-powered saws which have already been sent to the Government'. The threat which was distant and unreal in 1849 had now become close and overwhelming. An unreal threat could be tolerated. But one which augured nothing less than annihilation was more difficult to tolerate. This is what lies behind the changed opinion of the commission members. They said it themselves: 'The main reason why the members of the commission have now come to a different opinion than they had as members of the Forest and Sawmill Commission is the fear of the steam-power saw's influence ... which has now arisen because of the extremely strong development which utilization of steam power has recently shown in our country'.

Behind the about-turn of the commission's members we can glimpse a growing anxiety amongst the well-established. The question of the fate of the privilege system suddenly meant something real and fundamental. Thus the seeds of a new conflict were sowed, making it more difficult to unite the different viewpoints, which the Forest Commission of 1849 had thought it could do. Already in the 1849 commission's report we can discern the vague contours of an orientation of policy towards the established interests, in the commission's strong emphasis, for instance, that the abolition would not result in especially great structural changes in the trade. In the second commission's report this policy orientation is clearly put forward, so that the weight of this argument becomes apparent.

The threat which was perceived to be directed against the relatively small group of established sawmill interests made the commission members change their minds. In the sharpened conflict of interests the second commission stood on the side of the well-established sawmill interests. The members of the commission abandoned the vessel which was sailing safely with the wind and jumped over into the sinking ship. In its argument for keeping the privileges, the commission, not unexpectedly, launched the protection-of-capital argument. It pointed out that the large amount of capital which the water-powered sawmills represented would be lost if they were eradicated. Furthermore it argued that foreigners might invest in the steam mills and thus reap an advantage which had hitherto been reserved 'for the state's own sons'. The commission gave as an example 'the latest sign of the times' which showed that 'foreign capitalists' were interested in investing in Norway. A social argument was also provided with reference to the 2400 workers at the old privileged sawmills who would be out of work if they closed. Finally the commission pointed to what we have referred to as the dilemma of the transition from a mercantilist to a liberal system, arguing that the abolition of the system would mean the 'destruction of the wealth of private individuals', which according to the new liberal creed was outside the competence of the state.

The commission's main task comprised valuation of the privileges in case the parliament decided to abolish the system. From the experiences of 1850–51 it seemed that a relatively high assessment could be an argument for keeping the system intact. It is clear that the commission did not adopt this line. The valuation was generally amazingly low. In total, the compensation amounts were estimated at 140,000 *speciedalers*, not much more than one year's timber customs duty. During the consideration of the case in 1850–51, there had been very guarded estimates of what sums one had in mind. However, a crossed-out sentence in a draft of the Ministry's pronouncement named a sum of several hundred thousand *speciedalers*. The modest valuation indicates perhaps that the commission wanted to make the compensation proposal palatable in case their principal recommendation was not followed. In short, it seems as if the commission feared that their cause was a lost one.

The Ministry was very clear on its rejection of the commission's conclusion. It stated that it 'must declare itself determinedly in favour of release' inasmuch as the running of the sawmills must be 'opened up for competition'.[9] Expanding on this liberalistic argument, it focussed on the commission's economic arguments. It must, the Ministry said, be wrong to keep the opportunity to establish steam-powered sawmills 'barred because of fear of its productive and profitable force in relation to the methods that have been used to run the mills hitherto'. What is interesting, however, is that this argument came only after the Ministry had commented upon the commission's report to the effect that 'there is some exaggeration present in the fear that the steam power will supersede the water-saw'. In the Ministry one sat far removed from the world of reality. For those in the Ministry, the new threat which had made the commission's members change their standpoint was not so overwhelming. Or perhaps this was only a tactical sweetening of the Ministry's conclusion, which was that the system could be dismantled provided that there was compensation and a reasonable time limit.

The Parliamentary Committee which was the next to handle the case recommended a waiting period, but no compensation. It pulled out all the stops and turned the commission's argument against itself: 'it is a misguided sort of conservatism when, to preserve the *status quo*, one shuts out every kind of reform, however much one acknowledges its usefulness'.[10] The question of whether the steam-powered sawmills represented a threat was not discussed by the committee.

Resistance to the abolition of privileges was a lost cause before it came up in Parliament. The proposal to keep the privileges was not even put forward, and only a small minority voted for compensation. As we have seen, however, a waiting period of six years up to 1860 was passed. The defeat inflicted on the commission was nearly complete. The attempt to use the system of privileges as a bulwark against new technology was thus a failure.

CHANGING BOUNDARIES FOR THE BOUNDED RATIONALITY

Resistance to new technology is a well-documented phenomenon in economic history, and there can be many reasons for resistance of this kind.[11] What is particularly notable in this particular case is that apparently the holders of the privileges felt that it was in their interests to use the system of privileges as a bulwark against the new steam technology. The basis for this supposition is the stand taken by the representative of the established interests, Christopher Tostrup, along with the relative silence of the rest of the group of owners of large mills. It is worth noting that they had originally been willing to allow the principle of free competition. When they felt that there were bigger interests at stake, they changed their attitude. What is problematic is the point made by Mokyr, that apparently they did not have a rational reason for opposing the new technology, as they themselves had the best opportunity to buy into the new technology and reap the advantages.

This is indeed conspicuously illustrated by the example of Christopher Tostrup. He was the one among the small group of owners of large sawmills who rose in defence of the privileges in the decisive phase. His argument was that the old trade would be spoilt, and time showed that he was correct in his assertion, just as his clear emphasis in the commission's report on the competitiveness of the steam-powered saws was shown to be correct. He had a very clear understanding of the economic and technological, if not of the political, aspects of the actual situation. When his defence of the privileges did not succeed, he was one of the first to make use of the new situation and his own insights. Along with many of the old guard of the mill owners, he built a large new steam mill.[12] His investment was extremely successful. The business grew and yielded better profits than it had done earlier. About fifteen years after the privilege system had been abolished, his business had the highest tax assessment in Christiania (Oslo). What was his motive for resisting a development which was so clearly advantageous for him personally?

Tostrup identified himself with the old production system, or with the privileged group of sawmill owners. We must understand his objectives as a businessman within this social and cultural framework. Within the system, he pursued the maximization of profit in competition with the other mills. He had done well in this competition. However, when faced with a more fundamental threat against the system which gave him his success, he put aside the consideration of profit maximization and replaced it with the aim of preserving the system. In short, this is a beautiful example of how rationality somehow has to be bounded.

The reaction against the new system must also be partly interpreted as a reaction to uncertainty. The commission had referred to 'steam power's application in recent times'. Perhaps the most important event which had occurred between the two commissions' respective reports was the construction of the first railway in Norway, which was in operation, if not officially opened, when the recommendations of the latter report were issued. Christopher Tostrup joined the board of this first railway company in 1853, so he was able to observe the revolutionary properties of steam power at close hand. Originally the concept of the railway was supported with enthusiasm, but as time went by several people began to get cold feet. The railway was an example of what big shake-ups the new technique could bring. People hesitated when confronted with real drastic changes. But as soon as the old system of privileges had been abolished and the way was opened up for the new revolutionizing technology in the sawmill industry as well, Tostrup must have realized the impossibility of keeping on with the old production method. But he was not a 'happy warrior' or an 'irrational' seeker of novelty in the Schumpeterian sense. On the contrary, he was forced to take up the new technology. It was a rational choice to change the boundaries for his bounded rationality, which means that he behaved as a 'satisficer'.[13]

It is odd that the threat from foreign competition does not seem to have played any part in the debate. Somehow Tostrup must have felt that the natural advantages of Norway, of which closeness to the big markets was the most important, ruled out this argument, at least as a decisive one.

There is one final reason why Tostrup stood up for the old production system. In the middle of the nineteenth century there had not yet been any bureaucratization of private firms. This meant that most of the firms were still personal enterprises, in which the owner personally took care of the day-to-day operating as well as of the long-term investment decisions.[14] To change to a new production method meant – for Tostrup personally – night work, increased worries, perhaps also invidious tasks. The reference in the commission's report to the many workers at the old sawmills may have a concrete source in Tostrup's own deliberations about what he could possibly do with the workers when production had to be moved to another place. Inertia can also be rational.

CONCLUSION

The great technological revolution in the largest industry in Norway was not introduced by the established group of sawmill owners. On the contrary, they tried to use the political system to raise barriers to technological changes. The new technology was forced upon them by the state, using the market mechanism as an instrument.

Contrary to the common image of a passive state, the Norwegian state adapted, at least from the 1840s, what can be called a modernization programme. This means that the state introduced a coherent policy to further economic growth and material well-being. The market mechanism was used consciously, but if it did not live up to expectations, the state did not hesitate to introduce compensatory measures.[15] Our case is just one of many examples of how the state successfully acted as an agent of change. In the beginning the government hesitated to use the old system of privileges to further technological change, although the possibility was discussed. In the end, however, it overruled vested interests by letting loose the market, with the pronounced aim of generating a technological revolution.

NOTES

1. See J. Mokyr, 'The Political Economy of Technological Change: Resistance and Innovation in Economic History', Chapter 3 of this volume.
2. This article is a revised version of an article published in Norwegian in Francis Sejersted (1993), *Demokratisk Kapitalisme*, Oslo: Universitetsforlaget, and in an expanded version in Francis Sejersted (1979b), 'Da Sagskuren ble Frigitt', in Ottar Dahl et al. (eds), *Makt og Motiv. Festskrift til Jens Arup Seip*, Oslo: Gyldendal Norsk Forlag, pp. 94–109. It is originally a 'by-product' of a history of a firm. Cf. Francis Sejersted (1979a), *Fra Linderud til Eidsvold Værk*, Vol. III, Oslo: Dreyers forlag. For the final version I have benefitted from the discussions following the presentation at the seminar in May–June 1996.
3. See Sejersted, *Demokratisk Kapitalisme*, pp. 186–7.
4. *Indberetning og Indstilling fra den ved Kongelig Naadigst Resolution af 10de Januar 1849 Nedsatte Commission Angaaende Skov- og Saugbrugsvæsenet m.m.* (1850), Drammen. All quotations in the next four paragraphs are from this report, translated by the present author.
5. Regarding licences to operate steam powered sawmills, cf. August Schou (1940–41), 'Fra Dampsagens Gjennombruddstid', *Historisk Tidsskrift*, **32**, 258–65; and L. Alnæs (1973), 'Oppgave over Dissenser i Visse Næringspolitiske Spørsmål innenfor det Vogt-Løvenskioldske Ministerium', in Per Maurseth, *Prosjekt Dissens*, unpublished paper, pp. 14–22.
6. *Dept. Tidende* (Ministerial News) *1851*, pp. 717–25, and *Dept. Tidende 1853*, pp. 473–80.
7. The second commission's report is printed in its entirety in the Government bill, *St. Forh.* (Parliamentary Papers) *1854. D.4*, O. no. 39, pp. 1–50. All quotations in the next four paragraphs are from this report, translated by the present author. The commission's file has not been found.
8. Home Office minutes, State Archives, give the following figures:

	1847	48	49	50	51	52	53	54	55	56	57	58	59
Applications:	13	10	3	5	13	17	39	38	17	17	3	2	6
Licences:	8	7	3	4	13	11	31	25	14	15	3	1	5

9. All quotations in this paragraph are from the Government bill, *St. Forh. 1854, D.4*, O. no. 39, pp. 62–4, translated by the present author.
10. *St. Forh. 1854. D.8*, p. 339, translated by the present author.
11. Martin Bauer (ed.) (1995), *Resistance to New Technology*, Cambridge: Cambridge University Press, gives an overview and presents many examples mainly from resistance to modern technology.
12. The first three steam-powered sawmills to be set up in Fredrikstad (cf. above) were those of Gutzeit, Kiær and Jacobsen. Gutzeit and Kiær were well-established owners of water-powered sawmills whereas Jacobsen had worked in the trade for many years. The three firms were for a long time the most successful of the many new steam-powered sawmills in Fredrikstad. Cf. Sejersted, *Demokratisk Kapitalisme*.
13. On the concept of 'satisficing' see Sejersted, *Demokratisk Kapitalisme*, pp. 357–90.
14. See Alfred D. Chandler junior and Herman Daems, 'Introduction: The Rise of Managerial Capitalism and its Impact on Investment Strategy in the Western World and Japan', in Herman Daems and Herman van der Wee (eds) (1974), *The Rise of Managerial Capitalism*, Leuven, Belgium: Leuven University Press and The Hague, The Netherlands: Martinus Nijhoff, pp. 1–34, p. 5.
15. See Sejersted, *Demokratisk Kapitalisme*, pp. 47–105.

13. Technology, Trade and Wage Inequalities in the Late Nineteenth Century

Nick von Tunzelmann and Edward Anderson

INTRODUCTION

In recent years, the phenomenon often styled as 'globalization' has drawn the attention of many scholars to both national and international patterns of trade, technology and wage inequalities, and especially to their putative interactions. The widening gulf between 'skilled' and 'unskilled' wages in advanced Western industrial countries has been attributed by some to competition in trade from later industrializing countries,[1] by others to biassed technological change favouring skilled employees, and by others again to factors such as changes in political or corporate governance.[2]

Similar concerns have hovered in the background of studies of the late nineteenth century, a period which deserves consideration in these respects as one in which the leading industrial country (Great Britain) was for the first time being seriously challenged by subsequent industrializers, in the wake of its mid-century heyday as 'workshop of the world'. Now Germany and the United States were posing major threats to Britain's dominance of the world trade and technology scene. The period, of course, came to an end in 1914 when economic conflict among them gave way to military conflict.

Although such interrelationships appear to be of key importance, in the form enunciated here they have not been given explicit attention by economic historians. Instead, there are many studies, dating from the time itself, which look at the individual experience of each major country. There are also more scattered studies that undertake cross-country comparisons, but on the whole these tackle each of the subjects separately, and normally in discrete studies. The work of Jeffrey Williamson, especially, together with his colleagues such as Alan Taylor, Timothy Hatton, Kevin O'Rourke and others, has provided us with rich material and thought-provoking results in many of these areas, and we are indebted to them for both data and findings. Yet, though Williamson and colleagues sometimes explicitly relate their work to the current debates briefly outlined in our first paragraph, they tend to treat the issues in turn rather than conjointly. For example, they have produced invaluable studies of wage differentials (between skilled and unskilled, and so on) within countries, and more recently of differentials between unskilled labourers across countries. But the intra-national inequalities have not to our knowledge been extensively combined with international inequalities, to build up the full picture. And the relationships with trade and technology patterns remain very sketchy in this context, though again widely studied in themselves.

This chapter[3] therefore aims to integrate some of the data on these various fields in an international setting. So far as the data are concerned, we are simply collating extant material, though the limitations of readily accessible secondary material are soon exposed in this treatment. On the side of the findings, we are fully aware that at present they are at best merely exploratory.

Our basic quest here is to examine wage differentials between the broad classes of labour on a limited cross-country basis, and to relate these to underlying trade and technology patterns. The latter rotate around the growing competition to Britain's leadership, expressed not just through foreign trade as international movements of goods and services, but also through foreign investment and migration as international movements of factors. The latter were in many respects a more potent challenge to Britain than the goods trade itself. Although we consider just three countries in this chapter – the U.K., the U.S.A. and Germany, and generally in that order – they are sufficient to counterpose two themes: the rivalry between Britain as the first industrial nation and its new industrial challengers of the U.S. and Germany; and the competition between the 'Old World' of Britain and Germany as lower-wage sources of emigration, and the 'New World' epitomized by the U.S.A. as recipient.

Moreover, since the capital and labour migrations in part embody technology and skills, they modify the ability to draw simple conclusions about the wage inequality patterns. We shall give particular attention to the so-called 'American invasion of Europe' at the end of the nineteenth century as a key defining moment in the process of industrial overtaking: an invasion characterized by direct investment by Americans in Europe as well as by trade, while at the same time the U.S.A. continued to receive huge numbers of both skilled and unskilled immigrants from Europe.

CONCEPTUAL ISSUES: TECHNOLOGY AND SKILLS

In terms of labour and the wage inequalities, we must begin by defining the classes of labour that are relevant to the story. This is by no means a trivial exercise, and indeed the larger research enterprise to which this present study relates revolves around the issue of the relationship between technology and skills. Our ideas here are continually evolving. For present purposes, we take two cuts at the nature of skills.

First, we can conceptually divide skills into *skills of creativity* and *skills of operation*. This relates to concepts of the worker as being composed partly of brain and partly of hands. The creative rôle is one with which economists have not always been too happy, inasmuch as creativity is usually regarded as exogenous. Economists would prefer to relegate it, if indeed they ever need to consider it, to a final residual rôle, once everything more readily capable of conceptualization has been stripped out. However, it was much emphasized in comparisons between Britain and America in the mid-nineteenth century, for example by the British industrialist Joseph Whitworth in his celebrated 1854 Report on a visit to the U.S.A.[4] More recently, it has lain not far below the surface of the body of work in the so-called 'new growth theory', in which human capital formation via education has been given a special rôle.[5] It is worthy of a passing note that Whitworth, along with many others, drew attention to education at that time, but he also pointed out that it was not a sufficient condition for growth, as Prussia had had universal primary education for considerably longer than the U.S.A. without – at that stage – much development.[6]

Labour of skilled kinds, involving hands rather than brains, is the more customary notion of what skills involve. The two may of course be connected. The Japanese system of production is famous for making use of the fact that 'every pair of hands brings a free brain'. The substantive issue is how much the two depend on each other. The human capital argument implies a causal connection running from brain to hands, while equally fashionable notions of technological 'learning' (Arrow's learning by doing[7] and Rosenberg's learning by using) imply a reverse causal connection running from hands to brain. In fact, all three were succinctly summarized long ago in just two sentences of the *Wealth of Nations*:

> All the improvements in machinery, however, have by no means been the inventions of those who had occasion to use the machines. Many improvements have been made by the ingenuity of the makers of the machines, when to make them became the business of a peculiar trade; and some by that of those who are called philosophers [scientists] or men of speculation ...[8]

Such interdependencies are thus crucial to understanding technological achievement and performance. Mid-nineteenth-century commentators on the American success, such as George Wallis, believed that multiple skilling, encouraged by the social mobility of labour, helped to bring about the high degree of creativity of American skilled workers.[9] Similar views have been held about multiskilled Japanese workers in recent times.[10] Both arguments have appreciated the contrasting Smithian point about the advantages of specialization and division of labour; the trade-offs between specialization and variety remain controversial.

Leaving the interrelationships aside for the meantime, we can make some progress by adopting a structure similar to that used by Edward Wolff and colleagues in studies of wage differential movements in modern times.[11] They divide skills into three kinds: motor, cognitive and interactive. *Motor skills* are the traditional practical skills associated with working machinery, *cognitive skills* are those that are nowadays more in demand (for example, in software), and *interactive skills* are those which are linked to dealing with other people, for example, in management. Wolff establishes a pronounced decline in the wage patterns related to motor skills as compared with rewards for cognitive and interactive skills in the U.S.A., and there are indications of similar patterns in other countries.[12] Evidently these categories overlap with our division into creative and operational, but they also suggest a relationship with technological trends. While the mechanical 'paradigm' of technology was in the ascendant – roughly from the mid-eighteenth century to the mid-twentieth – motor skills were more handsomely rewarded. However, with the rise of the information technology paradigm in the last half-century, the cognitive skills have been in greater demand and increasingly well paid relative to motor skills. Skill differentials and technological trends are thus complementary. Yet technology is not the only factor underlying skill structures, as the case of services shows – salaries for, say, financial services were much higher than for domestic service long before technology boosted them further. Thus the creation/operation distinction retains validity even where technology is not crucial.

To adapt these notions to the second half of the nineteenth century, we need to consider a somewhat wider range of both skills and technologies. Operational skills associated with working with 'hands' need to be subdivided into those assisted by machinery and those not. Conventional handicrafts using hand tools come into the latter category, but these were being overtaken by mechanically skilled workers operating machinery. The low substitutability between craft skills and machine skills forms the backdrop to many of the labour struggles of the nineteenth and early twentieth centuries, from the hand-loom weavers of the early nineteenth century to the dispossessed coal hewers a century later.

These well-known examples point to a sectoral component to the technology aspect of determining skills. In turning to the data, we first consider, rather briefly, the case of agriculture. The situation of British agriculture in the second half of the nineteenth century has since the time itself been recognised as one in which trade – in the guise of free trade – had fundamental effects on output and returns.[13] The orthodox view, which we do not propose to challenge here, is that the agricultural sector was the sacrifice that the British economy made in order to achieve free trade and the status of being workshop of the world. This followed the acrimonious debates over the Corn Laws between the agricultural interests and the rising manufacturing interests from the late eighteenth century onwards – a debate too well-known to require much elucidation here. In the late nineteenth century and the so-called 'Great Depression', the impact of free trade was to depress prices and profitability in British agriculture, especially in English corn and wool, which suffered the most immediate competition from the 'regions of recent settlement'.

The case of mining is somewhat less clear-cut, because Britain continued to be a major exporter of coal, and indeed attained its highest levels of international competitiveness in coal-mining in the 'Great Depression' period.[14] The rôle of coal in international trade is less transparent than that of agriculture, because coal as a bulky item was customarily taken on as ballast for ships leaving Britain to collect imports.[15] Wages of coal hewers were about 20 per cent above those of all manual workers in the 1880s, rising to about 60 per cent above in the 'Edwardian' period before World War I,[16] though with some very sharp cyclical fluctuations that perhaps in part reflected the price-indexed wage systems adopted in some British coalfields from the 1890s.[17]

EMPIRICAL ISSUES: OUTPUT, TECHNOLOGY AND WAGES

The main thrust of this chapter is, however, directed towards manufacturing, where a much greater degree of diversity prevails. So far as British manufacturing is concerned, the late-nineteenth-century period of 'retardation' can be usefully divided into three main branches. The first consisted of traditional areas of British international strength, with textiles as the most obvious representative of this group. The second comprised newer product areas and industries emerging in this period, including new areas of mechanical engineering (sewing machines, later bicycles, and so on) as well as activities associated with new technological paradigms (chemicals, electricity, internal combustion engine, and so on). The third represented those industries which produced more perishable items, such as brewing. The last group is not considered here, as it was little involved in international trade – Britain was able to advance in some of these fields well into the twentieth century, if not beyond. So far as trade was concerned, consumer perishables in this period behaved more like services. The first group is also given less attention here. It was to face its most severe international competition shortly after the end of the period we are describing, when the challenge came increasingly from Third World countries like Japan and India, based on low-wage and comparatively unskilled labour. It should be noted that this group generated strengths in some branches of mechanical engineering, such as textile machinery, which continued to flourish in the late nineteenth century, although its markets were increasingly to be found abroad.[18] Our main concern is thus with the second group, the newer activities.

From an American perspective, the structure of industries looked slightly different. Daniel Nelson[19] suggests, when looking at the question from the standpoint of labour, a three-way division of manufacturing into 'precision engineering', 'heat-using production' (for example, iron and steel), and more conventional mechanical engineering as in textiles. The first of these is the most interesting for us, since it involved the so-called 'American system of manufactures', built around concepts such as interchangeable parts, in its starkest form. According to Nelson, factories of this kind employed perhaps only about 5 per cent of the labour force; however, it was these that turned out the products that were the spearhead of the 'American invasion of Europe'. The development of American competitiveness in particular products was thus anchored on the country's distinctive strengths in process technologies, most notably its lead in putting together technological 'systems'.[20] These systems in turn derived in part from the multiple skilling and labour mobility in the U.S.A. emphasized in the British Parliamentary Reports noted above. However, they were equally indebted to the distinctive American ability, via its management systems, to couple production processes with prowess in marketing.

A similar categorization might also suffice for Germany, but we should need to add the important contribution that was coming to be made by more science-based industries, such as chemicals. Ideally such industry classifications ought to meet our needs of tracing the linkages between wages, technology and trade. Unfortunately we do not have data structured according to these categories, particularly as regards wages. The assessment of the ensuing hypotheses must therefore remain very rough and ready at this stage. The data as presented overlap some boundaries that we would like to draw, but we have little option other than to accept them.

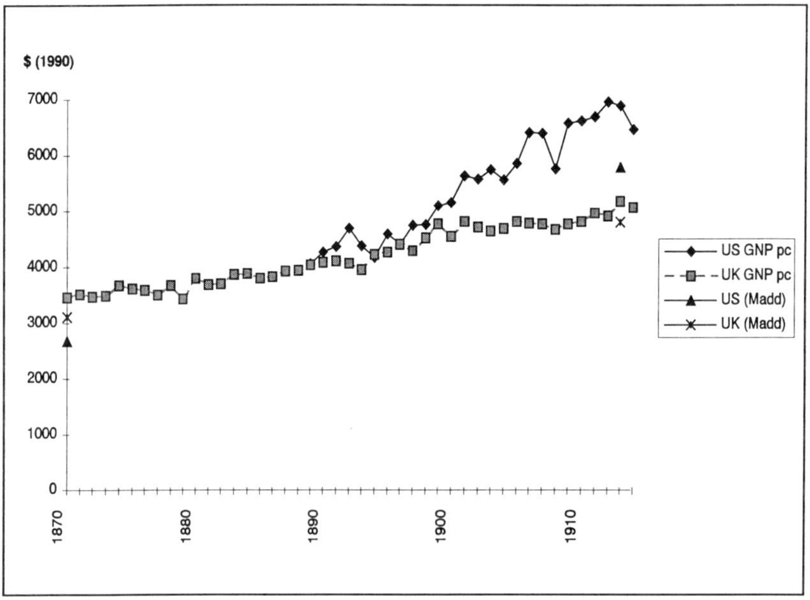

Sources: U.K.: Feinstein, *National Income*, Appendices;[21] Maddison, *Dynamic Forces*, p. 7.[22] U.S.A.: Department of Commerce, *Historical Statistics*, vol. I, p. 224;[23] Maddison, *Dynamic Forces*, p. 7.

Figure 13.1. G.N.P. per capita in the U.K. and U.S.A., 1870–1913

National Income

Annual data on G.N.P. *per capita* are shown in Figure 13.1 for the U.S.A. and the U.K. These indicate similar levels of G.N.P. per head *c.* 1890, but the U.S. opens up a substantial gap by the end of the period, and from the turn of the century there is a pronounced drift apart (during the so-called 'Edwardian climacteric' in Britain). Recent re-estimates from Maddison for the end years of 1870 and 1913 are shown for comparison, revised to our price base of 1990 (Maddison used 1985). Through many revisions, Maddison has over the years refined the crude estimates in an effort to put them on an internationally comparable basis. Some of these revisions appear justified, others perhaps less so. Evidently the figures in our U.K. series are somewhat higher than Maddison's throughout – about 11 per cent in 1870 and 8 per cent in 1913 (much of course depends on the precise choice of series for G.N.P., prices, and exchange rates) – while those in our U.S. series for the only available year of comparison (1913) are considerably higher. Nevertheless our data, like his, suggest a cross-over in G.N.P. *per capita* between the U.K. as the original leading industrial nation, and the

U.S.A., as its new rival, probably in the 1880s. We do not have annual series for Germany: the data of Maddison[24] suggest that German G.N.P. per head was only half that of the U.K. in 1870, rising to nearly two-thirds by 1913 (58 per cent of the U.S. in 1870 and 54 per cent of the U.S. in 1913).

Industrial Production

Data for industrial production shown in Figure 13.2 show some similarity with G.N.P. *per capita*. Since the figures are aggregates rather than per head, the relative growth of the U.S. is naturally much greater. The data, indexed to 1869, give a more adequate picture for Germany, which appears to grow unspectacularly until the mid-1880s, and then out-performs the U.K.[25]

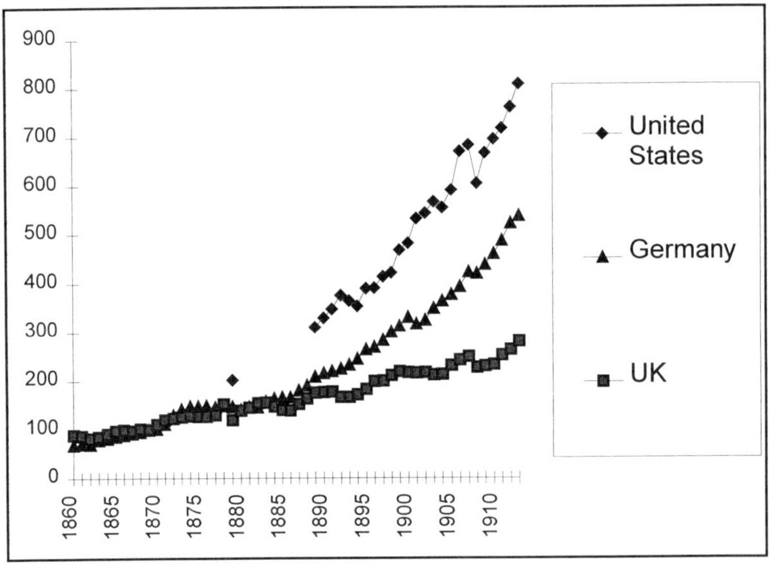

Sources: U.K.: Feinstein, *National Income*, Table 51; U.S.A.: Shaw, *Value of Commodity Output*, Table I-1;[26] Germany: Hoffmann et al., *Das Wachstum der Deutschen Wirtschaft*, Table 76.[27]

Figure 13.2. Industrial Production in the U.K., U.S.A. and Germany (1869 = 100)

For the U.K. the engineering industries, as computed by Feinstein, showed strong cyclical fluctuations. The metal industries and chemical industries show cyclical fluctuations that are synchronous with those in engineering industries, though not quite so sharp. Most areas of industrial production show a lower rate of growth after 1890 than before; the chemicals sector, however, shows a pronounced acceleration, especially in the later years of the period.[28] The main difference among the newer activities is that, while *engineering* and *metals* show a slight decline over the years 1899–1903, *chemicals* shows a slight rise, and thus pull away somewhat from the trends of the other two. *Metal manufactures* grows considerably more slowly than the other two after 1890. There is no indication in any of them of particularly rapid growth during most of the 1890s (though *other industries* does grow at above-average rates in this decade).

For the U.S.A. we do not have annual figures on a similar basis until 1889, but there are various compilations based on the U.S. Census of Manufactures. For the years that are available, the data are probably much more reliable than those for the U.K., which had no Census of Production until 1907. Shaw's data[29] on sectoral value-added (calculated as percentages of total value-added in manufacturing) indicate a surprising stability of the composition of U.S. manufacturing. 'Traditional' industries like food and beverages maintain their share over the period 1869–1914, a point implicitly made by Wright.[30] *Textiles* shrinks a little, but the fall occurs mainly in the decade from 1879 to 1889. These two together account for about 40 per cent of manufacturing value-added in 1869 and nearly the same in 1889–1914. *Iron and steel* accounts for about 15 per cent of the total in 1869 and again much the same throughout, with a slight fall in 1909–14. *Other-metal manufactures* approximately doubles, from 3 per cent or so to around 6 per cent, with all the increase coming in the decade 1889–99. *Chemicals* shows a slight increase throughout, from about 5 per cent to 8 per cent of total manufacturing. *Vehicles* is stable in its share, apart from a not unexpected rise in 1909–14, but even then is only a small part of the total. Overall, we can say that there are some structural shifts in U.S. manufacturing in expected directions, but they are much smaller and more restricted in time than one would probably have anticipated.

As regards the annual data, over the period 1869–1913 as a whole, *producer durables* and *consumer durables* grow at almost identical rates, but between 1899 and 1907 the former is considerably above the latter (indexed to 1869). Between 1899 and 1907, *producer durables* shows strong cyclical movements, presumably through the operation of the 'accelerator', ending in a heavy crash in 1907. It is again surprising that two of Shaw's other categories, namely *consumer semi-durables* and (especially) *consumer perishables*, more or less keep pace with *durables* until about 1907, so the long-term restructuring is mostly the result of the last few prewar years.

The results for Germany indicate a sustained growth throughout of the chemicals industries, for which the data begin in 1870. *Metals* grows at rates very little different from manufacturing as a whole over the years 1860–95, but then appears to jump on to a new track paralleling that of *chemicals*, though well below it if indexed to 1870. 'Traditional' sectors fall behind after the late 1890s, with a particularly marked decline in *textiles*.

The production indices as a whole thus suggest structural changes at the sectoral level in the directions expected, if perhaps less rapidly than expected. The U.K. shows some 'retardation' in the second half of the period, except for the acceleration in chemicals. The U.S.A. also slows down at this time, but from a much higher growth base. Germany shows some acceleration overall. The data confirm the expected sectoral differences across countries, including the earlier rise of chemicals in Germany and the expansion of producer durables (basically machinery and transport equipment except automobiles) in the U.S.A. from the late 1890s.

Technology

We do not as yet have robust indicators of technological strengths of the three countries to relate to these sectoral patterns. The results square with what is known about the growth of research and development laboratories in each case. The early development of research laboratories in German chemicals companies is well known, especially for their focus on specific industrial objectives such as producing synthetic indigo, and for their achievements in pharmaceuticals such as aspirin.[31] American laboratories, and especially those that took the important early rôle of contract research, undertook somewhat more routine tasks such as materials testing over these years.[32] The in-house central laboratories in electrical companies such as General Electric and chemicals companies mostly came around the turn of the century,[33] though some smaller operations preceded them.[34]

Data on patenting are not readily available at the sectoral level. Figure 13.3 charts total patenting in the U.S.A. from 1860. The U.S.A. in these figures is the 'home country', and naturally the extent of patenting from U.S. sources is much higher in level than that from foreign sources. We have divided the number of patents from U.S. sources by 10 to make the graph interpretable. The faster growth of foreign patenting from its low early base is to be expected for similar reasons – some considerable part of it is probably accounted for by general growth in trade, rather than by the spread of industrial production and technological development across countries. The figures are also distorted by differential fees charged by the U.S. Patent Office. In the mid-1850s, foreigners were charged ten times as much as U.S. citizens, and British patentees substantially more still, to exercise their intellectual property rights in the U.S. (as Whitworth noted). The disaggregated country data show a surge for both Britain and Germany in the later 1890s, which is consonant with the trade and technology patterns

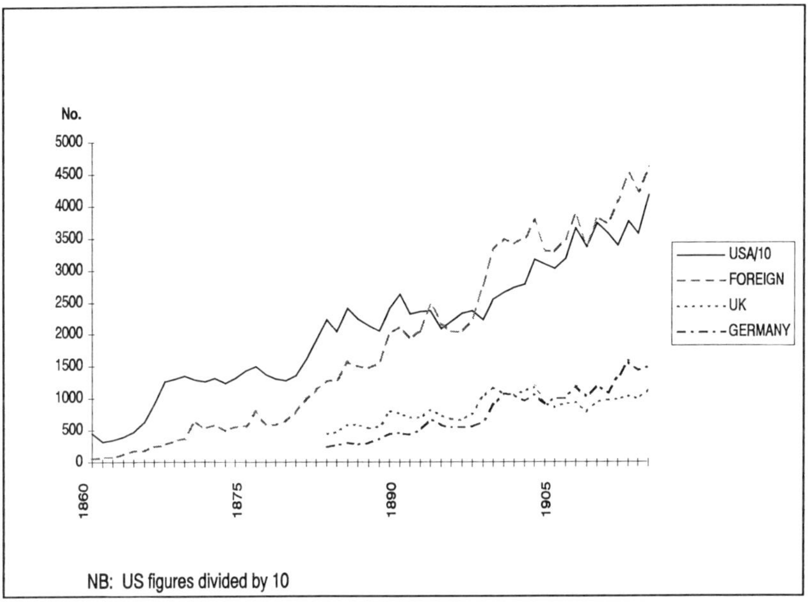

Source: U.S. Department of Commerce (1977), *Technology Assessment and Forecast: 7th Report*, Washington: Patent and Trademark Office, Tables A-1, A-3.

Figure 13.3. Patenting by U.S. Residents and Foreigners in the U.S.A., 1860–1914

observed elsewhere in this chapter. More telling, perhaps, is the rise of German sources relative to British, crossing over in the early twentieth century, especially as fee changes are likely to have benefitted the latter more than the former. The high patentability of chemicals, in which the Germans were so strong, probably accounts in part for this crossover.

Employment

Sectoral shares in employment partially reflected changes in the structure of production, with, for example, the share of agriculture in employment in the U.K. falling from 15 per cent in 1871 to 8 per cent in 1911, in the U.S.A. from 52 per cent in 1870 to 35 per cent in 1910, and in Germany from 42 per cent in 1882 to 34 per cent in 1907. Our concern here is more with the occupational than with the sectoral structure. The pattern in the U.K. from 1871 to 1911 shows a sustained rise throughout of the percentage of the occupied population working in commerce and finance, government and professional services.[35] In other respects we can obtain only proxies for the occupational structure. The data for 1911 collated by Routh[36] give the following shares: professional and managerial workers 14 per cent, clerical workers 5 per cent, foremen and supervisors 1 per cent, skilled manual workers 31 per cent, semi-skilled manual workers 39 per cent, unskilled manual workers 10 per cent. Data are available for the U.S.A. in 1900 and 1910, and show professional and technical workers at about 7 per cent in both years, and managerial and administrative workers at about 10 per cent. The largest growth occurs in the share of clerical workers, which rises from about 5 per cent to 8 per cent of the labour force between 1900 and 1910; there is also a substantial tranche of employment in sales, of about 7 per cent in both years. Shares of craftsmen (about 15 per cent) and operatives (about 21 per cent) are stable over this decade, but the share of other non-farm labourers falls sharply, as to a lesser extent does that of domestic employees. Finally, data for Germany are available for 1882, 1895 and 1907,[37] which include a component for unpaid family members (which rises considerably between 1895 and 1907). The most marked trend in these data is the fall in the proportions of independent proprietors and higher officers, matched by a commensurate rise in that of salary earners. The rise of the proportion of wage-earners in industry more or less offsets the fall of wage-earners in non-industry. As in the U.S.A., the share of domestic servants falls in the early twentieth century.

These patterns, sketchy as the data are, are suggestive of a number of important trends. The U.K. data point to a trend towards professionalization in British society.[38] This has been argued at greater length by one of us for the important case of engineering,[39] which in the latter nineteenth century – and indeed up to the present day[40] – was faced by a choice essentially between becoming commercial and becoming professional. In Britain, the high ground was taken by professionalization: engineers sought to raise their social status through membership of professional societies, introduction of specialized university degrees, exclusion of lower-status elements (for example, women), and the like. In the U.S.A., the commercial route was chosen, with a pronounced increase of employment of engineers in the emerging large corporations, the growth of business-oriented college education in engineer-ing, etc.[41] The reporting of a large number of salespersons in the U.S.A. is also indicative of a marketing-oriented society. The parallel in terms of perceived trends in Germany would lie in the growth of bureaucracy, modelled originally on the Prussian Army;[42] this too appears to be mirrored even in our limited data in the shift from entrepreneurs to salary earners in Germany.

Regrettably our data do not allow us to say much about differences in occupational structure among manual workers, including skill structures, but more can be learnt from analysing wage differentials, as we shall do next.

Skilled and Unskilled Wages

A recent exhaustive study by Williamson[43] has provided us with a clear picture of the contrasting patterns of 'real wages' across countries since 1830. The expression 'real wages' is here used in its standard sense to mean money wages adjusted for the cost of living, for unskilled labourers. The international relativities are computed by Williamson according to purchasing-power-parity (P.P.P.) assessments. He shows that the patterns and major underlying influences differ according to the sub-period chosen for study. For the sub-period which is of most interest here, the late nineteenth century, Williamson finds a period of substantial convergence in such real wages, more so than in data based on Maddison's estimates on G.N.P. per worker-hour: U.S. real wages halve their lead over the U.K. between 1855 and 1913, though Germany does not catch up on the U.K. between 1870 and 1900. Williamson leaves open the causes of this convergence, including the rôle of price movements rather than wage movements; O'Rourke and William-son[44] make a strong argument for convergence in factor prices being led by convergence in commodity prices, but Williamson's more recent Presidential Address to the Economic History Association[45] seems to play this down.

In 'The Evolution of Global Labor Markets', Williamson does not deal with differences in skilled labour or wage differentials. He has, however, with various colleagues, written extensively on the subject in other studies of particular countries.[46] These two issues can fruitfully be linked, which will be our main task here.

Figure 13.4 estimates the wage ratios of skilled to unskilled manual workers in the three countries from 1870 to 1913, estimated for what we referred to above as 'operational skills'. The data in this form suggested a much higher average wage ratio in the U.S.A. than in the other two countries. However, it was apparent that like was not being compared with like. In order to allow a more valid comparison of the cross-country differences, a benchmark comparison for manual workers in the engineering and building industries of these three countries in 1905 (1909 for the U.S.A.) was obtained from the Board of Trade data cited in Williamson.[47] The results in tabular form are given for selected years in Table 13.1, and the results for all years are incorporated into Figure 13.4. In this format, the skill differential for the U.S.A. is similar to that in the U.K. from the 1870s to the 1890s, whereupon that in the U.S. starts to climb sharply, up to 1913. The U.K. differential in 1913 is instead about the same as in 1900. The data for Germany suggest a higher gap from the 1880s to the early 1900s; thereafter the German wage ratio falls as steeply as the American one rises.

Table 13.1. Relative Skill Differentials, 1880–1913 (U.K. in 1890 = 100)

	1880	1890	1900	1910	1913
U.K.	102.66	100.00	93.35	100.61	98.06
Germany	104.61	109.80	106.73	94.01	93.75
U.S.	99.90	98.06	105.15	110.56	112.93

Sources: U.K.: Knowles and Robertson, 'Wages of Skilled and Unskilled Workers'; U.S.A.: Williamson and Lindert, *American Inequality*, Table D5; Germany: Bry, *Wages in Germany*, Table A4.[48] See also Williamson, 'The Evolution of Global Labor Markets'; and compare V. Zamagni (1989), 'An International Comparison of Real Industrial Wages 1890–1913: Methodological Issues and Results', in P. Scholliers (ed.), *Real Wages in 19th and 20th Century Europe: Historical and Comparative Perspectives*, New York: Berg, pp. 107–39.

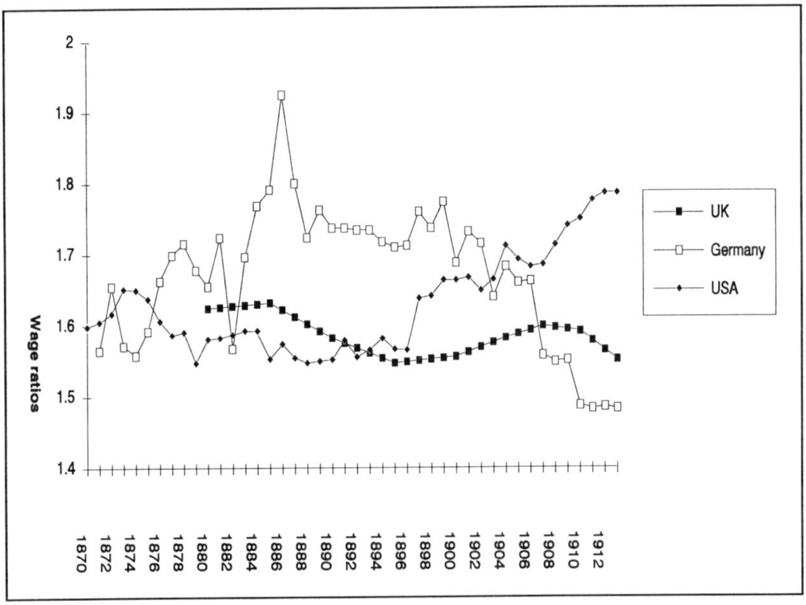

Note: The original data series for the U.K. includes observations at five-year intervals only; intervening observations have been interpolated using a simple linear trend.

Sources: As for Table 13.1.

Figure 13.4. Skill Differentials for Manual Workers in the U.K., Germany and the United States, 1870–1913

The trends in the skill differential suggested by Figure 13.4 can, with some modifications, be supported by evidence from a number of other sources. Williamson[49] provides average annual earnings data for the U.K. between 1871 and 1911 which suggest a steadily declining differential between the earnings of skilled and unskilled manual workers. The data for British shipbuilding given by Pollard and Robertson[50] are particularly interesting because of the careful definition of skills involved – they show a slight rise from 1880 to a peak (for engineers, and so on) in the late 1890s, then a rather sharp fall to 1910.[51] Long's study[52] of manufacturing wages in the U.S. between 1870 and 1890 shows that, after an initial decline in the early part of the period, skill differentials began to increase after the mid-1880s; this broadly confirms the data of Williamson and Lindert,[53] although here the increase started slightly earlier. Another commonly-cited source on historical skill differentials in U.S. manufacturing is Ozanne,[54] who presents a uniquely continuous picture of the skill differential, from 1858 to 1959, for one engineering firm – the important one, for our purposes, of McCormick (from 1901 the McCormick works of International Harvester). While the evidence for the whole period suggests a volatile series of recurring rises and falls, the period between the early 1890s and the outbreak of World War I is a period of prolonged and rapidly rising differentials.[55]

It is also possible to look at relative wages among different categories of skilled manual workers across the three countries.[56] In the U.K., hourly wage rate data for the period 1870–1920 suggest that the wages of bricklayers increased relative to those of both fitters and turners, in the engineering sector, and also to those of compositors.[57] The wages of fitters and turners, on the other hand, showed no clear trend relative to those of compositors over the majority of the period.[58]

The evidence in Long allows a similar analysis for the United States for the period 1870–90. There, like Britain, the wages of skilled building workers increased relative to those of both skilled engineering workers and skilled printing workers, most notably during the 1880s, while the gap between the wages of skilled engineering and skilled printing workers remained relatively constant.

For Germany, the ratio of skilled workers' wages as between builders (masons and carpenters) and printers rose from the early 1880s up to the First World War, while that between builders and metal workers rose more slowly and erratically over the years 1870–1913;[59] the data for metal workers, however, are of low reliability.

Generally speaking, one can thus consider that differentials for skills follow somewhat similar patterns over time across major sectors in the three countries. The predominant pattern is one of a relative rise for employees in the building trades, which we attribute to these activities being 'nontradeables'. Workers in 'tradeables' like metal manufactures were losers, relatively, despite the growing complexity of their work. An analogous result for the differing regions within the United States was obtained by Sundstrom and Rosenbloom,[60] who found greater wage convergence in occupations linked with 'tradeable' products than in building. Differences in the technologies accompanying the skills required in each occupation (craftsmanship, and so on) may also be influencing this result, but we believe that this impact is smaller.[61] But these explanations do not easily account for the overall difference in trends of skilled versus unskilled manual wages as between the three countries, which is our main concern at present.

Creative and Operational Skilled Earnings

As before, our present interest lies more in the occupational differences than in the sectoral differences in skilled earnings. The problems here with cross-country comparability are probably even greater than for the operational skills discussed above.

Figure 13.5 displays the data of Williamson for higher and lower non-manual workers in the U.K., at the Censuses from 1871 to 1911. Relative to average manual wages, both 'higher professional' and (to a lesser degree) 'clerks' decline, but below them again there is a relative increase in the earnings of 'lower professional' (including teachers). A more comprehensive picture, using annual data, can be taken from the data of Routh,[62] which are for the Civil Service. Relative to the wages of manual civil servants (Routh uses postmen as this indicator), those of the higher ranks (permanent heads, administrative class and executive class) showed a slow decline for most of the period from 1875 to 1914, while those of the lower ranks (including teachers) rose. Such convergence is not exhibited in engineering: the most interesting aspect of Routh's data is the *divergence* between senior engineers (inspectors and above), whose wages show a relative increase from the mid-1870s to the mid-1890s and then stability, and lower grades of engineers, whose pay shows a slight relative fall. This supports the notion of a wedge being driven between professional engineers and those grades regarded more as mechanics in the U.K., as suggested above.

A rather similar narrowing of differentials in the higher earnings brackets occurs in the U.S.A., using data from Williamson,[63] which, however, rely heavily on ministers, and from Douglas,[64] whose observations run only from 1890 (Douglas's figures indicate partial divergence in the 1890s until convergence resumes before the turn of the century). For Germany we have very little data. The one-off figures of Scitovsky[65] for professors (1897) and civil servants (1913), expressed as multiples of average *per capita* incomes, show high ratios compared with interwar and postwar data. It is perhaps

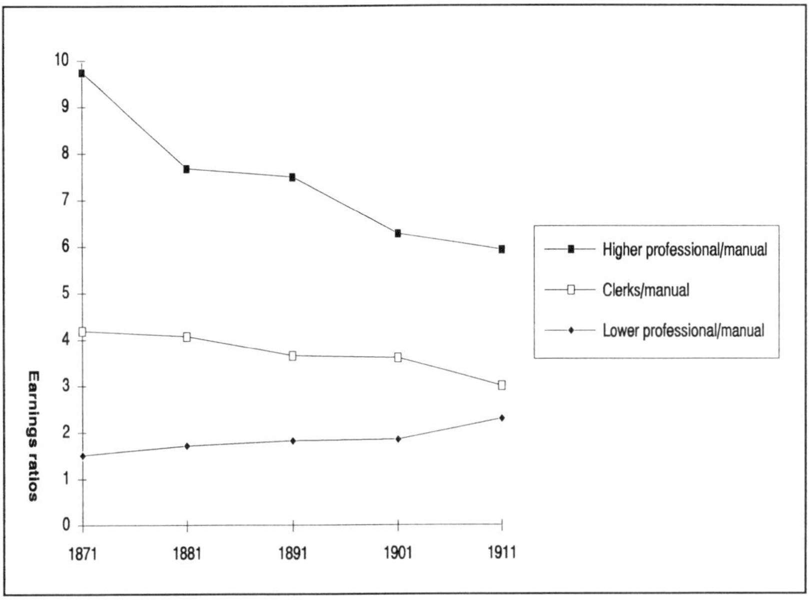

Notes: Earnings ratios are relative to (unweighted) average earnings of manual workers. Calculations are based on nominal annual earnings data. 'Higher professional' includes: government high-wage workers, clergy, solicitors and barristers, surgeons, medical officers, engineers and surveyors; 'lower professional' includes teachers; 'clerks' excludes government clerks.

Source: J.G. Williamson (1982), 'The Structure of Pay in Britain, 1710–1911', *Research in Economic History*, **7**, 1–54.

Figure 13.5. Relative Earnings of Higher and Lower Non-Manual Workers in the U.K., 1871–1911

stretching the data to or beyond its limits to say much about the relative differentials across countries, but an amalgam of the above sources would suggest that the earnings ratios for high salary groups were greatest in Germany and least in the U.S.A. Again there are many problems with comparing unlike occupations.

For more direct evidence we have to rely on more scattered data. One of the leading 'stylized facts' that came out of the debate over Anglo-American technology differences was that American wages for skilled workers were relatively higher than for Britain, hence encouraging technologies which 'saved' on skilled labour in the U.S.A.[66] The empirical evidence for or against this belief is weak, and there is general agreement about the lack of a quantitative basis for any strong assertions. Tweedale[67] has several observations for Sheffield steel, which had strong links with America; indeed many of its more skilled employees were sent across the Atlantic when the U.S. industry was in its infancy. In the 1860s, 'In two days work in a Pittsburgh furnace, a Sheffield melter could earn as much as his English counterpart earned in a week [presumably six days]'.[68] In the 1870s, Pittsburgh melters were earning four to five times as much in a week as in Sheffield, but by the 1880s they were earning little over double, and by the 1890s only about one-third more.[69] The same source additionally shows the great range of variation as between one 'skill' and another: for example, Pittsburgh melters in 1872 were paid about two-and-a-half times the wages of cokers or converters, who were also regarded as skilled in some sense.

In our view, the large gaps found especially in the earlier years of our period as between the U.S. and U.K. are partly accounted for by the American data of Tweedale and others, including an implicit element for creative as well as operational skills.[70] It may be useful to think in terms of 'capabilities' in this respect. The literature on newly industrializing countries finds it convenient to divide firms' abilities to absorb inward transfers of technology into operational, investment, and innovative.[71] The first level refers to capabilities of taking over the day-to-day operations of running production, the second to capabilities of constructing new plants on the same design and getting them started up, and the third to capabilities of adapting and improving upon existing designs. The Sheffield steel-workers brought into the United States were being induced in for their investment and innovative skills, in these senses, as much as for their operative skills, coming as they were to a country in the early stages of imitating Britain in this industrial segment.[72] Once operational skills were indigenized, there was less need to pay for the investment and innovative aspects, since these too were becoming indigenized (sometimes via the children of the British immigrants), and along somewhat different paths from those stressed in the U.K.[73] It was also the case that British skilled workers were coming to be denigrated for what was seen in the U.S. as antisocial behaviour: absenteeism, alcoholism, obstructivism, and so on,[74] a pattern which Bruland[75] also found for British skilled workers in Norway.

Some of the larger gaps occasionally observed in skilled earnings as between the U.K. and U.S.A. thus probably reflect these implicit differentials for 'learning' contributions, that is, paying for the brains as well as the (skilled) hands. It may be noted parenthetically that, in so far as this was the case, it contradicts the view widespread in the 'new growth theory' that learning from production is an unpaid-for externality.

For Germany we do not have sufficient indicators as yet of learning behaviour of these kinds. The major difference in production structures between Germany and the other two countries in any case derived from the early expansion of chemicals in Germany, which was in turn associated with the early development of formal research and development laboratories. Thus learning structures in Germany, to the extent that they differed from the U.K. and especially from the U.S.A., were built around more formal 'learning' in research and development and in the natural sciences. Professional management seems to have earned relatively highly in Germany, judging by the few observations we have, but that management was also deeply infused with technical and engineering talent.[76] In that respect, there was some fusion between cognitive and interactive skills, in the sense of Wolff, which helped to generate the high differentials for top bureaucrats observed in Germany.

EMPIRICAL ISSUES: TRADE

We are now in a position to offer a partial explanation of the wage structures observed, and especially of the various patterns of skill differential. Two rather overlapping suggestions have in fact emerged. One is the differing types of complementarity across the skill divisions we are employing (operational/creative, and so on), brought about partly for socio-economic reasons (professionalization in the U.K., commercialization in the U.S.A., bureaucratization in Germany, and so on). The second is the rôle of technological opportunity, including the differences in sectoral patterns of development (chemicals in Germany, and so on), and temporal differences in stages of development (catching up in the U.S.A. and Germany earlier in the period). To the latter we would be tempted to add differences in technological opportunity arising out of inherent possibilities of the technologies themselves,[77] which helped to cause the technological paths pursued by the three countries to drift in rather different directions. Already it is apparent that it is very difficult to say which is cause and which is effect in these interrelationships.

Such problems are compounded when we introduce the remaining factors we wish to discuss, which relate to international trade and factor flows, under which we subsume a few remarks about labour markets.[78]

Figure 13.6 highlights certain common trends between the three countries in the commodity composition of manufactured exports over this period. Machinery and transport equipment was taking a growing share, with the largest increase occurring in the U.S.A.[79] Germany and the United States increased their proportion of metals and metal manufactures, while for Britain they remained constant. Textile goods were accounting for a declining share in all three countries, but still remained the relatively largest sector in the U.K.

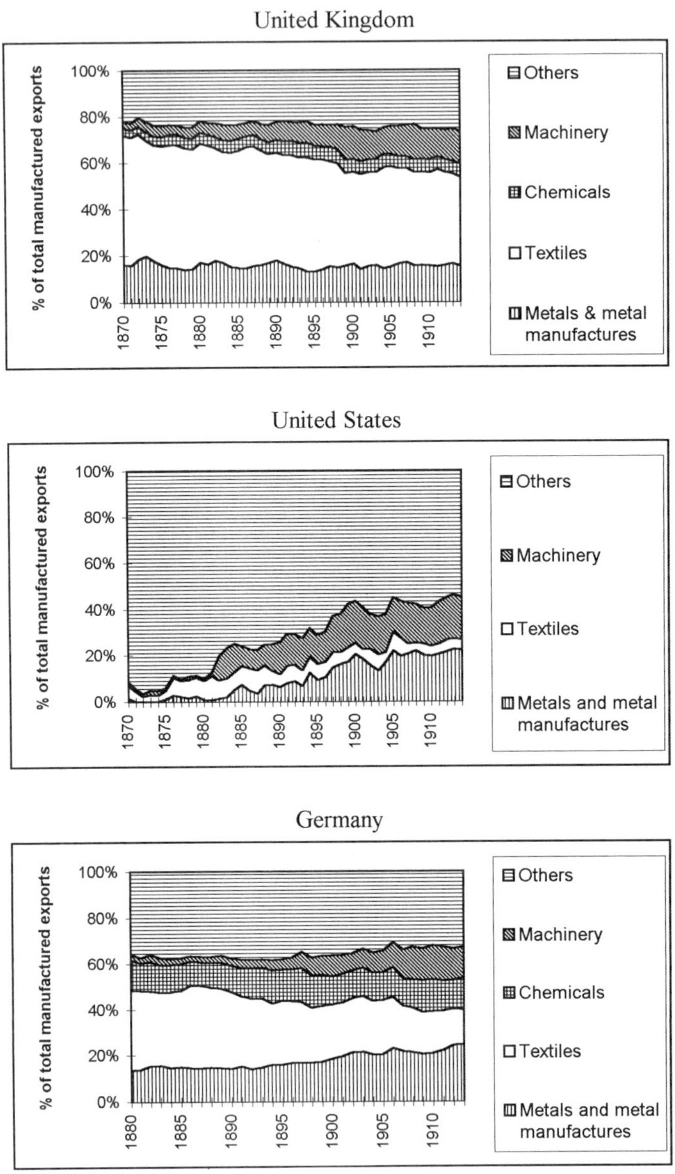

Sources: U.K.: Mitchell, *British Historical Statistics*, Table 9.11; U.S.A.: Department of Commerce, *Historical Statistics*, Table U274–294; Germany: Hoffmann et al., *Das Wachstum der Deutschen Wirtschaft*, Table 126.

Figure 13.6. Shares in Manufactured Exports, U.K., U.S.A. and Germany, 1870–1913

Figure 13.7 below portrays the absolute value of exports of the U.K., Germany and the United States in these commodity sectors, and highlights more clearly how the three countries were performing relative to one another. From this diagram, the 'revealed comparative advantage' of the three countries, and how it changed, can readily be deduced.[80]

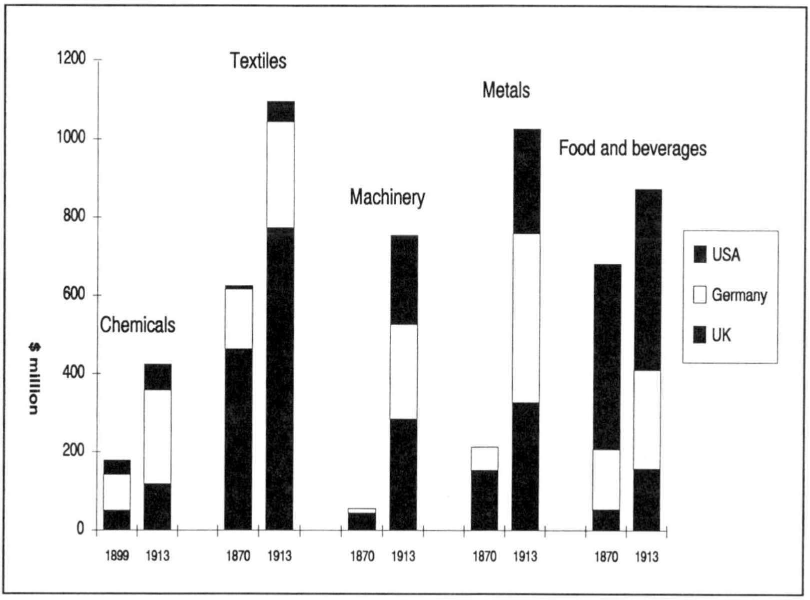

Sources: as for Figure 13.6, plus A. Maizels (1971), *Industrial Growth and World Trade: World Trends in Production, Consumption and Trade in Manufactures*, Cambridge: Cambridge University Press, Tables A51, A61, A65.

Figure 13.7. Exports of Manufactures by Commodity and by Country, 1870–1913

Figure 13.8 uses Maizels' data to note the corresponding changes in imports. Imports can be high or rising because of domestic deficiencies in production, or more often in recent times through growth of 'intra-trade' in similar items (indicative of demand factors as well as supply). In the latter case they need to be compared more directly with exports. Germany's substantial but stable import shares in machinery and chemicals probably reflects such intra-trade, at least over these late years. Brown[81] explains the persistence of German textile exports in this period as being due to imperfect competition in the context of intra-trade, notwithstanding the strong 'revealed comparative disadvantage' of German textiles.[82] The rising shares of machinery and transport equipment for the U.K., and of chemicals for the

U.S.A., equally probably reflect domestic supply weaknesses, at least in some segments of these industries. Annual data[83] show that the share of machinery and transport equipment in imports rose monotonically from 1896 (the data are not separated from iron and steel until 1896). More disaggregation of the data from Maizels indicates that the share of machinery in exports rose slightly more than did that of transport equipment between 1899 and 1913, but that the reverse was true in imports, which also casts doubt on the intra-trade explanation in this area. Much more detailed analysis of the machine tool segment by Floud[84] showed that Britain remained strong in heavier items but lost markets, including its own market, to the U.S.A. in lighter items, such as were associated with the 'American invasion of Europe'.

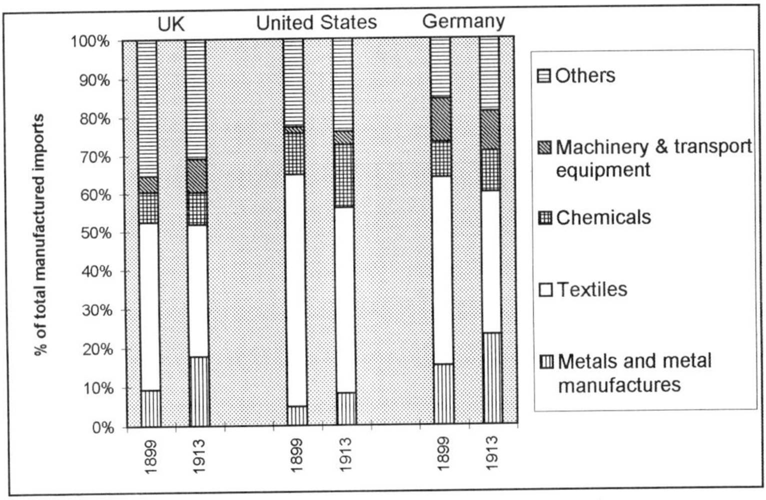

Source: Maizels, *Industrial Growth and World Trade*, Tables A17, A23, A25.

Figure 13.8. Imports of Manufactures by Commodity, 1899 and 1913

We have also examined data for bilateral trade between these three countries. As a percentage of total British exports, those to the U.S.A. fell throughout the period, but there was a partially offsetting rise in those to Germany from the 1890s.[85] But at the same time U.S. exports to Britain fell from about the 50 per cent share they had held in the 1880s to a one-third share by the turn of the century and a one-quarter share by the end of the period.[86] The lower share of Germany in U.S. exports, however, rose consistently between 1870 and 1913. Finally, there is a break in the series for German exports in 1889 which makes long-term comparisons awkward, but the data thereafter show slowly falling export shares to both the U.K. and U.S.A.[87] This suggests that the trade impacts were felt predominantly not in one another's markets, but in other markets. Germany's foreign trade was typically of the *Mitterstellung* (intermediary) kind – importing capital goods from advanced countries like Britain and exporting products based on them to less developed countries further south and east in Europe.[88] These conflicts in third markets are supported by some of the secondary literature on the subject.[89]

If the findings on trade are combined with the preceding ones on wage differentials, some intriguing contrasts appear. In the U.K., the general lowering of differentials tended to accompany the expansion of those exports that were sensitive to foreign competition, and since the newer goods were likely to have been skill-intensive in the ways relevant to the measure of differentials, it can be concluded that foreign competition eroded the margins for skills. An especially interesting case is that of shipbuilding – Pollard[90] demonstrated that British shipbuilding productivity up to 1914 was well above that of the rest of the world (U.S.A. included), and that the only serious competition came from Germany; but the data of Pollard and Robertson reported above[91] show the erosion of skill differentials from the late 1890s. However, there were periods when this relationship was weak or reversed, including the increase of machinery exports from about 1898 to 1908, which went alongside an apparent increase in skill differentials. For Germany, the erosion of skill margins was held off, perhaps through imperfect competition, until about 1906; after that, the skill margins fell sharply while machinery exports continued to rise. Possibly there was a connection here with growing political anxieties in Germany.

The most interesting results are, however, for the U.S.A. Figure 13.9 shows a marked and positive synchrony between skill margins and exports of manufactures. In other words, rising exports in skill-intensive products accompanied a *rise* in rewards for skills.[92] To the extent that the two are directly related, this – in conjunction with what has just been said about America's main rivals – suggests that cost factors were not crucial, and that either American exports were coming to dominate world markets on quality grounds, or demand factors predominated (and these two are probably linked). The detailed empirical survey of Ferguson[93] uses Consular reports and other contemporary material to argue that American imports into Europe won on product quality rather than cheapness.

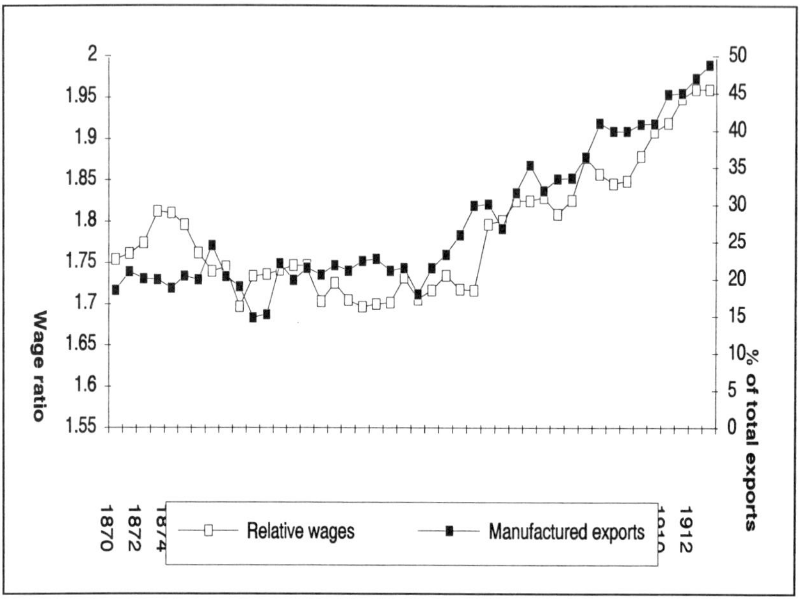

Sources: Williamson and Lindert, *American Inequality*, Table D5; Department of Commerce, *Historical Statistics*, Table U274–294.

Figure 13.9. Relative Wages and the Share of Manufactured Exports, U.S.A. 1870–1913

EMPIRICAL ISSUES: INTERNATIONAL FACTOR MOVEMENTS

While Figure 13.9 seems striking, it does not seem valid to deduce any simple causal relationship between wage differentials and skill-intensive exports. There is general agreement in the literature on U.S. differentials that the main driving force underlying their widening from the 1890s was the 'new' immigration of unskilled labour, particularly from southern and eastern Europe.[94] Emigration from Germany to America was notable from about 1880 to the early 1890s, but then fell to low levels. In absolute numbers, emigration from the U.K. to America remained higher than that for Germany, though also tending to fall until the late 1890s, but thereafter it rose again up to 1913.[95] The occupational background of all immigrants into the U.S.A. is shown in Figure 13.10; the rise in 'new immigration' is hinted at by the proportion recording agriculture as their occupation.

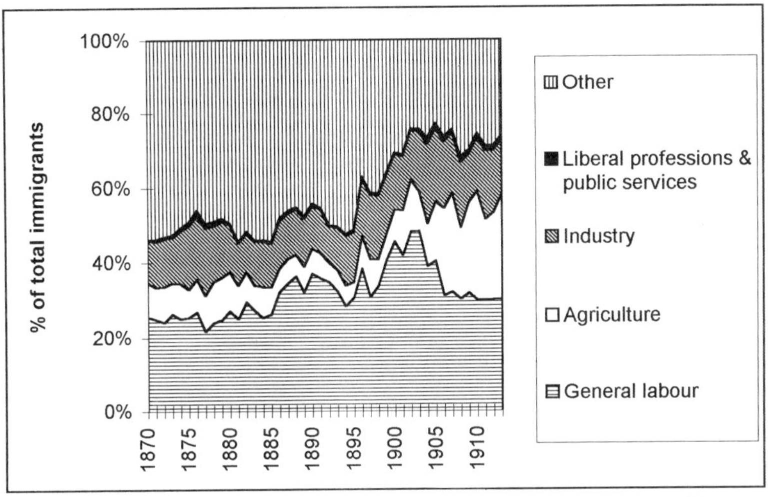

Source: Ferenczi and Wilson, *International Migrations*, vol. I, Table 8, pp. 399–400.[96]

Figure 13.10. Total Immigration into the U.S. 1870–1915, by Occupation or Economic Sector

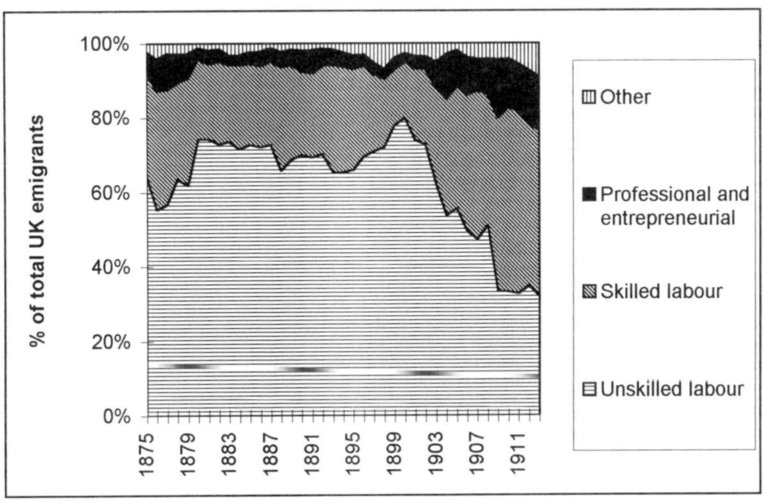

Source: Thomas, *Migration and Economic Growth*, Table 80, p. 382.[97]

Figure 13.11. U.S. Immigration Originating from the U.K. by Broad Skill Group, 1875–1913

But for migration from the U.K. to the U.S.A., the skill profile was very different, as Figure 13.11 shows. The initial decline in the proportion of skilled emigrants from the mid-1890s was quickly reversed, and it continued to rise in the Edwardian period, at the time when absolute numbers were also rising. This was the period in which American skill differentials were widening most rapidly; the British picture is more confused. These phenomena clearly involve a complex set of causal interrelationships of their own, of kinds which have been analysed in the huge literature on this migration, but we cannot hope to do justice to them here. It is evident that both demand factors (of the rising U.S. differentials) and reverse causation (e.g. the impact on both American and British differentials) operated.

The parallel between international movements of labour and of capital has long been drawn,[98] in a literature which has shown a strong synchrony in time but a great difference in region – flows of capital out of Europe to the New World, including the U.S.A., took place at about the same time and with similar violent fluctuations as the outward flows of labour, but as the labour increasingly came from southern and eastern Europe, the importance of north and west Europe, and especially the U.K., for capital remained undiminished. Thus flows of labour and capital altered the underlying factor proportions on which trade could be based and the factor returns established, via the complex set of causal interactions just hinted at.

These external flows of capital have also been examined at the corporate level, in particular the establishment overseas of branch plants of manufacturing enterprises, that is, nascent multinational companies.[99] This literature itself descends from a more polemical tradition that attempted to explain the origins of the First World War.[100] The results for American mechanical engineering firms are especially interesting to us.

> The American machinery companies successfully 'invaded' Europe because so few European enterprises were able to produce better-performing machines at comparable prices and to provide the necessary marketing services of demonstration, installation, after-sales repair and service, and consumer credit. Precisely because the market power of these American firms was based on such functional supremacy, they rarely felt the need to enter into agreements with European competitors either before or after World War I, unlike the leading American firms in oil, glass, chemicals, nonferrous metals and other materials, and electrical machinery.[101]

The question of the extent to which trade and foreign investment (or international migration) acted as substitutes for each other has recently been addressed by Collins et al.,[102] who argue very strongly that they did not. Ideally, for present purposes we would like to know more about the wage and skill patterns associated with this Foreign Direct Investment (F.D.I.), but little appears to have been collated on this question until a much later period. We shall have to leave this matter, like other concerns of this chapter, on the research agenda.

CONCLUSION

Strong parallels between the late nineteenth century and the modern era have been drawn by Williamson.[103] Both were periods of substantial convergence in 'real wages' among the industrializing countries, though the most important determinant a century ago, namely international migration, has been negligible in recent times.

In this chapter we have suggested, on the basis of the three leading industrial countries, that such convergence did not extend to skilled wages. In the U.S.A., the differential between skilled and unskilled wages widened from the 1890s, while in Britain it was narrowing from mid-century, and in Germany narrowing from the end of the century. The international migration factor clearly plays some part in the altering differentials, most obviously in regard to the American unskilled worker and perhaps to some degree to the British skilled worker. Our main concern has, however, been with the complex interplay of technology and trade factors associated with these patterns, which to some extent match the widening differentials in industrial leaders, and the narrowing of differentials in followers, as of recent years.

Two main themes related to technology run through this period. One is the balance between the initial leader (Britain in its workshop of the world phase) and those catching up, that is, the U.S.A. and Germany. A second cross-cutting theme is the balance between the New World and Europe, in which the links to technology run through resource bases and skills. The catching-up issue was still evident in the earlier part of the period studied here, when wages for the skilled in a catching-up country like the U.S.A. could embody a premium for creativity, which we believe was attributable to the knowledge potential of such workers. Initially many of these skilled employees were 'imported' from Britain, but as later generations grew up locally acquiring these skills, there was less need for 'imported skills' and the premia subsided. The balance of power in technology switched from Britain to the U.S.A. in engineering and to Germany in chemicals.

The pattern of skill differentials later in our period is, however, mainly a story of New World versus Old, as Britain and Germany behave rather similarly, and in opposite ways to the U.S.A. To explain this brings in the question of trade, and its interrelationships with technology. The growing power of the United States in these technologies did not seem to come from any massive scientifically-based technological leadership in these areas. As we and others have pointed out elsewhere,[104] America was still predominantly a borrower of radically new technologies, which of course is why the skills associated with them needed at first to be 'imported'. The U.S.A. instead excelled at working the new technologies into commercially powerful systems. This involved characteristics that are especially linked with the

U.S.A. in this period: the 'American system of manufactures' based on interchangeable parts, the vertical interlinkages between mechanical engineering and industrial users (and hence between 'learning by doing' and 'learning by using'), the rise of corporate organization, as stressed by Chandler, and not least the absorption of skilled engineers into the needs of 'big business'.

This American leadership in trade and technology was sector-specific, being dominated by certain branches of mechanical engineering. Essentially the American supremacy combined key product innovations with an underlying strength in process technologies (the aforesaid 'American system of manufactures'). The additional margins offered for 'creative' skills narrowed, except in so far as new opportunities arose through professionalization (for top British engineers) or management (in Germany). The gains in the U.S.A. were stronger for 'operational' skills associated with machine-based technologies.

The rôle of such skilled workers was enhanced through commercial success at home and abroad, including Foreign Direct Investment. Engineers acquired a new source of 'creative' skills in addition to 'operational' skills through their involvement in corporate management. In Germany, too, engineers were heavily involved in management, though commercial aspects in an increasingly competitive world were less well attended to. In the U.K., professional engineers gradually acquired such rôles, but the lower echelons of engineers were sidelined. The result was an erosion of their earnings differentials and, perhaps more profoundly, of the technological and commercial strength of the British economy.

NOTES

1. For example, A. Wood (1994), *North-South Trade, Employment and Inequality: Changing Fortunes in a Skill-Driven World*, Oxford: Clarendon Press.
2. For a partial review, see J.-P. Cotis, J.-M. German and A. Quinet (1996), 'Technical Progress, International Trade and Low-Skilled Labour', *O.E.C.D. S.T.I. Review*, **18**, 187–228.
3. This chapter is the first output from a project between the two named authors and Adrian Wood, on 'Technology, Trade and Inequality', substantially funded by the U.K. Economic and Social Research Council (E.S.R.C.), grant no. R000236878. We are indebted to Adrian for his constant advice and range of ideas.
4. Reprinted in N. Rosenberg (ed.) (1969), *The American System of Manufactures*, Edinburgh: Edinburgh University Press.
5. For example, R.E. Lucas junior (1988), 'On the Mechanics of Economic Development', *Journal of Monetary Economics*, **22**, 3–42; R.J. Barro (1991), 'Economic Growth in a Cross-Section of Countries', *Quarterly Journal of Economics*, **106**, 407–43.
6. Rosenberg, *The American System of Manufactures*, p. 389; see also L.G. Sandberg (1979), 'The Case of the Impoverished Sophisticate: Human Capital and Swedish Economic Growth before World War I', *Journal of Economic History*, **39**, 225–41. Sandberg notes the apparently paradoxical co-existence of excellent education and lack of growth in Sweden for most of the nineteenth century.

7. K.J. Arrow (1962), 'The Economic Implications of Learning by Doing', *Review of Economic Studies*, **29**, 155–73.
8. Adam Smith (1776), *An Inquiry into the Nature and Causes of the Wealth of Nations*, London: W. Strahan and T. Cadell; bicentennial edition (1976), ed. by R.H. Campbell, A.S. Skinner and W.B. Todd, Oxford: Clarendon Press, p. 21.
9. Rosenberg, *The American System of Manufactures*, p. 204.
10. For example, H.L. Carmichael and W.B. MacLeod (1993), 'Multiskilling, Technical Change and the Japanese Firm', *Economic Journal*, **103**, 142–60.
11. D.R. Howell and E.N. Wolff (1992), 'Technical Change and the Demand for Skills by U.S. Industries', *Cambridge Journal of Economics*, **16**, 127–46; E.N. Wolff (1996), 'Technology and the Demand for Skills', *S.T.I. Review*, **18**, 95–123.
12. See the special issue of the O.E.C.D.'s *S.T.I. Review* for 1996.
13. For example, D.N. McCloskey (1981), 'Magnanimous Albion: Free Trade and British National Income, 1841–1881', in D.N. McCloskey, *Enterprise and Trade in Victorian Britain: Essays in Historical Economics*, London: Allen & Unwin, pp. 155–72.
14. R.A. Church, A. Hall and J. Kanefsky (1986), *The History of the British Coal Industry, vol. 3: 1830–1913, Victorian Pre-Eminence*, Oxford: Clarendon Press.
15. C.K. Harley (1989), 'Coal Exports and British Shipping, 1850–1913', *Explorations in Economic History*, **26**, 311–38.
16. Data are from B.R. Mitchell (1987), *British Historical Statistics*, Cambridge: Cambridge University Press, p. 170.
17. J.G. Treble (1987), 'Sliding Scales and Conciliation Boards: Risk-Sharing in the late Nineteenth-Century Coal Industry', *Oxford Economic Papers*, new series, **39**, 679–98.
18. S.B. Saul (1967), 'The Market and Development of the Mechanical Engineering Industries in Britain, 1860–1914', *Economic History Review*, 2nd series, **20**, 111–30.
19. D. Nelson (1981), 'The American System and the American Worker', in O. Mayr and R.C. Post (eds), *Yankee Enterprise: the Rise of the American System of Manufactures*, Washington: Smithsonian Institution Press, pp. 171–88.
20. G.N. von Tunzelmann (1995), *Technology and Industrial Progress: the Foundations of Economic Growth*, Aldershot: Edward Elgar, chapter 6.
21. C.H. Feinstein (1972), *National Income, Expenditure and Output of the U.K., 1855–1965*, Cambridge: Cambridge University Press.
22. A. Maddison (1991), *Dynamic Forces in Capitalist Development: a long-run comparative view*, Oxford and New York: Oxford University Press, p. 7.
23. U.S. Department of Commerce (1975), *Historical Statistics of the United States: Colonial Times to Present*, vols. I and II, Washington: Bureau of the Census.
24. Maddison, *Dynamic Forces*, p. 6.
25. C.H. Feinstein (1990), 'What Really Happened to Real Wages?: Trends in Wages, Prices, and Productivity in the United Kingdom, 1880–1913', *Economic History Review*, 2nd series, **43**, 329–55. Feinstein's investigations conclude, through a process of elimination, that the U.K. figures are probably overstated for the years after 1900.
26. W.H. Shaw (1947), *Value of Commodity Output since 1869*, New York: E.L. Hildreth.
27. W.G. Hoffmann, F. Grumbach and H. Hesse (1965), *Das Wachstum der Deutschen Wirtschaft seit der Mitte des 19 Jahrhunderts*, Berlin: Springer.
28. Growth rates were calculated from regressions on logarithms of the production series, and Chow tests were used to establish the significance of breaks in trend, using 1889 as a base. The results were generally as expected, and to save space they are not reported here.
29. Shaw, *Value of Commodity Output*.
30. G. Wright (1990), 'The Origins of American Industrial Success, 1879–1940', *American Economic Review*, **80**, 651–68.
31. J. Liebenau (1987), *Medical Science and Medical Industry: the Formation of the American Pharmaceutical Industry*, Basingstoke and London: Macmillan; C. Freeman and L. Soete (1997), *The Economics of Industrial Innovation*, 3rd edition, London: Pinter, chapter 4.
32. D.C. Mowery and N. Rosenberg (1989), *Technology and the Pursuit of Economic Growth*, Cambridge: Cambridge University Press.
33. L.S. Reich (1985), *The Making of American Industrial Research: Science and Business at GE and Bell, 1876–1926*, Cambridge: Cambridge University Press; G. Wise (1985), *Willis R. Whitney, General Electric, and the Origins of U.S. Industrial Research*, New York: Columbia University Press; A.D. Chandler jr. (1990), *Scale and Scope: the Dynamics of Industrial Capitalism*, Cambridge, MA: Belknap Press, p. 84.

34. L. Galambos with J.E. Sewell (1995), *Networks of Innovation: Vaccine Development at Merck, Sharp and Dohme, and Mulford, 1895–1995*, Cambridge: Cambridge University Press.
35. U.K. Department of Employment (1971), *British Labour Statistics: Historical Abstract, 1886–1968*, London: H.M.S.O.
36. G. Routh (1965), *Occupation and Pay in Great Britain, 1906–60*, Cambridge: Cambridge University Press; G. Routh (1987), *Occupations of the People of Great Britain, 1801–1981*, Basingstoke and London: Macmillan.
37. G. Bry (1960), *Wages in Germany, 1871–1945*, Princeton: Princeton University Press.
38. See H. Perkin (1989), *The Rise of Professional Society: England since 1880*, London: Routledge.
39. G.N. von Tunzelmann (1997), 'Engineering and Innovation in the Industrial Revolutions', *Interdisciplinary Science Reviews*, **22**, 67–77.
40. M. Finniston (1980), *Engineering Our Future: Report of the Committee of Inquiry into the Engineering Profession*, London: H.M.S.O.
41. M.A. Calvert (1967), *The Mechanical Engineer in America, 1830–1910: Professional Cultures in Conflict*, Baltimore: Johns Hopkins Press.
42. See M. Weber (1927), *General Economic History*, (transl. F.H. Knight), New York: Greenberg; ed. by I.J. Cohen (1981), New Brunswick: Transaction Books; C.R. Littler (1982), *The Development of the Labour Process in Capitalist Societies: a Comparative Study of the Transformation of Work Organization in Britain, Japan and the USA*, London: Heinemann Educational.
43. J.G. Williamson (1995), 'The Evolution of Global Labor Markets since 1830: Background Evidence and Hypotheses', *Explorations in Economic History*, **32**, 141–96.
44. K. O'Rourke and J.G. Williamson (1994), 'Late Nineteenth-Century Anglo-American Factor-Price Convergence: were Heckscher and Ohlin right?', *Journal of Economic History*, **54**, 892–916.
45. J.G. Williamson (1996), 'Globalization, Convergence and History' (Presidential Address), *Journal of Economic History*, **56**, 277–306.
46. For example, J.G. Williamson and P.H. Lindert (1980), *American Inequality: a Macroeconomic History*, New York: Academic Press; J.G. Williamson (1985), *Did British Capitalism Breed Inequality?*, Boston: Allen and Unwin.
47. Williamson, 'The Evolution of Global Labor Markets', p. 184.
48. K.G.J.C. Knowles and D.J. Robertson (1951), 'Differences between the Wages of Skilled and Unskilled Workers, 1880–1950', *Bulletin of the Oxford University Institute of Statistics*, **13**, 109–27; Williamson and Lindert, *American Inequality*; and Bry, *Wages in Germany*.
49. J.G. Williamson (1982), 'The Structure of Pay in Britain, 1710–1911', *Research in Economic History*, **7**, 1–54.
50. S. Pollard and P. Robertson (1979), *The British Shipbuilding Industry, 1870–1914*, Cambridge, MA: Harvard University Press, p. 195.
51. Pollard and Robertson, ibidem. Their data show no great difference in pattern between machine-based skills like engineering and those of 'joiners machine-men', and craft skills like those of carpenters and blacksmiths, though the latter rose and fell less over the period 1880–1910.
52. C.D. Long (1960), *Wages and Earnings in the United States, 1860–1890*, Princeton: Princeton University Press.
53. Williamson and Lindert, *American Inequality*.
54. R. Ozanne (1962), 'A Century of Occupational Differentials in Manufacturing', *Review of Economics and Statistics*, **44**, 292–9.
55. Ozanne, ibidem, attributes this trend to a combination of growing unionism among skilled workers (1898–1904), and a changing immigration pattern from the north European industrial countries to the east and south European agricultural countries. See below.
56. The data available allow us to make cross-country comparisons of the relative wages of three 'categories' of skilled manual workers: skilled building workers (generally masons, carpenters and bricklayers); skilled engineering workers (fitters and turners in the U.K.; machinists, moulders, patternmakers and blacksmiths in the U.S.; unspecified in Germany); and skilled printing workers (generally compositors).
57. U.K. Department of Employment, *British Labour Statistics*.
58. Evidence from Williamson, 'The Structure of Pay', on the average annual earnings of skilled manual workers in the building, engineering, and printing sectors, generally supports

these findings. Although he finds the wages of skilled engineering workers to have increased relative to skilled building workers over the period 1871–1911, the increase is not large, and changes in the series are consistent with those of the DoE. The irregular data of Knowles and Robertson ('Wages of Skilled and Unskilled Workers') show a widening gap between the skilled:unskilled wage ratio for building and that for engineering between 1885 and 1900, narrowing thereafter.

59. Bry, *Wages in Germany*.
60. W.A. Sundstrom and J.L. Rosenbloom (1993), 'Occupational Differences in the Dispersion of Wages and Working Hours: Labor Market Integration in the United States, 1890–1903', *Explorations in Economic History*, **30**, 379–408.
61. Deskilling may also have been having an impact on skilled wages in printing, but we have not explored this possibility.
62. G. Routh (1954), 'Civil Service Pay, 1875–1950', *Economica*, **21**, 201–23.
63. Williamson and Lindert, *American Inequality*.
64. P.H. Douglas (1930), *Real Wages in the United States, 1890–1926*, repr. (1966), New York: Augustus M. Kelley.
65. T. Scitovsky (1966), 'An International Comparison of the Trend of Professional Earnings', *American Economic Review*, **66**, 25–42, p. 66.
66. H.J. Habakkuk (1962), *American and British Technology in the Nineteenth Century: the Search for Labour-Saving Inventions*, Cambridge: Cambridge University Press, pp. 22–4.
67. G. Tweedale (1987), *Sheffield Steel and America: a Century of Commercial and Technological Interdependence, 1830–1930*, Cambridge: Cambridge University Press.
68. Tweedale, ibidem, p. 17.
69. Tweedale, ibidem, p. 42 and Table 2.2.
70. See Ozanne, 'A Century of Occupational Differentials', who interprets his data for McCormick as showing no long-run convergence or divergence of skill differentials. Unfortunately for us, Ozanne has to change his definition of skills, albeit just once, using the mean hourly rate of the upper sextile of day-rate *production employees* other than supervisors for 1858–97, but the mean hourly rate of pattern *shop craftsmen* for the overlapping period 1888–1941. While the two move similarly during the years of overlap, they are different in terms of our attempt to classify skill structures, and may have moved apart over the longer period.
71. For example, J.M. Katz (ed.) (1987), *Technology Generation in Latin American Manufacturing Industry*, New York: St Martin's Press.
72. The almost complete absence of science in high-quality steel making, and the extreme importance of 'art', is stressed by Tweedale, *Sheffield Steel and America*, chapter 2.
73. In the case of high-quality steel, Pittsburgh more rapidly adopted the Siemens gas melting furnace, and later replaced crucible steel with open-hearth (Tweedale, *Sheffield Steel and America*, pp. 19, 28). For more general differences in technological trends between the U.S.A. and U.K., see von Tunzelmann, *Technology and Industrial Progress*, chapter 6.
74. Tweedale, *Sheffield Steel and America*, p. 18.
75. K. Bruland (1989), *British Technology and European Industrialization: the Norwegian Textile Industry in the mid Nineteenth Century*, Cambridge: Cambridge University Press.
76. B. Dornseifer and J. Kocka (1993), 'The Impact of the Preindustrial Heritage: Reconsiderations on the German Pattern of Corporate Development in the late 19th and early 20th Centuries', *Industrial and Corporate Change*, **2**, 233–48.
77. von Tunzelmann, *Technology and Industrial Progress*, p. 198.
78. Variations in economic conditions as a determinant of wage differentials were largely rejected by Ozanne ('A Century of Occupational Differentials') in his study of McCormick. To the extent that these worked through changing labour supplies and demands, they will be touched on below. Unionization is another factor that we cannot do justice to here, except inasmuch as unionization itself is usually regarded as fluctuating according to wider economic conditions. Unionization also had inadvertent effects on trade patterns, which are briefly considered below.
79. The apparent surge in this category from the early 1880s is an artefact brought about by their not being adequately distinguished before that time.
80. Cf. N.F.R. Crafts (1989), 'Revealed Comparative Advantage in Manufacturing, 1899–1950', *Journal of European Economic History*, **18**, 127–37.
81. J.C. Brown (1995), 'Imperfect Competition and Anglo-German Trade Rivalry: Markets for Cotton Textiles before 1914', *Journal of Economic History*, **55**, 494–527.
82. For the adverse comparative advantage, see Crafts, 'Revealed Comparative Advantage'.

83. From Mitchell, *British Historical Statistics*, pp. 478–9.
84. R. Floud (1976), *The British Machine Tool Industry, 1850–1914*, Cambridge: Cambridge University Press.
85. B.R. Mitchell (1992), *International Historical Statistics: Europe, 1750–1988*, Basingstoke and London: Macmillan, pp. 644–5.
86. U.S. Department of Commerce (1975), *Historical Statistics of the United States: Colonial Times to Present*, vols. I and II, Washington: Bureau of the Census.
87. Mitchell, *International Historical Statistics: Europe*, p. 593.
88. S. Pollard (1981), *Peaceful Conquest: the Industrialization of Europe 1760–1970*, Oxford: Oxford University Press, pp. 177, 226.
89. For example, Brown, 'Imperfect Competition'. Tariffs or other hindrances to trade could prevent cost competition in international trade from having its full impact on internal or third markets. These have probably not been given their full due for a period conventionally linked to free trade and 'globalization'. This was also the period of Bismarck, McKinley and the like, and studies of individual industries do support the importance of tariff structures (for example, Floud, *The British Machine Tool Industry*). For an assessment of the U.S. 'effective rate of protection' after McKinley, see G.R. Hawke (1975), 'The United States Tariff and Industrial Protection in the late Nineteenth Century', *Economic History Review*, 2nd series, **28**, 84–99.
90. S. Pollard (1957), 'British and World Shipbuilding, 1890–1914: a Study in Comparative Costs', *Journal of Economic History*, **17**, 426–44.
91. Pollard and Robertson, *The British Shipbuilding Industry*.
92. This raises the issue of whether American technology saved on skilled labour and was thus deskilling in orientation. This point is raised by Habakkuk, *American and British Technology*, and is developed much further in the literature on 'labour process', for example H. Braverman (1974), *Labor and Monopoly Capital: the Degradation of Work in the Twentieth Century*, New York and London: Monthly Review Press. Our findings elsewhere (von Tunzelmann, *Technology and Industrial Progress*) suggest that the main orientation of American technology was towards saving time rather than skilled labour, a result which we believe the bulk of detailed studies support.
93. E.S. Ferguson (1981), 'History and Historiography', in O. Mayr and R.C. Post (eds), *Yankee Enterprise: the Rise of the American System of Manufactures*, Washington: Smithsonian Institution Press, pp. 1–24, especially p. 12.
94. For example, Douglas, *Real Wages*; Ozanne, 'A Century of Occupational Differentials'.
95. Data are from Mitchell, *International Historical Statistics: Europe*, pp. 126–32.
96. I. Ferenczi and W. Willcox (1929), *International Migrations: Statistics, vol. I*, New York: National Bureau of Economic Research/International Labour Organization. The shortcomings in these data are well known, e.g. their reliance on passenger lists.
97. B. Thomas (1973), *Migration and Economic Growth*, Cambridge: Cambridge University Press.
98. E.g. J.G. Williamson (1964), *American Growth and the Balance of Payments: a Study of the Long Swing*, Chapel Hill, New Carolina: University of North Carolina Press.
99. See M. Wilkins (1970), *The Emergence of Multinational Enterprise: American Business Abroad from the Colonial Era to 1914*, Cambridge, MA: Harvard University Press; L.G. Franko (1976), *The European Multinationals: a Renewed Challenge to American and British Big Business*, London: Harper and Row; Chandler, *Scale and Scope*; G. Jones (1996), *The Evolution of International Business: an Introduction*, London and New York: Routledge.
100. V.I. Lenin (1917), *Imperialism: The Highest Stage of Capitalism: a Popular Outline*, Petrograd, repr. (1977) New York: International Publishers.
101. Chandler, *Scale and Scope*, pp. 157–61, 199–201, especially p. 201. Chandler points out that tariffs, often regarded as an important determinant for moving investment rather than goods abroad, could not have been very important here, since the majority of branch plants were set up in the U.K. (under free trade) or in circumstances where tariffs appear not to have been the issue.
102. W.J. Collins, K.H. O'Rourke and J.G. Williamson (1997), 'Were Trade and Factor Mobility Substitutes in History?', National Bureau for Economic Research Working Paper 6059, Cambridge, MA.
103. Williamson, 'Globalization, Convergence and History'.
104. N. Rosenberg (1972), *Technology and American Economic Growth*, New York: Harper and Row; von Tunzelmann, *Technology and Industrial Progress*.

PART 5

Law and Institutions

14. The Birth of a Network Technology: the First French Railway System

François Caron

This chapter[1] examines the nature of the technological developments leading to the establishment of a major technical network, the railway, which synthesized and to some extent completed the technologies of the first industrial revolution. The processes integrating and co-ordinating the technologies defining the network need to be analysed. The problem is not specific to French railways, but is of a universal nature. Nevertheless, every railway network has its own technical configuration. From the start, the technology involved was subject to a process of differentiation, and the French network was simply one of the varieties of a European system contrasting sharply with its American counterpart. Taking the French example as a basis, we need to understand what factors determined the differences between systems. This particular network reached a first state of maturity and co-ordination during the Second Empire. The various technical options chosen were more or less successful in creating an equilibrium that was, however, to be gradually undermined from the 1870s onwards.

First of all, we should call to mind the essential features of the history of railways in France. The total length of the track in use jumped from 38 kilometres in 1830 to 3870 in 1852, 9500 in 1860 and 17000 in 1869, a growth arising from the constant pressure from local élites for new lines. After a long period of hesitation and argument between the state and the concessionary companies, the legal status was first firmly established in 1859, when the *conventions* (agreements between the private companies and the state to ensure finance for new lines) were signed. The general features of the system may be defined as follows:

The companies were at once private-law undertakings responsible for running a set of concessionary lines, and bodies subject to strict supervision by the state by virtue of either its sovereign statutory right to ensure public order (the law of 1842 and the *ordonnances* of 15 November 1846), or the responsibilities laid down in the agreements, or the financial clauses forming part of them.

From 1852, the concessions were valid for a uniform 99 years, as opposed to a much shorter period in the 1840s.

The six companies established between 1852 and 1859[2] achieved a *de facto* monopoly in their areas. They were also granted, in exchange for new concessions, a system of guaranteed interest for bonds issued for building new lines, but state payments under the guarantee had to be reimbursed, and the companies were in fact operating within very tight financial limits.

As the financial régime became stable, the technical system achieved a degree of consistency. Engineers created a railway science combining a mainly economics-based theory of utility, and a largely technically-based science of systems.

A three-stage approach seems appropriate here. The first will involve the factors determining the initial choices made by the engineers, the second a brief consideration of some of the particular technical advances in certain components of the system, and the third an attempt both to define the general features of the new system and to point out the features specific to the French variant of it.

THE FACTORS DETERMINING TECHNOLOGICAL CHOICES

In the early stages of the process, the railway engineers inherited original technical cultures which they put to good use in designing installations and plant. Faced with an uncertain situation, they transposed equally uncertain foreign models into a French context and were immediately confronted with serious dysfunctions. Nowadays it is hard to imagine just how starkly inadequate their barely-mastered technology proved to be. Although it has theoretical aspects, railway science was built up on the basis of experience and is the fruit of a critical reading of its lessons. It was a learning-by-doing process, born of initially inadequate forecasting on quite a large scale, and in each department and on each network solutions had to be improvised and a totally new technology put into practice. The result was a collection of very dissimilar bodies of knowledge. Finally engineers adopted commercial policies that directly influenced the technical choices they had to make.

The Legacy of the Past

Railway technology is rooted in a threefold inheritance: the techniques of (mainly road) transport that had grown up in France since the 1740s around the projects initiated by the *Corps des Ponts et Chaussées*; the technologies involving steam, controlled by the *Corps des Mines*; and the industrial basis provided by the mechanical and metallurgical industries. The knowledge acquired in road-building was passed on to railway science essentially but not exclusively by the engineers of the *Ponts et Chaussées*, who very soon took charge of the fate of the big companies. The engineers of the *Mines* had been made responsible for both monitoring and improving steam engines, and were quite naturally put in charge of supervising the quality of the railway companies' plant. Within those companies, their rôle was complementary to that of the engineers of the *Ponts et Chaussées*. The other element in the inheritance was that provided by the iron and steel, mechanical, and woodworking industries, transmitted through both the skills of workers and the knowledge accumulated by the group of civil engineers currently in training. The railways were being built at the same time as the first generations of students were graduating from the *Ecole Centrale*, founded in 1829. Taking their acquired skills with them, they too set out to conquer the companies, and several of them reached management positions. Thus some companies were dominated by the culture of the *Ponts et Chaussées* and other by that of the 'civil engineers'.

Road-Building Technologies and Practices

There were four chief elements in road-building culture as it had developed since 1740.

It was based on a close dovetailing of the economic and technical approaches to problem-solving. Certain engineers of the *Ponts et Chaussées* had begun to establish what Antoine Picon has called 'a new relationship between economic calculations and the realities of production'.[3]

Like the roads, the railway network was clearly a major instrument in the creation of the national territory. Such unification was the result of both the increasingly close-knit nature of the network and its star-shaped form. The hierarchy of railway lines reproduced that of roads.

The railway network inherited from road-building culture the idea of designing alignments based on large-radius bends and gentle gradients.[4] It was a decisive choice, entailing high construction costs as a result of the major work involved in carrying out such a programme. The design and creation of railway infrastructures were directly derived from the practices used in the road and canal systems. In addition, road maintenance had created, within the administration, a corps of road-menders, which served as a model for incorporating the departmental structure of the *Voies et Bâtiments* into the various companies.

The development of road transport since the end of the eighteenth century had radically transformed both the commercial aspects and the technologies involved, with several new factors playing a part. These included high-speed travel, at the gallop for those favoured passengers able to afford the mail coach, and faster stagecoaches in general. But, in the engineers' view, speed was a luxury, and remained so for a long time. A further new factor was a timetable based on punctuality and regularity: 'The roads from Paris to Le Havre are superb, with stagecoaches available at fixed times', wrote Stendhal in 1811.[5] With regard to goods, the development of faster haulage (as opposed to normal haulage or water-borne transport) for valuable articles heralded the new distinction between high and lower speeds. Finally, a new division of labour was copied by the railway companies. Postilions become coachmen and were the forerunners of engine drivers, and coach drivers those of guards. Their respective rôles were codified. Mail-coach staging-posts were the precursors of railway stations.

Three features of the technology of horse-drawn carriages and stage-coaches were transferred directly to permanent-way rolling stock: the hierarchical allocation of space, which was to lead to the three classes on the railways; the way the bodywork was constructed, which was to lead to wooden railway carriages with separate compartments and no corridor; and in particular technical devices such as wheels, suspension, axles and (astonishingly primitive) braking.

The Industrial Legacy

French industry might seem to have been in a position to meet the needs of the emerging transport system fairly quickly. The iron and steel industry could offer a diverse range of products. The first engineers sought to identify their needs in terms of the various uses they had in mind and to find the most suitable metal for each of them from that point of view. But in fact none of the metals currently available had the qualities necessary for making rails, as they did not stand up sufficiently well to the loads and shocks involved.

The appearance of a real locomotive industry in France by the early 1840s is an indication of the maturity of the transformation of the metals and mechanical engineering sector. The French locomotive industry was born around 1842. A contemporary report by a mechanical engineering contractor[6] listed eight manufacturers, of whom only two, Le Creusot and the Alsatian, Koechlin, were to survive. The others were either to disappear or to give up making locomotives. The original proliferation does, however, bear witness to the dynamism of the mechanical engineering sector, based principally in Alsace, Paris and Lyons.

Rolling-stock manufacturing grew directly out of the carriage industry, which had been thriving since the beginning of the century, and had developed considerably throughout the first half of it in order to meet an increasing demand for security and speed. Initially, the greatest progress was in the area of the materials used. Axles were henceforth built from a single piece of forged metal, and tyres consisted of a continuous laminated welded iron strip. Despite such advances, transferring the techniques to railway equipment created serious difficulties.

On the whole there was little justification for the engineers' initial optimism. The technologies which the railways were anxious to use were ill-adapted to their requirements. Engineers were put into a doubly uncomfortable position, since, as a result of the serious dysfunctions for which they had to find an immediate remedy, nothing was sure and everything was urgent.

Uncertainties and Hazardous Choices

The railway system did not spring ready-made from the creative imagination of the engineers, but grew up gradually in a situation of uncertainty about the best solutions. With regard to locomotive engineering, François Arago was still speaking about 'the currently very imperfect state of the art' in 1838.[7] He was a member of the official commission which in 1844 recommended the construction of the St. Germain atmospheric railway that was to cross the slope separating le Pecq, the terminus of the line built in 1837, from St. Germain. It was judged to be 'incomparably more economical and rapid than the locomotive system, and perfectly safe'.[8] Carriage design provides a further example. In 1870 Charles Couche, discussing the first stages in the field, wrote that 'the problems of convergence have been greatly exaggerated and there have been attempts to solve imaginary difficulties'.[9] The 'imaginary difficulties' had serious consequences for the initial choices made by European engineers, who adopted the solution of rigid parallel axles, interdependent with the wheels, on short carriages.

It is therefore not unreasonable to talk about hazardous developments, even though initially the French engineers took the British model as their inspiration. British influence was decisive, and definitively established certain features of the French network. This was the case with Stephenson's gauge (4' 8½", or 1.44 metres) about which Charles Couche, who was an authoritative figure at the end of the Empire, said in 1868, 'there would have been some advantages and very few disadvantages in increasing it a little'.[10]

Both the sizes and the shapes of rails were largely influenced by British practices. The Paris–Rouen line, which was built in the 1840s, was designed and constructed by British engineers and contractors.

Locomotive technology was also directly based on British models and in the early stages slavishly imitated them. C.E. Jullien, who had been an engineer in the Le Creusot construction shops, spoke in favour of that approach in 1841. Comparing the speed with which manufacture in the field had spread in France with the slowness of the process in the case of fixed steam engines, he explained the difference by the fact that 'the normal practices of manufacturing on the basis of one's own inspiration had been abandoned in favour of examining what others had done before getting down to work'.[11] It was not until the late 1840s that French industry began to offer its own models, and even then British influence remained important, especially with regard to high-speed locomotives. Some British models were more successful in France than they had been in the United Kingdom, the best-known example being the Crampton.

Couche had some hard things to say about Anglomania in the technological field. By the early 1850s, he was foremost amongst those advocating German solutions in the fields of both track and locomotives. He was not systematically opposed to tradition and could see that the 'first manufacturers' had sometimes made judicious choices. Yet he would only accept them on an experimental basis, and urged a procedure based on rationally-conceived trials always conducted in identical weather, route and track conditions. This 'experimental' approach was adopted to a limited degree only. It was rather the day-to-day experience of running networks that determined which choices were made. The way the system developed was therefore determined by the lessons to be learnt from its dysfunctions.

Inadequate Forecasting and Dysfunctions

The rôle of accidents

Accidents have occurred regularly in the history of railways, but some of them have had much more serious repercussions than others. There were 55 deaths in the Meudon disaster of May 1842, for instance, mostly of people burnt alive, which stimulated serious thinking about the ways of running passenger services, and inspired many of the articles in the *ordonnances* of 1846. Railway employees had a particularly dangerous job. A committee was therefore set up in 1853 to 'consider ways of ensuring the regularity and safety of railways'.[12] It recorded 513 accidents between 1835 and 1854, of which 274 were derailments and 239 collisions. Safety was therefore viewed as the basis of the technical culture of railways. The majority of railway regulations were drawn up taking into account the experience of accidents.

An accident is simply an incident to the *n*th degree. Incidents – delays, congestion on the lines, difficulties with locomotives because of bad weather as well as of failures in the system, points and signalling errors, poor distribution and loss of plant, to name but a few – were a daily occurrence. A further part of the 1853 commission's remit was to establish the best ways of ensuring a regular service, which was all the more difficult as it was growing ever more complex. As the network became more tightly-knit, the number of potential accident black-spots increased.

The lack of adequate financial forecasting

The difficulties encountered in construction had been very widely underestimated. Auguste Perdonnet, who had had experience of such work, noted in 1859[13] that earthworks called for much more powerful plant than that needed for ordinary road building, particularly where carrying earth for long distances was involved. The most serious unforeseen problem was landslides. Estimates of building expenditure were always surpassed.

The growth in the volume of traffic was satisfactory and indeed, in the case of goods, higher than forecast, but the uneven temporal and spatial distribution of flows made them particularly difficult to manage, creating more bottlenecks and increasing the risk of congestion. The engineers therefore discovered, to their surprise, that there was no end to start-up costs, in the sense that they were always having to enlarge their installations to adapt them to constantly-increasing traffic, and to engage in 'supplementary work'.

The lack of adequate financial forecasting was also evident with regard to running costs. In 1868 François Jacqmin, the director of the Eastern company, wrote as follows: 'It is very certain that no-one had foreseen how much would have to be spent each year on renewing or repairing tracks, enlarging stations or the volume of rolling stock ... or dreamed that tracks would have to be repaired in their entirety every ten or twelve years on major lines and every four or five years on certain sections'.[14] Rails wore out much more quickly and trains were derailed more often than had been expected, braking and signalling were notoriously inadequate, and increasingly more powerful locomotives were launched on a network that could not properly control flows.

Commercial Policies

Both the boom in goods traffic and the lesser growth in passenger traffic were the result of strategic choices. The development of long-distance heavy haulage in direct competition with water-borne transport and the creation of high-speed expresses for élite first-class passengers were the result of deliberate choices. As a matter of fact, the view of price-setting developed by the engineers of the *Ponts et Chaussées* was one of the sources of marginalism, and opened up a debate on differential costs based on taking technical data into account. The decisions concerning railway rates arose from this debate, in accordance with marginalist views, as developed in the 1840s by the *Ponts et Chaussées* engineer Jules Dupuit. They were, however, also a consequence of a clear perception of the opportunities offered by technology. The engineers soon realized that there were hidden reserves of capacity in the networks and looked forward to the kind of performance the locomotives were capable of giving.

On the other hand, the systematic application of rational economic theory did lead to some decisions that, paradoxically, compromised the efficiency of the system. The most striking example has to do with setting passenger prices. To their astonishment, the engineers discovered that carriages were simply not being filled. In the 1860s, three-quarters of them were travelling empty. And yet the companies were stubbornly opposed to making third-class travel more comfortable. As the 1863 commission presided over by Michel Chevalier pointed out, the directors judged that 'it was in the interest of the companies to make second-class travel inconvenient and unpleasant and to remove any trace of comfort or convenience from third-class carriages'.[15] Discomfort was the result of a rational economic choice.

Strategic decisions of that kind had a considerable effect on the development of the technical system, determining the make-up of trains and the design of carriages as well as the direction taken by locomotive or track technology.

THE DEVELOPMENT OF SPECIALIZED AREAS

It is not possible to analyse here in any great detail the way in which the different fields combined to make up the technical system of the railways. I shall merely define the major axes of development of some of them, under the separate headings of man-management, track techniques, traction, rolling stock, and the regulation of train movements on the open track.

Man-Management

From the very beginning, man-management seemed to the whole body of engineers not only the major need, but also the major problem. At first they hired large numbers of men, preferring an unskilled and low-paid labour force, and companies very quickly decided to provide their own in-service training for their employees. This decision was justified by the new kind of qualifications needed and the uncertainty as to what they really were. This policy could not, of course, be applied to complicated jobs, like those of engine-drivers, but even in such cases the companies quickly adopted a method of recruitment based for the most part on hiring unskilled shop-floor workers and on internal promotion.

This made it possible to hire hands for low wages, but did have certain disadvantages. From this point of view, the conclusions of the 1853 commission on safety are typical. It saw accidents as resulting chiefly from 'insufficient regard for regulations and a lack of assiduity and intelligence on the part of employees' rather than 'the mindless power of locomotives or the instability and fragility of the materials used in the construction of the permanent way'. The commission therefore concluded that 'the men of science are capable of the new tasks facing them and it behoves supervisory and safety staff to prove by their choice of employees and their constant vigilance that they are capable of theirs'.[16]

That statement provides the key to the organizational model adopted by the companies. It was multi-functional and very hierarchical, depending for its working on the principle of obedience and the mastery of written material. The division into three departments (track and buildings, traction and equipment, and commercial operations) was largely based on the *Ponts et Chaussées* administration and that of the economy of the road system. The 1853 commission was of the view that 'a unified authority' was the prime requirement for an efficient service, insisting that 'there should be no division of authority, the orders given should be carried out without discussion, departments should work towards the same end without at least any visible friction or strain, which can only be achieved by concentrating power intelligently'.[17] Obeying orders to the letter was seen as the cardinal virtue of the railway employee, which explains why a complex system of rewards and penalties was introduced.

Overall, the range of organizational principles followed from the constraints imposed by the technical system itself, with the division into departments reflecting the complexity of the tasks, and the hierarchical structure that of qualifications and powers. An authoritarian approach was seen as the only way of ensuring that the undertaking was regular and safe.

Permanent-Way Techniques

We shall stress three features of the techniques used between 1850 and 1860. In the first place, uncertainty meant that for each problem there was an astonishing proliferation of possible solutions. As the engineer E. Baude wrote in 1882, 'there is no field in which there have been more inventions and the number of really practical solutions has been more limited'.[18] The most graphic illustration of the situation was the 'rail profile war' which began in the 1850s. Initially, the companies had used the British solution of a double-headed rail on a chair. In the 1850s, the Eastern and Northern companies, and later the P.L.M. (Paris–Lyons–Méditerranée), all decided to adopt the Vignole T-headed rail, which had already been tried out on the St. Germain line. Although this type had obvious advantages, the P.O. (Paris–Orléans), Midi and Western companies kept the double-headed rail, since it meant that trains could be reversed. Thus there were two Frances, each with its own kind of rail. There is certainly no shortage of similar examples.

Secondly, as has been pointed out, materials were singularly ill-adapted to their purpose, particularly in the case of rails. In 1859, Perdonnet wrote that manufacturing rails was still 'under discussion'.[19] The real way out of the impasse was not found until the development of manufacturing steel rails from the 1860s. This example illustrates the difficulties the materials industries encountered in meeting their customers' needs. In that sector, technical change was the result of pressures on producers from users rather than of the logic of supply, as in the twentieth century.

Thirdly, supervising and maintaining the permanent way involved a considerable labour force. Maintenance was carried out on a piecemeal or 'as and when necessary' basis which, in the words of one engineer trying to defend traditional methods in 1893, relied on 'the sharp eye and practical sense of junior employees'.[20] But by the 1860s some company engineers were once again following the German example and advocating periodic overhauls. The fact that so many engineers did not wholeheartedly support this shows the strength of the belief in the superiority of the weight of experience over any other type of regulation.

Traction

Locomotives, like the permanent way, appealed to the imagination of inventors. This technological creativity, however, was a great deal more organized, and the differentiation was functional rather than proliferating. Every engine was designed in terms of the uses it would be put to. The very idea of standardizing models seemed opposed to the functionality which plant needed and to the demands of technical progress, which every engineer thought it would have the disadvantage of hampering.

From the initial British models, two main technological branches, high-speed and goods locomotives, developed alongside each other. With the former, the British influence remained predominant, as has been noted. In the case of the latter, the Northern company started the race to greater power as early as 1851. When it decided to develop full coal trains, it adopted, 'as a matter of course', in the words of Charles Couche, 'more powerful engines than six-wheelers'.[21] This was the beginning of the line of locomotives derived from the Engerths, and the strategic and technological decisions were inextricably linked.

Although locomotive technology was a special field of application for the experimental scientific method, the research process was discontinuous in nature. Even at that early stage it was almost exclusively directed towards the reduction in fuel consumption necessitated by the high price of coal. In the 1850s, an innovation as simple as improved grids in fireboxes provided one of the most striking examples of this. The experiments carried out by Nozo and Geoffroy in the early 1860s to establish the best sizes for boiler tubes directly influenced the decisions companies made, and were only the first of a long series of such tests.

The system was still very far from perfect, and braking provides a typical example of this. From the start, the administration was opposed to its being operated by locomotives, fearing that over-abrupt stops might completely destroy the train. Yet the engineers all thought that, in the words of an 1857 report, 'the means of stopping the train are not adequate, given the speeds reached'.[22] Despite the reluctance of the administration, the companies experimented with several locomotive braking procedures. No satisfactory solution was found, however. Reversing steam, proposed by Le Chatelier in the 1860s, was only a partial one, and the engine-drivers claimed that it damaged the devices on their locomotives. The state of the art, therefore, was still imperfect.

Rolling Stock

The same varying level of technological achievement was also a feature of rolling stock. Real progress had been made in the design of parts, and as a result carriages were considerably safer. Springs were subjected to intensive research from the point of view of both the materials used and their shape, which in theory was formalized in 1852. There were remarkable improvements in tyres, such as the introduction of non-welded ones. The fear of broken axles encouraged a great deal of research and consequent improvements in both form and material. On the other hand, certain companies remained faithful to wooden underframes, and no really satisfactory solution was found to the problem of lubrication.

Body design was the outstanding area of deadlock and conservatism. The shape still largely followed the design of stagecoaches. French engineers, like most of their European colleagues, rejected the American system of long, central-corridor carriages resting on double-axled flat trucks. To justify this point of view, Charles Couche used both the psychological argument of

European individualism and the technical argument that 'adopting American rolling stock on European lines, where very considerable sacrifices have been made to avoid short-radius bends and steep gradients, would be quite illogical'.[23] In fact, there was a fear of bogies. Separate compartments and the lack of a corridor certainly had disadvantages, making it impossible to install even the most elementary conveniences and raising the insoluble problem of how to communicate with passengers. Certain companies tried out electrical communications systems, but the commonest approach was to use footboards and a handrail along the whole length of the train, to enable the guard to move from one compartment to another while it was in motion. The engineers maintained that accidents were less common than might be thought!

Traffic Control on the Line

The increase in both traffic density and the number of sensitive points made flow control more and more delicate and dangerous. Initially, traffic was controlled by precise regulations governing the time-lapse between trains leaving stations. In addition the Western network saw, in the 1840s, the introduction and subsequent widespread use of traffic charts enabling the guard and driver to visualize both the progress of their train along its whole route and overall movement on the line. Finally the electric telegraph became a very common method of transmitting information and orders from one station to another in case of incidents or of the irregular presence of a train on the line, or quite simply to inform of the departure of trains.

The weakness of the system lay in its signalling faults. Controlling traffic actually on the line had originally been the task of day and night guards. The former were abolished after 1848, the latter from the 1860s, as they were not much used and their presence on the line was dangerous. It was track staff, that is, teams of permanent-way men, level-crossing keepers and pointsmen at junctions, who were made responsible for signalling. The purpose of fixed signals was to protect points beyond which it was impossible to proceed without special precautions, such as the approaches to stations and halts, intersections, branch lines, turning points and dangerous stretches of the line. The increasing complexity of junctions and track systems meant that more devices for operating signals at a distance were needed. The engineer Brame advised a maximum of 800 metres, given the fragile nature of the wires, although in fact certain devices extended over two kilometres. Electricity was used to warn of broken wires by means of trembler bells.

The marriage of electricity and the railways had begun in the 1840s with the electric telegraph, which was brought fully into use in the following decade. But the overall reaction of the commission of 1853 to the numerous inventions based on the telegraph and electricity brought to its attention was rather reticent. It expressed doubts about their usefulness, declaring that 'for the most part, the inventors had gone too far ... in trying to make batteries solely responsible for safety and to use them for both near and distant operations, on stations, on the permanent way, on signals and locomotives', a programme which it found unrealistic and impracticable.[24]

Yet it was not to be long before certain factors made it necessary to use such apparatus. The increasing number of trains travelling at various speeds made the problem of spacing them an urgent one. All the engineers were in agreement that doing so on a temporal basis had disadvantages, and that what was needed was a spatial separation, that is, the introduction of the block system. To make it practicable, it had to be possible to announce instantly the departure of trains at the beginning of each section and their arrival at the other end, which could only be done by using electricity. However, with the first devices for this purpose (Tiers' on the P.L.M. and Régnault's on the Western network) the electric signal sent to the employee in the box had to be confirmed manually by means of a semaphore disc. This meant that 'the block system lost many of its theoretical advantages', as the inventor of the electric semaphore pointed out in 1877.[25] It was not until the late 1870s that visual and electric signalling devices were successfully combined.

Despite such achievements, electrical technology was still playing no more than a marginal rôle in the railway system around 1870, a fact which is partly explained by a distrust of automatic apparatus. Nothing brings this home more clearly than what Charles Couche had to say in 1870: 'Although in theory, on paper, a good signalling system can cope with any eventuality, that is not the case in reality. There is nothing to match the train driver's direct sight when he can satisfy himself on an alignment with long straight stretches and long-radius bends that the way ahead is clear ... and that he can proceed calmly'.[26] The sentence presents the whole thinking of the French railway system in a nutshell.

THE ORIGINAL FEATURES OF THE FIRST RAILWAY SYSTEM

In France as elsewhere, the technical system of the railways was formed by the process of combining several existing technical fields. In our view, the influence of this legacy is a major factor to be taken into consideration if we are to understand how the systems developed. The railway engineers, who were working against time, had nothing but earlier practices to base their work on. In short, they synthesized the technologies of the first industrial revolution, and that synthesis went hand in hand with commercial strategies precisely defined in terms of economic science. This happened because controlling major technical systems demands not only mobilizing and harmonizing infinitely varied technical resources, but also defining the basic economic principles of their commercial exploitation, which are major determinants of progress in such systems.

The engineers were working in a situation of technological uncertainty. Some initial decisions were risky, which meant that the networks became increasingly different from one other. It was as if each one was aware that only second-best or stop-gap solutions had been adopted, which is why the refusal to standardize became a dogma. Each sought the solution best fitted to its own routes, commercial policy and traffic. The diversity, however, was not merely functional, but also a guarantee of technical progress and a virtue.

Nevertheless, the process of formalization, which was sometimes inevitable, began to become apparent. It was a matter of quite specific aspects of the system (points, springs, one or two devices on locomotives) and was closely linked to the development of an experimental method based on comparative trials carried out in exactly identical weather, route and traffic conditions. The approach was part of the tradition of the *Ponts et Chaussées*, but was mostly inspired by German practice. Such efforts reduced the diversity only marginally, however, and the research process remained discontinuous.

In its earliest form, railway technology was not only labour-intensive, but also labour-dependent. It was not simply a case of directing the productive system towards labour-intensive, low-capital technologies, but also of an approach that moved choices towards solutions which made the proper working of the system depend on its employees. Their 'intelligence' and 'assiduity', their ability to pay attention and react to very varied circumstances, and their obedience were sought within the framework of the written regulations that were their guides. The management of complex situations was completely entrusted to them.

The distrust of automatic devices was justified because of the imperfect and unreliable way in which they worked, but went beyond that particular perception. It was an essential component of a culture based on a respect for acquired or created skills, and was to last for a long time. The engineers found it much easier to accept the idea of mechanical transmission for movement at a distance than that of transmitting an electric force to initiate it. It took the increased pressure of traffic in the 1860s, and the greater demands of the administration and the customers, to bring a second railway system, dominated by automated regulatory processes, into being.

The French model was almost a caricature of some of the features (a labour-dependent technology and an uncontrolled diversification of systems) mentioned here. In 1905, Clément Colson spoke of 'an extremely ingenious way of dealing with matters of detail'[27] as the feature distinguishing the French from the German railway network, which required large inputs of capital. The financial clauses in the agreements were so stringent that they obliged companies to adopt a 'parsimonious' management policy with regard to complementary investments, and increased the trend towards choosing labour-intensive technologies. In contrast, a further feature of the French model, inherited from the tradition of the *Ponts et Chaussées* administration, was a search for perfect alignments and infrastructures. This preoccupation went beyond purely economic concerns.

In France, the technical diversification was taken to its extreme limit. The process of differentiation between the six networks was already becoming evident under the Second Empire. The desire to be different became almost a knee-jerk reaction amongst the engineers. Even the administration thought that diversity was necessary for technical progress, and only changed its behaviour very gradually. On the other hand, the search for economic rationality played a major part in decision-making. For both the administration and the engineers, running the networks was an exercise in applied economics. In France, theories of utility were the sources of inspiration, and had been the subject of intense debates in the *Ponts et Chaussées* for a century.

CONCLUSION

Although the railways were the culmination of the first industrial revolution, they were also the starting-point for processes that were to challenge its technological bases, for it soon became clear that the techniques used had their limits. The other components of the system were inadequate for the ever-growing power of the locomotive. Plant was badly controlled, flow control was embryonic, permanent-way materials were insufficiently hard-wearing. There were enough such challenges to demand a radical transformation of all the fields now forming part of the system. A new materials technology appeared, and new ways of using electricity emerged. Automatic devices and signalling technologies began to be developed. All these signs heralded the creation of a new technical system that was to undermine the very foundations of the dominant technologies of the 1840s. What has been said about carriage design, the way of understanding accidents at work, and the hesitation with regard to automatic devices clearly shows, however, the difficulties which the engineers experienced in trying to break out of their inherited intellectual framework.

NOTES

1. This chapter, including all quotations, has been translated from the original French by Dr. J.C. Whitehouse of the Department of Modern Languages at the University of Bradford, to whom I am most grateful.
2. The six companies were: the Northern, the Western, the Eastern, the P.L.M., the P.O. and the Midi. See François Caron (1973), *Histoire de l'Exploitation d'un Grand Réseau: la Compagnie du Chemin de Fer du Nord, 1846–1937*, Paris, La Haye: Mouton; and François Caron (1997), *Histoire des Chemins de Fer en France 1740–1883*, Paris: Fayard.
3. Antoine Picon (1992), *L'Invention de l'Ingénieur Moderne, l'Ecole des Ponts et Chaussées, 1747–1851*, Paris: Presses de l'E.N.C.P., p. 345.
4. The first to apply the principle was Marc Seguin, on the Lyons to St. Etienne lines from 1829 to 1832, thus falling in with the thinking of the Ponts et Chaussées.
5. Stendhal (1811), 'Journal de Voyage à la Mer de Paris au Havre, du 29 Avril au 3 Mai 1811, tenu par Stendhal, Félix Faure et Louis Crozet (Récit de Louis Crozet)', in Stendhal (1961), *Oeuvres Intimes*, ed. Henri Martineau, Paris: Gallimard, pp. 1080–87.
6. François Calla (1842), 'Rapport sur Plusieurs Etablissements affectés à la Construction des Grandes Machines à Vapeur et des Machines Locomotives', *Bulletin de la Société d'Encouragement*, **41**, 467–84.
7. François Arago (1838), 'Rapport à la Chambre des Députés sur le Projet de Loi de Classement du 8 Février 1838 au Nom de la Commission chargée de l'Examen du Projet', *Moniteur Universel*, 26 April 1838.
8. 'Rapport de Mallet, Inspecteur Divisionnaire des Ponts et Chaussées sur le Chemin de Fer Atmosphérique', *Le Journal des Débats*, 25 May 1844, Archives Nationales, F.14.11171; François Arago (1844), 'Communication sur le Chemin de Fer Atmosphérique', *Comptes Rendus des Séances de l'Académie des Sciences*, **6**, February 1844.
9. Charles Couche (1867–74), *Voie, Matériel Roulant et Exploitation Technique des Chemins de Fer*, Paris: Dunod (3 vols), vol. 2, p. 237.
10. Couche, ibidem, vol. 1, p. 5.
11. C.E. Jullien (1841), *Nouveau Manuel Complet du Constructeur de Machines Locomotives*, Paris, pp. v–vi.
12. Ministère de l'Agriculture, du Commerce et des Travaux Publics (1858), *Enquête sur les Moyens d'assurer la Régularité et la Sûreté de l'Exploitation sur les Chemins de Fer*, Paris: Imprimerie Nationale.
13. Auguste Perdonnet (1859–60), *Traité Elémentaire des Chemins de Fer*, 2nd edition, 2 vols, Paris: Langlois et Leclercq (vol. 1), Garnier Frères (vol. 2), vol. 1, p. 24.
14. François Jacqmin (1868), *De l'Exploitation des Chemins de Fer*, Paris: Garnier frères (2 vols), vol. 2, p. 55.
15. Ministère de l'Agriculture, du Commerce et des Travaux Publics (1863), *Enquête sur l'Exploitation et la Construction des Chemins de Fer*, Paris: Imprimerie Nationale, p. xxxvi.
16. Ministère de l'Agriculture, du Commerce et des Travaux Publics (1858), *Enquête sur les Moyens d'assurer*, p. viii.
17. Ministère de l'Agriculture, du Commerce et des Travaux Publics (1858), ibidem, p. vi.
18. E. Baude (1882) quoted in Charles Laboulaye, 'Chemins de Fer', in *Dictionnaire des Arts et Manufactures*, vol. 1, p. 4, col. 1.
19. Perdonnet, *Traité Elémentaire*, vol. 1, p. 120.
20. 'Rapport sur l'Entretien des Voies', *Congrès International des Chemins de Fer, 4ème Session, St. Petersburg* (1893), vol. 1.
21. Couche, *Voie, Matériel Roulant et Exploitation Technique*, vol. 2, p. 357.
22. Archives du Monde du Travail Roubaix, 48.AQ.3752–32.3 (1857). See François Caron, 'Histoire de l'Exploitation d'un Grand Réseau Français: la Compagnie du Chemin de Fer du Nord de 1846 à 1976', doctorate thesis, Faculté des Lettres de Nanterre, Paris, 2 vols, vol. 1, p. 102.
23. Couche, *Voie, Matériel Roulant et Exploitation Technique*, vol. 2, p. 25.
24. Ministère de l'Agriculture, du Commerce et des Travaux Publics (1858), *Enquête sur les Moyens d'assurer*, p. xxi.

25. M.H. Lartigue (1877), *Note sur les Electro-Sémaphores pour le Block-System, leur Application sur les Lignes à Double et Simple Voie*, Paris: E. Capiomont et V. Renault, p. 6.
26. Couche, *Voie, Matériel Roulant et Exploitation Technique*, vol. 2, p. 230.
27. Clément Colson (1905), 'Revue des Questions de Transport', *Revue Politique et Parlementaire*, 1905, p. 299.

15. Patent Institutions, Industrial Organization and Early Technological Change: Britain and the United States, 1790–1850

B. Zorina Khan and Kenneth L. Sokoloff

Comparisons of the patterns of technological change in Britain and the United States during early industrialization have long generated controversy. Nowhere has this been more evident in recent years than in the debate about manufacturing productivity growth over the course of the Industrial Revolution in Britain. The traditional view, whose classic statement is generally credited to T.S. Ashton, held that the pace of technological progress in Britain markedly accelerated late in the eighteenth century, and that this acceleration extended across a broad range of industries.[1] This position has been challenged, however, by N.F.R. Crafts, C.K. Harley, and others, who have argued that technological change up to 1860 was much slower than has been thought, especially in labour-intensive industries.[2] Their work implies that the British path of early industrialization, characterized by modest and unbalanced rates of productivity advance, was quite unlike that of the better-documented U.S. case. Scholars of early American industrialization generally agree that productivity grew at an impressive pace and was relatively balanced across manufacturing industries between 1820 and 1860, which helps to explain why U.S. economic historians are well represented amongst those most sceptical of the revisionist case.[3]

The central problem, of course, is whether and, if so, why these first two industrial economies appear to have had such divergent patterns of early industrialization. Because of a common cultural heritage, and of the general coincidence in the timing of when economic growth got under way in the two countries, one might reasonably have expected their paths of development to be similar. Indeed, some might hold that the appearance of stark contrast is an artefact of estimates based on inadequate data. It is not clear, however, that the most constructive reaction is to reject the implications of the findings out of hand. As economic historians are now rediscovering, our understanding of the sources and processes of growth can be much improved by considering the sources of variation in experience across nations. Some, like Douglass North, argue that such a perspective highlights the crucial importance of institutions to growth and the path of technological change.[4] Other scholars follow H.J. Habakkuk in emphasizing how comparative history enhances appreciation of the significance of factor endowments.[5] This chapter employs such a comparative approach by examining two salient institutional differences between the first two industrial nations, which may have substantially affected their paths of technological development and helped to account for why Britain had slower and less balanced manufacturing productivity growth than the United States during early industrialization. First, we contrast the British and U.S. patent systems, which differed significantly in ways that influenced the nature and direction of inventive activity. Second, we suggest how alternative forms of manufacturing organization in the two countries provide a potentially important source of divergence in their paths of technological change.

In selecting the patent system for such attention, we are inspired by scholars like North and Fritz Machlup who pointed long ago to the patent system as an outstanding example of an institution that had a major impact on the course of technological change and economic development.[6] Although the framers of the U.S. Constitution and statutes were certainly familiar with, and influenced by, the British patent system, they chose to make important departures in the ways in which property rights in technology were defined and awarded. More specifically, they sought to make such property rights available to a much broader class of the population, to ensure that details of new discoveries diffused more rapidly, and to develop trade and commercialization of technology to a greater extent than did the British system. These ends were achieved through innovations such as reserving patent rights to the first and true inventor, the administrative processing of applications, and fees that were set at a low level. Moreover, the fact that the public had ready access to patent specifications promoted the diffusion of inventions and facilitated extensive trade in patented technologies. If the design of the patent system mattered during this era of the onset of economic growth, then these institutional innovations of the U.S. should have led to a broader extension of property rights in new technological information, across both inventors and inventions, than did the British system, and should have affected the relative rate and direction of technological change.[7]

Although our research is still at an early stage, the evidence we have examined thus far is consistent with the predicted effects of the differences between the two patent systems. First, the social and occupational background of U.S. patentees was much more diverse than in Britain. Second, patenting was more concentrated (unequally distributed) across individuals in Britain than in the United States. Third, when the patent law in Britain was revised in the direction of the U.S. law, the levels of patenting activity increased significantly. Finally, patenting in Britain appears to have been more directed at capital-intensive industries than was the case in the United States. This pattern parallels the contrast between the two countries in the degree of balance across manufacturing industries in productivity growth.

Our second major focus is on alternative organizations of manufacturing in Britain and the United States. The basic point we make is that, relative to the United States, cottage manufacture was much more common in Britain than manufactories operating with centralized production. If cottage manufacturing was less conducive to invention than manufacturing in centralized plants, as we contend, this contrast could help to explain why productivity growth in labour-intensive industries was slower in Britain than in the U.S. Our analysis draws on the findings by Kenneth Sokoloff and David Dollar that the greater prevalence of cottage manufacture in Britain during early industrialization was due to differences in labour market seasonality and factor endowments, but our emphasis here is on the question of whether the difference in the organization of manufacturing production affected the paths of technological development in the two countries.[8]

We hope that this essay makes three types of contributions to the comparative study of early industrialization in Britain and the United States. First, our proposal of several reasons why these two industrial economies could have followed different paths of technological development implies that the revisionist estimates of the patterns of productivity growth during the British Industrial Revolution should be taken as a serious possibility, and highlights the need for further study. Second, by identifying several relevant conditions that could have influenced the path of technological change, we seek to suggest new ways of using the limited sources of information available to obtain a fuller understanding of the records of growth in the two early industrial economies. Finally, by linking the rate and direction of technological change to the patent system and the organization of economic enterprise, we seek to broaden the study of the processes of economic growth, and of the rôle of institutions in particular.

THE DEVELOPMENT OF PATENT INSTITUTIONS IN BRITAIN AND THE UNITED STATES

The granting of exclusive property rights vested in patents developed from medieval guild practices in Europe. English monarchs frequently used patents to reward favourites with privileges, such as monopolies over trade that increased retail prices of commodities. It was not until the seventeenth century that English patents were associated entirely with awards to inventors, when Section 6 of the Statute of Monopolies (1624) repealed the practice of royal monopoly grants to all except inventors.[9] Although the petition to the Crown was itself a formality, the British patent system retained many features that reflected its origins in royal privilege. Indeed, the evidence presented below suggests that the system restricted access to property rights in inventions in ways that had the consequence of limiting the class of inventors primarily to those who had substantial technical knowledge or were wealthy and well-connected. The structure of the system served to raise the average value of patents, and favoured the invention of high-valued physical capital inputs, such as textile machinery and steam engines.

Patent grants were regarded as a favour from the Crown, and as monopolies that restricted community rights, and thus as rights to be carefully monitored and narrowly construed.[10] This approach was also evident in the legal system, whose protection of property rights in patents was qualified by the hostile attitude of many judges.[11] At the same time, once the patentee had paid the required patent fees, the patent was sealed without any examination into the technical merits of the invention. Patent rights were not limited to original inventors, since importers of foreign inventions were also given exclusive rights. Patents were granted for a term of fourteen years, and could be extended only by a private Act of Parliament.

Towards the 1780s nation-wide lobbies of manufacturers and patentees expressed dissatisfaction with the operation of the patent system, and from 1829 onwards a series of Select Parliamentary Committees explored proposals for reform. However, it was not until 1852 that Parliament approved the Patent Law Amendment Act, which authorized a major overhaul of the patent system.[12] Among other features, the law established a renewal system that required the payment of fees in instalments if the patentee wished to maintain the patent for the full term. Two significant changes in the direction of the American system consisted of lower fees and costs, and the rationalization of patent administration.[13]

The system of patenting in the United States provides a striking contrast to that of Britain. The American patent system was created in accordance with the Commerce Clause of the Constitution, 'to promote the progress of science and useful arts by securing for limited times to authors and inventors the exclusive right to their respective writings and discoveries'.[14] Patents were viewed more as the natural right of the inventor than as monopoly privileges, and the consistent goal of those who shaped the system was to encourage domestic ingenuity, whatever the social class of the inventor. The judiciary recognized the importance of secure property rights in a market economy and were especially concerned about the enforcement of patents as 'the dearest and most valuable' of property rights.[15]

In keeping with the consciously utilitarian purpose of early economic and social policy in America, the patent application process was straightforward, and involved impersonal, routine administrative procedures. For the first few years after the Patent Act of 1790 was passed, patent applications were examined by a tribunal comprising the Secretaries of State (Thomas Jefferson) and War (Henry Knox), and the Attorney General (Edmund Randolph). This practice proved unwieldy and was replaced by a registration system in 1793. Concerns about overlapping patents led a committee headed by Senator John Ruggles to conduct an inquiry into the patent system. As a result, the Patent Act of 1836 instituted significant changes, and in July of that year the U.S. adopted the examination system that is still in use today. Under the new system, each application was scrutinized by technically trained examiners to ensure that the invention conformed to the law, and constituted an original advance in the state of the art.

The American patent system might well be regarded as the world's first modern patent institution. First, its objective was not to grant or limit monopoly rights and privileges, but to promote invention and domestic ingenuity, and to ensure the diffusion of information and innovation. Second, it largely ignored social class and privilege, and routinely granted all inventors the right to property in their discoveries. Later, when the patenting process became more stringent, the criteria for granting the patent depended on the merits of the application and not of the applicant. Third, the administration was straightforward and uncomplicated, and employees of the Patent Office – especially after 1836 – were skilled, specialized professionals. Fourth, the parameters of the patent institution were established by statute, rather than depending on individual discretion, and this led to more predictable rules.[16]

The drafters of the American Constitution and of its patent laws were familiar with European practice, so it might reasonably be inferred that departures from British precedent were self-conscious and deliberate attempts to establish a different system and pursue an alternative path of development.[17] In particular, we suggest that the framers of the U.S. system believed that inventors benefited society, and that inventive efforts were significantly influenced by prospects for material gain. An examination of the legislative history of the first U.S. Patent Act supports the view that early policy makers attempted to establish a system that would enhance rewards to inventors – whatever their social status – and accordingly stimulate inventive activity and economic growth.

Early in 1790 George Washington recommended that the legislature give 'effectual encouragement, as well to the introduction of new and useful inventions from abroad, as to the exertions of skill and genius in producing them at home'.[18] Congress responded with a bill, HR-41, that was designed to meet these objectives, but before the bill was passed, it was subject to several amendments. The most minor of these amendments is suggestive: the phrase 'great seal of the United States', which emulated the British custom, was altered to read simply 'seal of the United States'. Indeed, the patent bill that was approved by Washington in April, 1790, differed significantly from the British system, in ways that favoured the rights of inventors. First, the House deleted Section 6, which had imitated the English policy of granting patents for imported inventions. Second, the Senate extended the initial definition of novelty: the laws still employed the language of the English statute in granting patents to 'the first and true inventor', but unlike in England, the phrase was used literally to grant patents for inventions that were original in the world, not simply within U.S. borders. Third, a section regarding interferences (or conflicting applications) was replaced by a clause which ensured that information about prior inventions was readily available to potential patentees. Fourth, the Senate suggested forcing patentees to work the patent or else to license others to do so, but the House rejected this as an infringement of the patentee's rights. Also, small reductions were made to the fee schedule, which was modest to begin with.[19]

It is perhaps not surprising that American legislators, who wanted to create a system that indeed 'promoted the progress of ... useful arts', chose features which differed from the British system. Whereas British institutions retained a bias towards wealth and privilege that limited access and participation, the establishment of a patent system that encouraged broad participation was in line with the generally more democratic orientation of U.S. institutions. The ability of individuals with rather ordinary backgrounds to secure property rights in incremental inventions, and in small improvements in design and technique, meant that when markets expanded in the U.S. during early industrialization, a broad spectrum of the population was in a position to take advantage of the opportunities that were emerging throughout the economy.[20]

THE EFFECTS OF INSTITUTIONAL DIFFERENCES ON PATTERNS OF PATENTING

The previous section has described the broad differences in policy and philosophy that characterized the beginnings of the two countries' patent institutions. We focus here on the specific features that were most likely to have influenced the path of technological progress, such as the cost of securing a patent grant, criteria for patentability, public access to the specifications of patented inventions, and restrictions on the transfer of patent rights and on commercial development. The first of these features, the patent fee, highlights one of the most striking contrasts between the British and American systems. After 1793 American applicants paid a fee of $30 ($35 after 1861) – a sum that was substantial, but only a fraction of *per capita* income. The application procedure in England, however, was notoriously costly and complex. Patent fees for England alone amounted to £100–£120 or approximately four times *per capita* income in 1860. The fee for a patent that also covered Scotland and Ireland could be as much as £350. Adding a co-inventor was likely to increase the costs by another £24.[21] Complicated administrative procedures added further to the costs: patent applications for England alone had to pass through seven offices, from the Home Secretary to the Lord Chancellor, and twice required the signature of the Sovereign.[22] These much higher costs of obtaining property rights to inventions in England tended to restrict the use of the patent system to inventions of high value, and to individuals who could raise the capital to apply for the patent and had access to information and other privileges that reduced the bureaucratic and political costs. Some might question how sensitive potential patentees were to the costs of securing property rights to inventions, but Figure 15.1 suggests that English inventors were significantly influenced by the administrative and monetary costs.[23] Under the renewal system, registration fees for an English patent fell to £25 in the first year; patentees who extended the term of the patent in the third year paid an additional £50, and £100 more to maintain the patent past the seventh year. Furthermore, under the new system the application procedures were administered by a single 'Great Seal Patent Office'. These changes reduced the costs of obtaining a patent, not least because consolidating the reviews into one office simplified the process considerably. As a result, the number of patents issued jumped from 455 in 1851 to 1384 in 1852 and to 2187 in the following year. In 1883, the initial fee was lowered to £4, although the cost was approximately the same if the patent were taken to term, and applications again surged from 5993 in 1883 to 17100 in 1884.[24] Thus the evidence supports the view that inventors were sensitive to the cost of obtaining patent protection.

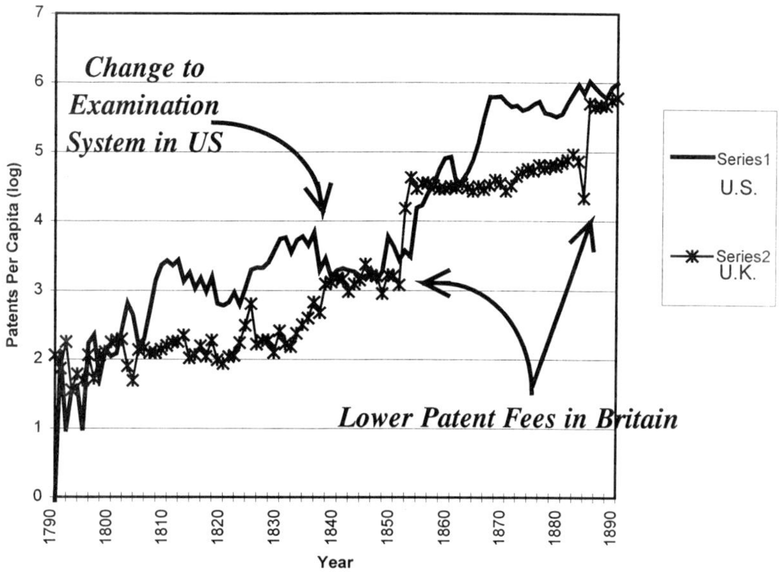

Note: In the United States, an examination system was implemented in July 1836, to replace what had previously been a straightforward registration system. In Britain, patent fees were reduced in 1852 and 1883 (see text for details). The figure presents the plot of the log of annual patents issued per million people by year for the two countries. Although the figures prior to 1852 should be regarded as English patents, after that year patents applied to all of Britain.

Sources: For the numbers of patents in the two countries, see U.S. Bureau of the Census (1975), Series W 99; Dutton, *The Patent System*; and *British Parliamentary Papers* (various years).

Figure 15.1. Patents Per Capita in Britain and America, 1790–1890

From the patentee's perspective, the high fees and procedural costs of the British system may have been offset by the lack of any further examination of applications for novelty or utility. It is true, for example, that in the U.S. the introduction of the examination system in 1836 raised the costs that patentees incurred in preparing their patent applications. What is essential, however, is that although a registration system with high patent fees and an examination system both apply filters to the population of potentially patentable inventions, the samples of inventions and inventors that survive the filter are likely to differ in important respects. In particular, the U.S. examination system selected inventions on the basis of novelty and contribution to knowledge, whereas the English registration system favoured applicants who already had access to the substantial capital outlay required to obtain a patent. In markets with complete information, an inventor with a valuable idea would be able to raise sufficient capital, but in general it might be expected that uncertainty about the value of a new invention, coupled with asymmetric information, was likely to pose significant obstacles. The bias in the English system also affected the distribution of patents, which was more likely to be skewed towards high-value, capital-intensive inventions. The effective result was that a broader segment of the population in the United States was able to secure property rights in a wider array of inventions, which probably generated greater incentives for inventive activity.

As discussed above, the two patent systems also differed in their interpretation of the requirements for novelty in the patent grant and in their policies towards imported inventions. The United States system awarded patents only to the original inventors, who were advised to obtain patent protection as soon as they had reduced their idea to a practical form. In England, patents were granted to importers of inventions from overseas as well as to original inventors, a policy which certainly must have favoured members of the commercial and élite classes, who were more likely to be familiar with developments in other countries. Moreover, this practice implied that individuals with sufficient capital to obtain a patent for foreign technology could legally constrain competitors.[25]

Another effect of the complex English system was to enhance the rôle of middlemen who were positioned to reduce transaction costs for potential patentees. The overwhelming majority of British patentees employed patent agents who provided information and advice, and channelled the invention through the bureaucracy for a charge of £40 to £100 above the patent fees.[26] This industry was oligopolistic in structure, for a small number of agents such as the Newton family, Moses Poole, and Pierre Fontainmoreau dominated the patent agency business. Some of the more active patent agents were at the same time employed by the Patent Office, such as Poole, who obtained over 100 patents in his own name, and allegedly granted his own clients preferential treatment. Between 1816 and 1852, English patent agents obtained 537 patents in their own names, mostly on behalf of their clients. Significantly, patent agents and Patent Office officials were among the most vociferous opponents of the 1852 patent reforms.[27]

One of the reasons why patent agents were more necessary for British patent applicants lay in their access to information. Before the middle of the nineteenth century, inventors could not readily obtain copies, nor see the descriptions and specification of patents that had previously been granted.[28] The patent documents were stored haphazardly in three separate offices, to which an admittance fee was charged. The fee for simply reading a patent was two shillings and sixpence, and the cost of a copy of the patent varied between two and forty guineas in 1829.[29] One rationale for this practice was to prevent foreign competitors from acquiring British technology, but, whatever the objective, the result was limited access to information and slower diffusion of technology.[30] This policy tended to increase the value of the property rights of those who were able to obtain a patent, and to reduce the likelihood of further developments in the area by other inventors.

The difficulties of gaining access to information and the general complexity of the British system probably influenced the geographical distribution of inventive activity, in addition to favouring the wealthy. The specifications of U.S. patents were quickly and widely diffused through publications, and inventors throughout the country could apply for patents without significant obstacles, because applications could be submitted by mail free of postage. It is, accordingly, not surprising that much of the initial surge in patenting during early American industrialization occurred in rural areas. The English pattern was quite different, for patents were awarded predominantly to residents of cities, particularly to patentees with addresses in London.[31] This extreme geographical concentration partially reflected the difficulty of negotiating from outlying areas the complex procedures required to apply for a patent. It may be expected that the wealthy, who could afford to travel to supervise the process, or to retain agents to do so, were less disadvantaged by their distance from the capital.

Policies that affected the means of appropriating returns from the patent constitute another important feature of the patent institution whose impact might vary across social class, and affect the overall incentives for inventive effort. After obtaining a patent, the patentee typically attempted to pursue profit opportunities, but his ability to derive returns from these strategies partially depended on the patent laws and other aspects of the legal system.[32] These too varied between the two countries. For example, English law limited the number of individuals who could share in the patent rights to a rather small group, whose size could only be increased by a private Act of Parliament. This restriction made it more difficult for inventors to raise capital from outside investors to cover the patent application costs, or to exploit the invention commercially, and one would suppose that it constrained the working class to a greater extent than other groups.[33] In contrast, by the middle of the nineteenth century U.S. inventors were routinely trading patent rights that were divided and subdivided among assignees. This enhanced their ability to raise capital for support of their inventive activity, and to extract material returns from their efforts.[34]

The American legal system further encouraged the evolution of trade in patent rights by protecting assignees and investors against fraudulent patents – unlike the system in England, where payments for invalid patents could not be recovered.[35] The ability to profit thus depended on the legal system and the attitude of the courts, which influenced investors as well as current and potential infringers.[36] Partly because of the registration system, British judges claimed broad jurisdiction over the determination of the validity of patents, and this may have increased uncertainty regarding property rights, and made it more costly to enforce the patent.[37] In contrast, American judges consistently attempted to implement the spirit of the Constitution and subsequent patent statutes, by protecting the rights of inventors. For example, they refused to interpret the clause that patents should be for 'new and useful inventions' as implying that the courts should determine whether inventions were useful, and instead left the question of utility for the market to decide.[38] In short, in this regard as in others, American patentees submitted their inventions to an arena in which all participants were provided with relatively equal, low-cost access.

THE COMPOSITION OF INVENTORS AND INVENTIONS

The previous sections have presented descriptions of the historical development of the patent system in Britain and the United States, and discussed specific features that may have affected the levels and direction of patenting, as well as the composition of patentees. In this section, we test this hypothesis for consistency with the evidence. The simplest approach is to compare the rates of patenting *per capita* in the two countries. If one does so, as in Figure 15.1, the data indicate that once the United States government and patent system had had ten years to get organized, patenting *per capita* was markedly and consistently higher in the new republic. The three short intervals during which England rivalled the U.S. in patenting were at the beginning, just after the former had lowered fees, and just after the latter introduced its examination system. Overall, this comparison of patenting *per capita* in the two countries supports our contention that the U.S. patent system led to a higher proportion of inventions being patented than in Britain, if not higher rates of inventive activity as well. Another perspective on the impact of the more restrictive granting of patents in England is provided in Table 15.1. The table presents the distributions of patents in both countries for various sub-periods between 1750 and 1850, classified by the total number of patents that patentees ultimately received over their careers. The distributions are intended to convey a sense of the contrast between the two countries in the degree to which patents were concentrated among a small fraction of the population – whether professional inventors or a class with disproportionate access to the patent system. Indeed, the figures do suggest that patent holding was much more concentrated in England than in the U.S., even among patentees alone. For example, between 1812 and 1829, only 42.9 per cent of English patents were granted to individuals who received a single patent over their career, compared to 57.5 per cent in the United States. If one considers the much lower patenting rates in England, this disparity with respect to the breadth of participation in the patent system seems all the greater.

Table 15.1. Distribution of Patents by Patentee Career Total, England and United States, 1750–1850

Period	1 Patent n	row%	2 Patents n	row%	3 Patents n	row%	4–5 Patents n	row%	6–9 Patents n	row%	> 10 Patents n	row%
1750–1769												
England	181	71.0	51	20.0	10	3.9	9	3.5	4	0.6	0	0.0
U.S.	–	–	–	–	–	–	–	–	–	–	–	–
1790–1811												
England	144	52.2	65	23.6	27	9.8	23	8.3	13	4.7	4	1.4
U.S.	263	51.0	98	19.0	62	12.0	39	7.6	36	7.0	18	3.5
1812–1829												
England	75	42.9	33	18.9	23	13.1	19	10.9	14	8.0	11	6.3
U.S.	823	57.5	249	17.4	102	7.1	109	7.6	78	5.5	70	4.9
1830–1842												
England	83	46.1	37	20.6	18	10.0	20	11.1	7	3.9	15	8.3
U.S.	1102	57.4	317	16.5	156	8.1	153	8.0	108	5.6	85	4.4
1843–1850												
England	100	51.8	28	14.5	21	10.9	21	10.9	10	5.2	13	6.7
U.S.	329	60.5	96	17.7	48	8.8	39	7.2	13	2.4	19	3.5
ALL YEARS												
England	583	49.1	214	18.2	99	10.9	82	10.4	48	6.5	43	6.1
U.S.	2517	57.1	760	17.2	368	8.3	340	7.7	235	5.3	192	4.4

Sources: Sokoloff and Khan (1990), 'The Democratization of Invention During Early Industrialization: Evidence from the United States, 1790–1846', *Journal of Economic History*, **50**, 363–78, for the U.S. sample. The English sample was drawn from Woodcroft, *Titles of Patents of Invention*.

Moreover, another sharp contrast between the economies is apparent if one looks at how the distributions change over time. In the U.S., the share of all patents accounted for by patentees with only one career patent grows during the initial phase of industrialization (from 46.1 per cent in 1790–1804 to 58.2 per cent in 1823–29), while this share shrinks dramatically in England during the analogous stage (71.0 per cent in 1750–69 to 42.9 per cent in 1812–29). The divergence in experience may reflect differences across countries both in how broadly the commercial opportunities created by economic growth extended across social classes, and in the capital requirements for securing a patent. Given the severe difficulties in obtaining property rights to their inventions that ordinary citizens in England faced relative to their wealthy countrymen, it is not surprising that such property rights were more concentrated there.

Table 15.2. Distribution of Patents by Patentee Occupation: All English Patentees and U.S. Urban[a] Patentees, 1750–1850

	1750–1769		1790–1804		1805–1822		1823–1836[b]		1836–1850	
	n	col%	n	col%	n	col%	n	col%	n	col%
Commerce and Professional										
England	131	54.8	110	41.8	74	40.9	89	47.7	70	39.1
U.S.	–	–	13	50.0	60	38.7	59	24.6	43	18.6
Engineers/Machinists										
England	7	2.9	28	10.6	26	14.4	37	20.7	47	26.3
U.S.	–	–	1	3.9	17	11.0	34	14.2	40	17.3
Artisans										
England	76	31.8	95	35.1	63	34.8	33	18.4	54	30.2
U.S.	–	–	9	34.6	48	31.0	80	33.4	67	29.1
Manufacturers/Metal Dealers										
England	17	7.1	26	9.9	17	9.4	17	9.5	5	2.8
U.S.	–	–	2	7.7	17	11.0	40	16.7	49	21.2
Other Occupations/None listed										
England	8	3.4	4	1.5	1	0.6	3	1.7	3	1.7
U.S.	–	–	1	3.9	13	8.4	27	11.3	32	13.9

Notes:
a. The U.S. figures show the occupations of patentees that were traced from city directories, and therefore refer to urban patentees.
b. The English figures in the 1823–1836 column pertain to patents filed in 1840, whereas the U.S. figures are based on patents filed through 3 July 1836, when the new U.S. patent law took effect. The occupational category for 'artisans' includes manufacturers of non-metal products, and 'commerce and professional' includes merchants and gentlemen.

Sources: As for Table 15.1.

The same qualitative pattern is evident in Table 15.2, which provides the distributions of patents by the occupation of the patentee for early industrial sub-periods. Occupation data are available for most English patentees, because it was required on applications. For the U.S., however, we are confined to reporting the distribution of occupations for urban patentees alone, because the information had to be retrieved from city directories.

Working with this subset of U.S. patentees underestimates differences in the occupational composition of patentees, because patentees from rural areas and small cities, who accounted for a substantial proportion of all patentees in the United States, were less likely to be from professional occupations. Even so, the distributions reveal that English patentees were more likely to be from the relatively élite classes ('gentlemen', merchants and professionals) than their U.S. counterparts, while the latter were more likely to be artisans, manufacturers, or from a miscellaneous category. The contrast in how this composition changes over time is especially interesting. In the U.S. the professional or élite share of patents falls sharply over the early nineteenth century as other segments of the population take advantage of the expansion of economic opportunities associated with early industrialization; not coincidentally, rates of patenting *per capita* jump during the same period. In England, however, both the occupational composition of patentees and the patenting rates change much more slowly over time. These patterns seem consistent with our view that the operation of the patent system in England discouraged many inventors or would-be inventors from participation in the system. Some of the contrast between England and the U.S. in the distribution of patentees may be attributed to differences in industrial composition, but it is difficult to avoid the conclusion that differences in effective access to the use of the patent system and to commercial opportunities play a rôle in the explanation. The substantial share of English patents going to 'gentlemen' (generally between 20 and 30 per cent over the period) is particularly relevant on this point.

Perhaps the most important question is whether the design of the patent system had an impact on the direction of inventive activity in the two early industrial economies. Given that patented inventions are not generally representative of all inventions, a definitive answer is beyond the scope of this essay. Nevertheless, it is clear that the composition of patented inventions was so affected. One would expect that the much higher costs of obtaining a patent in England would lead to a relatively greater amount of patenting of inventions in capital-intensive industries, both because patents would have been easier to enforce and extract a high return from if they covered substantial capital equipment, and because individuals employed in labour-intensive industries and making incremental advances would have been less able on average to raise the resources needed to file a patent. Indeed, this expectation is borne out in striking fashion. When the patents are classified by final use from 1790 through the middle of the nineteenth century (September 1852 in England and 1846 in the U.S.), only 4.8 and 10.0 per cent respectively of patents in England were in the relatively labour-intensive agriculture and construction sectors, whereas the corresponding figures for the U.S. were 22.3 and 16.7 per cent respectively.

At the other extreme, the relatively capital-intensive manufacturing and transportation sectors garnered 52.4 and 20.9 per cent respectively of English patents, but only 40.1 and 12.8 per cent of those in the U.S.[39]

The contrast between the inventions patented in the two countries is equally apparent at a more disaggregated level. The bias in England toward capital-intensive inventions and industries is reflected in 7.3 per cent of all of its patents between 1750 and 1851 having been for steam and gas engines, as opposed to 3.8 per cent in the U.S. (between 1790 and 1846). Similarly, patents for the textiles sector accounted for just over 15 per cent of English patents, but less than 7 per cent of U.S. patents. When one looks at a labour-intensive industry like footwear, tanning, and general leather goods, however, the relative shares are reversed: 3.2 per cent of all U.S. patents (8 per cent of those in manufacturing), as compared to 2.5 per cent in England (5 per cent of manufacturing patents).[40]

Overall, comparisons of the patterns of patenting and of the composition of patentees yield results that are in accord with our assessment of the effect of differences in British and American patent institutions. It is clear that the British patent system favoured the wealthy and inventors who focused on technologies that were highly capital-intensive or otherwise perceived *ex ante* to be more valuable, while the U.S. system offered property rights in new technology in a manner that was more democratic. Inventions that were incremental, aimed at labour-intensive industries, and discovered by ordinary workers who were confronting problems that were related to their occupation, were accordingly more likely to be patented in the U.S. than in England.[41] The difference in institutions meant, at the least, that commercial opportunities for invention and innovation were less available to undistinguished individuals in England than in the U.S., with possible implications for relative degrees of equality and social mobility.

At a more global level, the British patent system may have led to lower rates of invention (or to different types of inventions), and not just of patenting, if the more restricted provision of property rights meant that the expected returns to inventive activity were lower for a significant group of potential inventors, or for certain types of inventions. There may be yet other reasons why the system was less conducive to technical change than was its U.S. counterpart. For example, the lack of property rights in technological information could have discouraged inventors from publicly promoting their discoveries. Similarly, the limited public access to patent specifications in the British system undoubtedly slowed further the diffusion of new technical knowledge. Slower diffusion of technology would have slowed productivity growth not only directly, but indirectly as well if familiarity with that new knowledge was an input into current and future inventive activity.

THE RELEVANCE OF DIFFERENT ORGANIZATIONS OF MANUFACTURING

Another institution-based hypothesis for the different records of productivity growth in Britain and the United States focuses on the significance of how firms or production were organized for the generation of technological change. A number of economic historians have recently shown that seasonality in manufacturing was quite pervasive in early industrial England and France, and that forms of organization conducive to part-time or irregular operations, such as putting-out or cottage manufacture, were especially prevalent in many labour-intensive industries well into the nineteenth century. Scholars of the U.S. have pointed out, however, that such forms of manufacturing organization were quite rare in early industrial America, where production was typically carried out in centralized plants which operated full-time throughout the year even before the rise of the mechanized factory.[42] Cottage manufacture has been linked on both sides of the Atlantic to variation over time in the opportunity cost of labour, with seasonal fluctuation in agricultural demand for labour and the household responsibilities of women highlighted as particularly significant sources of this variability. On the basis of this and other evidence, Sokoloff and Dollar have argued that the difference between the two countries in the prevalence of cottage manufacture, as compared to the manufactory organization of production, was at least partially due to the greater importance of crops with highly seasonal labour requirements (like wheat) in England.[43] The logic of this perspective is that cottage manufacture could survive in competition with technically more efficient manufactories, because the flexibility of the former mode of production made it more conducive to the effective harnessing of an offpeak or part-time workforce whose opportunity costs (wages) were low and who divided their labour time between different activities.

Regardless of the reasons for the difference in industrial organization, however, the contrast could have contributed to differences in the two countries' paths of technological change. Indeed, there is a strong theoretical case for why manufacturing based on a centralized plant organization would generate more rapid technical change over time. First, in an era in which much progress was made through improved organization of labour and other inputs, the larger scale of production associated with centralized plants might well have made it easier to identify or implement ways of raising productivity such as increasing the division or intensity of labour or exploiting indivisibilities in capital equipment.[44] Second, the rate of investment in inventive activity might reasonably have been influenced by the organization of enterprise. The centralized plant's greater degree of specialization, as well as its larger scale, would be expected to boost invention by raising the return to that activity, because the return to an improvement in technology generally increases with the level of output. Third, large centralized plants facilitated the extraction of private returns from property rights in inventions by reducing the costs of monitoring the use of the technology and protecting against infringement. Higher expected returns would presumably induce higher rates of investment in inventive activity. Finally, studies of technological advance during early industrialization suggest that the greater division of labour within centralized plants may have helped to focus attention on how machinery or other changes in production processes could be designed to improve productivity.[45] In short, it is entirely reasonable to expect that cottage production, where workers laboured part-time, on a very small scale, and with relatively simple capital equipment, made for a less favourable environment for invention than centralized production did.

It is obvious that this sort of framework for thinking about how the organization of manufacturing could influence the course of technological change would work, as did our analysis of the difference in patent institutions, in the direction of helping to explain the disparity between the estimates of the records of manufacturing productivity growth for Britain and the U.S. The hypothesis applies with greatest force to the labour-intensive manufacturing industries, where virtually all the difference between the two economies in industrial organization was located, and where Britain's pace of technological advance appears to have been relatively slow.

CONCLUSION

In comparing the paths of technological change in early industrial Britain and the United States, we could have chosen to discuss the considerable similarities between the circumstances and experiences of these first two economies to industrialize. Focussing on contrasts, as we have done instead, runs the risk of sacrificing the broad perspective, but it also allows us to highlight the stark difference in the records of productivity growth implied by the most recent estimates in the literature. Economic historians have not yet fully grappled with the questions of whether and how these estimates can be improved, and how the divergence between the two economies is to be explained if the qualitative finding of slower and much less balanced productivity growth in Britain than in the U.S. holds up to scrutiny. We believe that the study of the processes of early economic growth has been hampered by the lack of attention to this comparative dimension, and that reconciling or accounting for the evidently divergent patterns is of fundamental importance.

Although we hope to stimulate scholars to undertake the formidable work in the archives which is probably necessary to address the puzzle effectively, the contribution of this essay towards resolving the issues is quite modest. Our intent was to explore how two specific differences in institutions may have had significant influences on the course of technological change in the respective countries. Most of our efforts have been devoted to specifying differences between the British and American patent systems, and to assessing what impact they could have had on the respective courses of technical change. We argue that these differences influenced the compositions of both patentees and patented inventions: the British system favoured the wealthy and highly capital-intensive technologies in the award of patents, while the Americans extended property rights in invention to broader ranges of both new technologies and the population. At the very least, the more limited access of ordinary people in Britain to property rights in their inventions must have served to preserve inequality and reduce social mobility. Moreover, although there were certainly other ways of extracting a return from invention than relying on property rights vested in patents, there are reasons to believe that American patent institutions made for more extensive trade in technologies, higher expected returns to (and levels of) inventive activity, as well as more rapid diffusion of technological information. Given the difficulties of drawing inferences about the volume of invention from data on patent counts alone, this interpretation has yet to be subjected to a systematic empirical test. Nevertheless, it is interesting that one would expect the effects of the different patent institutions to be most pronounced in the labour-intensive industries, and indeed, the productivity growth estimates we have do exhibit the greatest discrepancy in such industries.

The other difference in institutions we discuss concerns how early manufacturing was organized. The contrast between early industrial Britain and the United States in the prevalence of cottage manufacture, as opposed to manufactories or non-mechanized factories, is quite dramatic, and again there is good reason to believe that the organization of production had some effect on the rate or direction of inventive activity. Although this issue also awaits empirical investigation, researchers should be able to take advantage of systematic regional variation in the organization of manufacturing within Britain by examining whether a relationship between the predominant form of organization and the generation of inventions in the relevant industries held across regions.

Although our knowledge of the processes by which technological change accelerated in early industrial Britain and the U.S. has been much improved in recent decades, there is still much that we do not understand. Whether or not the two economies are determined to have diverged substantially in their records of growth, the influence of institutions like patent systems and industrial organization on the rate and direction of inventive activity in this and other contexts deserves more attention and study by scholars.

NOTES

1. See T.S. Ashton (1968), *The Industrial Revolution, 1760–1830*, London: Oxford University Press; D.S. Landes (1969), *Unbound Prometheus: Technological Change and Industrial Development in Western Europe from 1750 to the Present*, London: Cambridge University Press; and M. Berg (1994), *The Age of Manufactures, 1700–1820: Industry, Innovation and Work in Britain*, London: Routledge.
2. N.F.R. Crafts and C.K. Harley (1992), 'Output Growth and the British Industrial Revolution: A Restatement of the Crafts-Harley View', *Economic History Review*, **45**, 703–30.
3. For examples, see K.L. Sokoloff (1992), 'Invention, Innovation, and Manufacturing Productivity Growth in the Antebellum Northeast', in R.E. Gallman and J.J. Wallis (eds), *American Economic Growth and Standards Of Living Before the Civil War*, Chicago: University of Chicago Press; and P. Temin (1997), 'Two Views of the Industrial Revolution', *Journal of Economic History*, **57**, 63–82.
4. D.C. North (1981), *Structure and Change in Economic History*, New York: Norton.
5. In addition to H.J. Habakkuk (1962), *American and British Technology in the Nineteenth Century: the Search for Labour-Saving Inventions*, Cambridge: Cambridge University Press, see S.L. Engerman and K.L. Sokoloff (1997), 'Factor Endowments, Institutions, and Differential Paths of Growth Among New World Economies', in S. Haber (ed.), *How Latin America Fell Behind*, Stanford: Stanford University Press.
6. See North, *Structure and Change*, and F. Machlup (1958), *An Economic Review of the Patent System*, Washington: Government Printing Office.
7. For evidence that inventors were responsive to prospects for material gain, consistent with patent systems making a difference, see K.L. Sokoloff (1988), 'Inventive Activity in Early Industrial America: Evidence from Patent Records, 1790–1846', *Journal of Economic History*, **48**, 813–50; B.Z. Khan and K.L. Sokoloff (1993), '"Schemes of Practical Utility": Entrepreneurship and Innovation Among "Great Inventors" in the United States, 1790–1865', *Journal of Economic History*, **53**, 289–307.
8. K.L. Sokoloff and D. Dollar (1997), 'Agricultural Seasonality and the Organization of Manufacturing in Early Industrial Societies: The Contrast Between England and the United States', *Journal of Economic History*, **57**, forthcoming.
9. The Statute of Monopolies, 21 Jac. c., 1623, was implemented in 1624.
10. Blanchard v. Sprague, 3 F. Cas. 648, Mass. (1839), compared the laws toward patents in Britain and America.
11. H.I. Dutton (1984), *The Patent System and Inventive Activity During the Industrial Revolution, 1750–1852*, Manchester: Manchester University Press, chapter 4.
12. Patent Law Amendment Act, 15 & 16 Vict., Chapter 83, 1852.
13. Some had recommended the replacement of the registration system by the American examination system, but this proposal was not accepted, and the British retained the registration system until early in the twentieth century. One of the witnesses before the 1851 Select Committee argued against an examination system because it would 'erode the royal prerogative'. However, the Select Committee decided not to adopt this aspect of the American system more because they felt it would not be possible to find sufficiently skilled examiners. Law Officers did have the right to examine the technical merits of patent applications at their discretion, but there is no evidence that this right was ever exercised. See Dutton, *The Patent System*, p. 61, and *British Parliamentary Papers. Inventions* (1968–70), Shannon: Irish University Press.
14. Article 1, sec. 1, cl. 8 of the United States Constitution authorized Congress 'to promote the progress of science and useful arts, by securing to authors and inventors the exclusive rights to their respective writings and discoveries'. Accordingly, the first Patent Act of April 1790 stipulated that 'upon the petition of any person or persons that he, she, or they,

hath invented or discovered any useful art, ... it shall be lawful ... to cause letters patent to be made out in the name of the United States'.

15. The phrase is from the landmark Supreme Court case, *ex parte* Wood v. Brundage, 22 U.S. 603 (1824). These sentiments were apparent in the entire body of American law reports: Ames v. Howard, 1 F. Cas. 755 (1833) declared that patents were granted 'to promote the progress of science and the useful arts, an object as truly national, and meritorious, and well founded in public policy, as any which can possibly be within the scope of national protection'. Whitney v. Emmett et al., 29 F. Cas 1074 (1831): 'In England a patent is granted as a favor ... Here a patent is a matter of right.' Pennock v. Dialogue, 27 U.S. 1 (1829), another landmark case, contrasts England and America: 'In the courts of the United States, a more just view had been taken of the rights of inventors. The laws of the United States were intended to protect those rights, and to confer benefits; while the provisions in the statute of England ... are exceptions to the law prohibiting monopolies. Hence, the construction of the British statute had been exceedingly straight and narrow, and different from the more liberal interpretation of our laws.'

16. Note that Britain passed no major patent statutes between 1624 and 1852 (two minor changes were recorded in 5 & 6 Will. IV, (1835) and 2 & 3 Vict. (1839)). Patent policies were therefore unpredictable, effected on an *ad hoc* basis by the courts, Law Officers, and the Lord Chancellor (H. Harding (1953), *Patent Office Centenary*, London: H.M.S.O., p. 4). As U.S. Supreme Court Justice Baldwin declared in Whitney v. Emmett, 29 F. Cas. 1074 (1831): 'The silence of the [English] law left a wide field open to the discretion of the courts ... But in this country the law is more explicit. The Constitution ... is a declaration of the supreme law of the land ... which leaves no discretion to the judges to assign or presume any other.' In keeping with the consciousness that patents were critical to national policy, reforms in the antebellum period were accomplished and incorporated into U.S. law fairly rapidly, by means of the Patent Law Acts of 1790, 1793, 1800, 1819, 1832, 1836, 1839 and 1842. The 1836 statute set in place the basic system that operated in the twentieth century.

17. See the discussion in B. Bugbee (1967), *The Genesis of American Patent and Copyright Law*, Washington: Public Affairs Press, p. 145, of the resemblance between the copyright law of 1790 and the Statute of Anne (1710). Note also his conclusion that (p. 157), 'The national patent and copyright systems created in 1790 under the Constitution were founded not only upon English precedents ... but also ... upon a century and a half of distinctive provincial tradition'.

18. Bugbee, *The Genesis*, p. 137.

19. See L.G. de Pauw (1977), *Documentary History of the First Federal Congress: vol. 3,: House of Representatives Journal*, Baltimore: Johns Hopkins University Press, pp. 1631–7. Also see Bugbee, *The Genesis*, p. 147: 'It will be observed that certain other provisions which had appeared at one time or another in American intellectual property enactments were omitted from the Federal patent and copyright statutes of 1790 – in favor of the inventor or author ... During the term of a patent or copyright legally acquired from the United States, the property of the creative individual in his production was to be absolute.'

20. See K.L. Sokoloff and B.Z. Khan (1990), 'The Democratization of Invention During Early Industrialization: Evidence from the United States, 1790–1846', *Journal of Economic History*, **50**, 363–78.

21. See Attorney General (1851), *Hansard's Parliamentary Debates*, vol. cxviii, July 1–August 8, London: T.C. Hansard; Dutton, *The Patent System*; and J. Ruggles (1836), *Senate Committee Report*, reprinted in *Journal of Patent Office Society* (1936), **18(12)**, 853–63. In 1836, according to Ruggles, the fee for a patent in England alone amounted to $585, or nearly twenty times that for a patent in the U.S.

22. If the patent were extended to Scotland and Ireland it was necessary to negotiate another five offices in each country. For descriptions of the patent procedure, see Dutton, *The Patent System*, and C. MacLeod (1988), *Inventing the Industrial Revolution*, Cambridge: Cambridge University Press. Harding, *Patent Office Centenary*, p. 5, calls the system 'quite fantastic'. According to E. Holroyd (1830), *A Practical Treatise of the Law of Patents for Inventions*, London: A. Strahan, pp. 66–7, patent petitions had to be submitted to: the office of the Secretary of State; the Attorney General; office of the Secretary of State for King's Warrant; warrant of the Attorney General; patent office for signature of the Crown; office of the Signet; Lord Keeper of the Privy Seal; office of the Lord Chancellor; Great Seal Patent Office. Officials were not required to justify their decisions. The vast majority of

patents before 1852 were filed in England – only 16 per cent were extended to Scotland and Ireland as well. Hence the figures for England closely approximate the total amount of patents until 1852, when patents grants began automatically to cover all three kingdoms (Dutton, *The Patent System*, p. 35).

23. According to the Attorney General, a renewal system would 'give poor men an opportunity which they did not at present possess, of protecting inventions till such time as they might be able to derive advantage from them, and at a cost adapted to their means'. See the Attorney General, *Hansard*, p. 1537.

24. Fees were £4 for the first application; £50 to maintain the patent after the fourth year; and £100 due before the end of seven years. See Patents, Designs and Trademarks Act, 46 & 47 Vict. Chapter 57 (1883).

25. Under certain conditions, involving a high cost of importing technology from abroad and a low cost of it diffusing once imported, an argument can be made for why the English practice (which was common throughout Europe and probably reflected a tradition of granting royal privileges to first-time investors) would enhance economic growth. In general, however, we think that such cases would have been relatively rare, and that the detrimental effects of restraining competition more than offset the positive ones overall. More research is needed to resolve the issue.

26. 'The great majority of patents are still however procured through the instrumentality of Patent Agents ... The proportion obtained in the year 1852–53, by Patentees or Solicitors, to that by Patent Agents, was one to fifty' according to J. Coryton (1855), *A Treatise on the Law of Letters Patent*, London: H. Sweet, p. 147. Most British patent agents were located in London, in closer proximity to the patent offices (Dutton, *The Patent System*). In the United States, the Patent Office itself disseminated information and forms. Copies of prior patents could be obtained by writing to the Patent Office, which charged 10 cents per hundred words. The Secretary of State published lists of expired patents in the Washington newspapers. Patent agents could be found in most major cities in the United States, from San Francisco to Boston. Competition among numerous patent agents kept fees low: for instance, in the 1860s Munn and Co., the largest patent agency, charged $5 for a patent search and $1 to copy patent claims. Rather than negotiating with a bureaucracy, American patent agents offered advice on the technical merits of the invention, drafted technical drawings, and served as mediators between buyers and sellers of technology. For, as *Scientific American* (November 13 1842, p. 62) – published by Munn & Co – noted, 'we advise every inventor who is able, to make application for himself, and thereby save some expense. There are forms and rules that will require study, but you can soon master them'.

27. See Harding, *Patent Office Centenary*, and Dutton, *The Patent System*.

28. Specifications were not generally published in England, according to Holroyd, *A Practical Treatise*, p. 160, until 1852. The various patent offices concerned prohibited extracts of key aspects of the invention (Select Committee, 1829, in *Parliamentary Papers. Inventions*). Trade journals such as the *Repertory of Arts, Manufactures and Agriculture* did publish patent descriptions. However, these published lists tended to be incomplete and the length of time between patent grant and publication varied between five months and five years (D.J. Jeremy (1981), *Transatlantic Industrial Revolution: The Diffusion of Textile Technologies Between Britain and America, 1790–1830s*, Cambridge: Cambridge University Press, pp. 47–9).

29. Jeremy, *Transatlantic Industrial Revolution*, pp. 46–7. Jeremy also notes that descriptions were filed haphazardly and that searches could take hours. In the 1820s an employee of the Patent Office and a patent agent on his own account, had compiled an index, but it remained 'under Poole's personal control', and probably contributed to the value of his services as a patent agent.

30. See MacLeod, *Inventing the Industrial Revolution*, p. 51, for a discussion of the development of the specification requirements for patentability and of the basis for limiting access to the specifications of patents.

31. See MacLeod, *Inventing the Industrial Revolution*, and Dutton, *The Patent System*. Between 1617 and 1852, Londoners consistently received the majority of all patents granted. See Harding, *Patent Office Centenary*, p. 13.

32. Alternative strategies for extracting material returns from a patented invention included assigning (selling) the property rights in total or in part, licensing the technology, exploiting the patent directly as a sole proprietor or partner of an enterprise that used the invention, or some combination of these approaches.

33. In the 1860s, assignments in England varied between 600 and 700 *per annum*. The law limited the number of assignees to five (twelve after 1832). At least until 1830, it was not clear whether this limit extended to licensees (Holroyd, *A Practical Treatise*, p. 145). This law in effect also restricted financing for inventions, as Duvergier v. Fellows, 10 C.B. 826 (1830), indicates.
34. For example, it is noted in B.Z. Khan (1995), 'Property Rights and Patent Litigation in Early Nineteenth-Century America', *Journal of Economic History*, **55**, 58–97, that the U.S. Patent Office recorded some 2108 assignments in 1845. See N.R. Lamoreaux and K.L. Sokoloff (1996), 'Long-Term Change in the Organization of Inventive Activity', *Proceedings of the National Academy of Sciences*, **93** (12 November), 12686–92, for an analysis of patent assignments during a later period.
35. If a British patent was useless, the purchaser could not recover his payment: Hall v. Conder, 2 C.B., N.S. 22 (1857). Assignees or licensees of American patents that lacked utility were entitled to refunds of their payments. For details see Khan, 'Property Rights'.
36. See MacLeod, *Inventing the Industrial Revolution*, and Dutton, *The Patent System*.
37. According to Lord Coke, an invention should 'not be mischievous to the State' nor 'hurt trade' (Coryton, *A Treatise on the Law*, pp. 60–61). Patents could be invalidated because of inutility, and according to Hill v. Thompson, Web. P.C. 237 (1817) 'the utility of an invention is a question for the jury'. Importers of inventions were at times deemed to be of lower status than original inventors, as In re. Soames' Patent, 1 Web. P.C. 733 (1843).
38. See Khan, 'Property Rights'.
39. These figures were computed from random samples of English and U.S. patents. The English patents were drawn from B. Woodcroft (1854), *Titles of Patents of Invention: Chronologically Arranged from March 2, 1617 to October 1, 1852*, London: Patent Office, Great Britain, comprising 26.7 per cent of all patents issued between 1790 and October 1, 1851. The U.S. sample includes 29.6 per cent of all patents issued between 1790 and 1846, and is described in Sokoloff, 'Inventive Activity'. The classifications were done by us, but similar qualitative results are implied by the classifications done by the respective patent offices. See Dutton, *The Patent System*, Appendix A, and E. Burke (1847), *List of Patents for Inventions and Designs Issued by the United States From 1790 to 1847*, Washington: U.S. Patent Office.
40. The figures for England were drawn from Dutton, *The Patent System*, Appendix A, while those for the U.S. were computed by the authors from the individual patent descriptions in Burke, *List of Patents*.
41. Indeed, in eighteenth century England 'patenting of mere improvements was frowned upon', as pointed out by MacLeod, *Inventing the Industrial Revolution*, p. 13.
42. For examples of work on England and France, see D. Bythell (1978), *The Sweated Trades*, London: St. Martin's; R.D.M. Snell (1985), *Annals of the Labouring Poor*, Cambridge: Cambridge University Press; G.L. Gullickson (1983), 'Agriculture and Cottage Industry: Redefining the Causes of Proto-Industrialization', *Journal of Economic History*, **43**, 831–50; and G. Postel-Vinay (1994), 'The Dis-integration of Traditional Labour Markets in France: From Agriculture and Industry to Agriculture or Industry', in G. Grantham and M. McKinnon (eds), *Labour Market Evolution: The Economic History of Market Integration, Wage Flexibility, and the Employment Relation*, London: Routledge. For a discussion of the experience in the U.S., see Sokoloff and Dollar, 'Agricultural Seasonality'.
43. See Sokoloff and Dollar, 'Agricultural Seasonality'. Gullickson, 'Agriculture and Cottage Industry', makes a similar argument for France.
44. See Sokoloff, 'Invention, Innovation', and Berg, *Age of Manufactures*, for discussion.
45. For example, see Landes, *Unbound Prometheus*.

Index